Studies in Infrastructure and Control

The book series aims to publish top-quality state-of-the-art textbooks, research monographs, edited volumes and selected conference proceedings related to infrastructure, innovation, control, and related fields. Additionally, established and emerging applications related to applied areas like smart cities, internet of things, machine learning, artificial intelligence, etc., are developed and utilized in an effort to demonstrate recent innovations in infrastructure and the possible implications of control theory therein. The study also includes areas like transportation infrastructure, building infrastructure management and seismic vibration control, and also spans a gamut of areas from renewable energy infrastructure like solar parks, wind farms, biomass power plants and related technologies, to the associated policies and related innovations and control methodologies involved.

Ashutosh K. Giri · Sabha Raj Arya ·
Josep M. Guerrero · Shailendra Kumar
Editors

Power Quality: Infrastructures and Control

Springer

Editors
Ashutosh K. Giri
Department of Electrical Engineering
Government Engineering College
Bharuch, India

Sabha Raj Arya
Department of Electrical Engineering
Sardar Vallabhbhai National Institute of Technology
Surat, Gujarat, India

Josep M. Guerrero
Centre of Research on Microgrids (CROM)
AAU Energy
Aalborg University
Aalborg East, Denmark

Shailendra Kumar
Department of Electrical Engineering
Maulana Azad National Institute of Technology
Bhopal, Madhya Pradesh, India

ISSN 2730-6453 ISSN 2730-6461 (electronic)
Studies in Infrastructure and Control
ISBN 978-981-19-7955-2 ISBN 978-981-19-7956-9 (eBook)
https://doi.org/10.1007/978-981-19-7956-9

This Springer imprint is published by the registered company Springer Nature Singapore Pte Ltd.
The registered company address is: 152 Beach Road, #21-01/04 Gateway East, Singapore 189721, Singapore

Preface

The main aim of the proposed book is to present novel idea and concepts developed by the researchers/academia and practicing engineers working in the domain of the power sector infrastructures where power electronics infrastructures are used for improving the system reliability and efficiency in the on-grid and off-grid system. The infrastructures of distributed power generation based on wind, solar, hydro, and many other renewable energy sources have increased manifold since last decade due to availability of efficient power converters and small rating generators. The application of power electronics switching devices has made job much easier to make such system infrastructures more reliable and controllable. The power quality (PQ) issues in infrastructures of distributed power generation system are a major concern for customers. The custom power device such as voltage source converter is used to mitigate the PQ and other issues such as voltage and frequency control under different loading conditions on the supply system. The operation of these custom power devices or other power converters is dependent on the duty cycles generated by control algorithm. The scope of control algorithm may be varying from model predictive control techniques, machine learning techniques to other artificial intelligence-based techniques. In addition to it, some classical control algorithm and adaptive type control algorithm may also be used for power converter operation. Therefore, this book creates awareness among teachers, research students, and industry persons about better utilization of infrastructures of distributed power generation system by making it more efficient with the use of power electronics and its control.

The main features of this book are given as follows:

1. This is intended to cover all the traditional as well as advanced control algorithms for power control and other power quality issues such as neutral current compensation in advanced power infrastructures.
2. The application of various infrastructural converter topologies in solar, wind, hydro and diesel engine and others power generation and energy conversion system.

3. The least mean square principal-based adaptive control, model predictive control, neuro/fuzzy-based controllers, passivity and fractional order PI-based controllers and optimization techniques application for estimating the error are the focus area for estimating the reference current.

The readers will be able to utilize the concepts presented in this book for doing research further in this domain. This book will provide the insight of various topology as well as control algorithm used for power control in the infrastructures of distributed energy conversion systems such as solar, wind, and other power sources.

Simulation results based on various topologies of solar and wind energy system will create the good interest among readers.

Mathematical modeling of control algorithm and its implementation for power control will also create good impact on the reader's mind.

The coverage of extensive circuit design and its testing for solar and wind-based system will be interesting for readers.

UG/PG/Research students can take some insight for their work from this book. This book provides extensive coverage of recent development in the area of distributed power generation based on wind, solar, hydro and many others, generation and conversion system, control and power conversion.

Bharuch, India Ashutosh K. Giri
Surat, India Sabha Raj Arya
Aalborg East, Denmark Josep M. Guerrero
Bhopal, India Shailendra Kumar

Contents

Editors and Contributors

About the Editors

Ashutosh K. Giri received his Ph.D. from the Department of Electrical Engineering, Sardar Vallabhbhai National Institute of Technology, Surat, India, in 2020. In May 2011, he joined the Department of Electrical Engineering, Government Engineering College, Bharuch, as an assistant professor. He has authored and co-authored more than 30 research papers in reputed journals such as IEEE TRANSACTIONS, IET Journal, Wiley, Tailor Francis, Springer. He has also co-authored one book which is published in Springer Publication, Singapore. He has received the sponsored research grant of Rs. 25 Lacs from the director of technical education, Gujarat state, under STEM Scheme. He is working as a principal investigator in this project. He has coordinated 5 STTP including recently concluded ATAL FDP. He has delivered national- and international-level expert talk at various reputed universities across India. He had received Research Opportunity Week (ROW) scholarship and visited Technical University of Munich, Germany, in March–April 2019. He has also served as a technical program committee member in conferences organized at Scientific Federation of Dallas, USA, and Kenitra, Morocco. He has also chaired as technical session in IEEE sponsored international conference at IIITDM Kancheepuram, Chennai. He had organized special session in world most reputed IEEEECCE2020 Conference at Detroit, Michigan, USA, and in ECCE-ASIA 2021 at Singapore. He is also serving as a potential reviewer in various reputed journals such as IEEE TRANSACTIONS OF POWER ELECTRONICS, IEEE TRANSACTIONS ON INDUSTRIAL ELECTRONICS, IET, Wiley. He was nominated the best researcher of the year 2019 at electrical department, SVNIT, Surat, in 2019. He is also serving as an honorary ad visor (Technical Services) in GMDT Marine and Industrial Engineering Pvt. Ltd, Ahmedabad, India. He is also serving as an honorary advisor (Technical Services) in CUBATICS INDUSTRIES Pvt. Ltd, Surat. India. He is a life member of Indian Society of Technical Education. Recently, he was included in Advisory Board as a member in Society of Power Engineers, Vadodara Chapter.

Sabha Raj Arya (M'1 2, SM'15) received Bachelor of Engineering degree in Electrical Engineering from Government Engineering College, Jabalpur, in 2002; Master of Technology in Power Electronics from Motilal National Institute of Technology, Allahabad, in 2004; and Ph.D. degree in Electrical Engineering from Indian Institute of Technology (IIT) Delhi, New Delhi, India, in 2014. He is joined as an assistant professor of the Department of Electrical Engineering, Sardar Vallabhbhai National Institute of Technology, Surat. January 2019, he is promoted as an associate professor in same institute. His fields of interest include power electronics, power quality, design of power filters, and distributed power generation. He received Two National Awards, namely INAE Young Engineer Award from Indian National Academy of Engineering, and POSOCO Power System Award from Power Grid Corporation of India in the year of 2014 for his research work. He is also received Amit Garg Memorial Research Award-2014 from IIT Delhi from the high impact publication in a quality journal during the session 2013–2014. At present, he has published more than hundred research papers in international Journals and conferences in the field of electrical power quality. He also serves as an associate editor for the IET Renewable Power Generation.

Josep M. Guerrero (S'0 1-M'04- SM'0 8-FM'1 5) received the B.S. degree in telecommunications engineering, the M.S. degree in electronics engineering, and the Ph.D. degree in power electronics from the Technical University of Catalonia, Barcelona, in 1997, 2000, and 2003, respectively. Since 2011, he has been a full professor with the Department of Energy Technology, Aalborg University, Denmark, where he is responsible for the Microgrid Research Program. From 2014, he is a chair professor in Shandong University; from 2015, he is a distinguished guest professor in Hunan University; and from 2016, he is a visiting professor fellow at Aston University, UK, and a guest professor at the Nanjing University of Posts and Telecommunications. From 2019, he became a Villum investigator by the Villum Fonden, which supports the Center for Research on Microgrids (CROM) at Aalborg University, being, the founder and director of the same center (www.crom.et.aau.dk). His research interests are oriented to different microgrid aspects, including power electronics, distributed energy storage systems, hierarchical and cooperative control, energy management systems, smart metering, and the Internet of things for AC/DC microgrid clusters and islanded minigrids. Specially focused on microgrid technologies applied to offshore wind, maritime microgrids for electrical ships, vessels, ferries and seaports, and space microgrids applied to nanosatellites and spacecrafts. Prof. Guerrero is an associate editor for a number of IEEE TRANSACTIONS. He has published more than 600 journal papers in the fields of microgrids and renewable energy systems, which are cited more than 50,000 times. He received the best paper award of the IEEE Transact ions on Energy Conversion for the period 2014–2015 and the best paper prize of IEEE-PES in 2015. As well, he received the best paper award of the Journal of Power Electronics in 2016. During seven consecutive years, from 2014 to 2020, he was awarded by Clarivate Analytics (a former Thomson Reuters) as Highly Cited Researcher with 50 highly cited papers. In 2015, he was elevated as an IEEE fellow for his contributions on "distributed power systems and microgrids."

Shailendra Kumar (Senior Member, IEEE) was born in Mahoba, UP, India, in 1988. He received the B.Tech. degree in electrical and electronics engineering from Uttar Pradesh Technical University, Lucknow, India, in 2010, and the M.Tech. degree in power electronics and electrical machine and drives from the Indian Institute of Technology, New Delhi, India, in 2015. He received the Ph.D. degree in electrical engineering from the Indian Institute of Technology Delhi, New Delhi, India, in May 2019. He is currently working as an assistant professor with the Maulana Azad National Institute of Technology, Bhopal. His research interests include power quality, grid integration, custom power devices, microgrid, and renewable energy. Dr. Kumar is a recipient of prestigious POSOCO Power System Award (in master's as well as in doctoral categories) in 2016 and 2019. He is also a recipient of Prof. Som Nath Mahendra Student Travel Award for the IEEE PEDES 2018 Conference and the IEEE UPCON Best Paper Award in 2016 and 2018, respectively.

Contributors

Mohsen Alizadeh Bidgoli Department of Electrical Engineering, Yadegar-E-Imam Khomeini (RAH) Shahre-Rey Branch, Islamic Azad University, Tehran, Iran

Ramanjaneyulu Alla Department of Electrical Engineering, Sardar Vallabhbhai National Institute of Technology, Surat, India

Sabha Raj Arya Department of Electrical Engineering, Sardar Vallabhbhai National Institute of Technology Surat, Gujarat, India

Vinay Kumar Awaar Gokaraju Rangaraju Institute of Engineering and Technology, Hyderabad, India

Sauvik Biswas Indian Institute of Technology Delhi, Delhi, India

Ambrish Chandra École de Technologie supérieure (ETS), Montréal, Canada

Anandita Chowdhury Department of Electrical Engineering, Sardar Vallabhbhai National Institute of Technology, Surat, India

Mohsen Eskandari University of New South Wales, Sydney, Australia

Fadoul École de Technologie supérieure (ETS), Montréal, Canada

Abdelhamid Hamadi GSE Solution, Montréal, Canada

Abdeldjebar Hazzab University of Bechar, Algeria and ETS, Béchar, Algeria

Hussein Ibrahim CR2Ie, Sept-Îles, QC, Canada

Praveen Jugge Gokaraju Rangaraju Institute of Engineering and Technology, Hyderabad, India

S. Tara Kalyani Jawaharlal Nehru Technological University Hyderabad, Hyderabad, India

Nishant Kumar Indian Institute of Technology Jodhpur, Karwar, Rajasthan, India

Shailendra Kumar Department of Electrical Engineering, Maulana Azad National Institute of Technology Bhopal, Bhopal, India

Khyati D. Mistry Department of Electrical Engineering, Sardar Vallabhbhai National Institute of Technology Surat, Gujarat, India

Paresh Kumar Nayak Indian Institute of Technology (ISM) Dhanbad, Dhanbad, India

Bijaya Ketan Panigrahi Indian Institute of Technology Delhi, Delhi, India

Pemendra Kumar Pardhi Department of Electrical Engineering, Shri G. S. Institute of Technology and Science Indore, Indore, Madhya Pradesh, India

Haritha Dhanlalji Parmar MVSR Engineering College, Nadergul, Hyderabad, India

Veramalla Rajagopal Department of Electrical and Electronics Engineering, Kakatiya Institute of Technology and Science, Warangal, India

More Raju Department of Electrical Engineering, Maulana Azad National Institute of Technology Bhopal, Bhopal, India

Ankur Singh Rana Department of Electrical and Electronics Engineering, National Institute of Technology Tiruchirappalli, Tiruchirappalli, Tamil Nadu, India

Miloud Rezkallah CR2Ie, Sept-Îles, QC, Canada;
École de Technologie supérieure (ETS), Montréal, Canada

Aliva Routray Department of Electrical Engineering, Sardar Vallabhbhai National Institute of Technology Surat, Gujarat, India

Priyank Shah Intelligent Control and Smart Energy (ICSE) Research Group, School of Engineering, The University of Warwick, Coventry, UK

Shailendra Kumar Sharma Department of Electrical Engineering, Shri G. S. Institute of Technology and Science Indore, Indore, Madhya Pradesh, India

Bhim Singh Department of Electrical Engineering, Indian Institute of Technology, Delhi, New, India

Vedantham Lakshmi Srinivas Indian Institute of Technology (Indian School of Mines), Dhanbad, Jharkhand, India

Salkuti Surender Reddy Department of Railroad and Electrical Engineering, Woosong University, Daejeon, Korea

Xiaowei Zhao Intelligent Control and Smart Energy (ICSE) Research Group, School of Engineering, The University of Warwick, Coventry, UK

Chapter 1
Infrastructures for Wind Energy-Based Power Generation System—Modelling and Control

Vedantham Lakshmi Srinivas and Nishant Kumar

Abstract In this chapter, the different configurations of wind energy conversion system (WECS) are discussed. The permanent magnet synchronous generator (PMSG)-based WECS control is elaborated in detailed. The comprehensive modelling of wind turbine and permanent magnet synchronous generator is studied. The detailed control of machine side converter and grid side converter is explored to extract maximum power from wind turbine and to improve the power quality of the distribution grid, respectively. Various advanced control strategies are analysed including proportional-integral (PI) controller, phase-locked loop (PLL)-based technique, fuzzy logic-based controller, Kalman filter-based control, etc. The detailed implementation of these control strategies is discussed to adhere to the IEEE standard 519 and IEC standard 61,727. In addition, the robust algorithms are discussed to deal with abnormal grid scenarios and illustrative examples with MATLAB/Simulink models (version: 2020b) are also provided with this book.

1 Wind Power Scenario

Wind energy utilization has increased dramatically in recent years across the world. Wind energy technology continues to advance in efficiency, dependability and cost performance, resulting in extraordinary growth, making wind the world's fastest-growing source of electric generating [1, 2]. In 1980, the total installed wind capacity in the globe was over 13 MW (5 MW and 8 MW installed wind capacity of Denmark and US, respectively). By the end of 2005, 59.024 GW of wind power had been installed worldwide. The total deployment of wind turbines had reached 74.150 GW

V. L. Srinivas
Indian Institute of Technology (Indian School of Mines), Dhanbad, Jharkhand 826004, India
e-mail: vlsrinivas@iitism.ac.in

N. Kumar (✉)
Indian Institute of Technology Jodhpur, Karwar, Rajasthan 342030, India
e-mail: nishantkumar@iitj.ac.in

A. K. Giri et al. (eds.), *Power Quality: Infrastructures and Control*,
Studies in Infrastructure and Control,
https://doi.org/10.1007/978-981-19-7956-9_1

by the end of 2006, with an annual growth rate of around 23.8 percent. This installed capacity expanded by 93.926 GW and 121.188 GW at the end of 2007 and 2008, respectively, representing an annual growth rate of about 26% [3, 4]. As of 2020, 743 GW of wind power capacity worldwide, helping to avoid over 1.1 billion tonnes of CO2 globally [5]. When compared to other renewable technologies available in the globe, grid-connected wind power production has garnered a high level of attention and popularity.

2 Types of Wind Energy Conversion System

There are various topologies [6–8] for wind energy conversion system. There are four types of system configurations for wind energy conversion systems namely, type-1, type-2, type-3 and type-4. The detailed schematic of each type is depicted in Figs. 1, 2, 3 and 4.

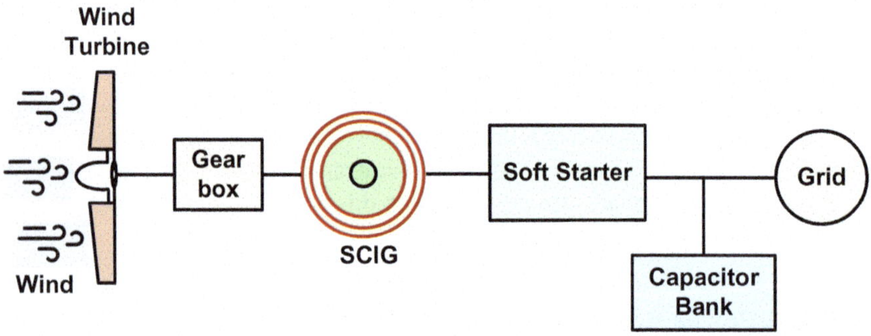

Fig. 1 *Type-1* wind energy conversion system

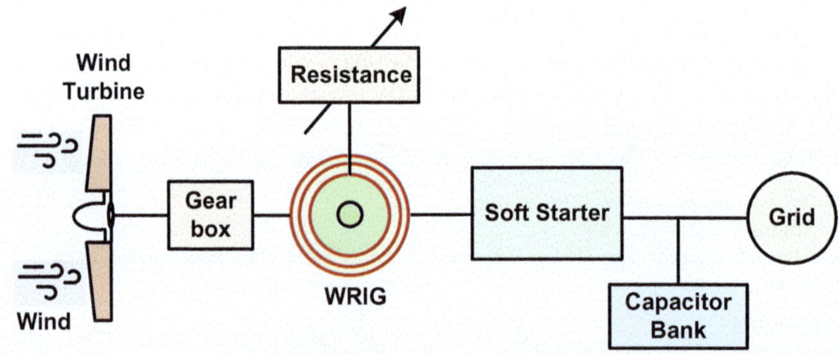

Fig. 2 *Type-2* wind energy conversion system

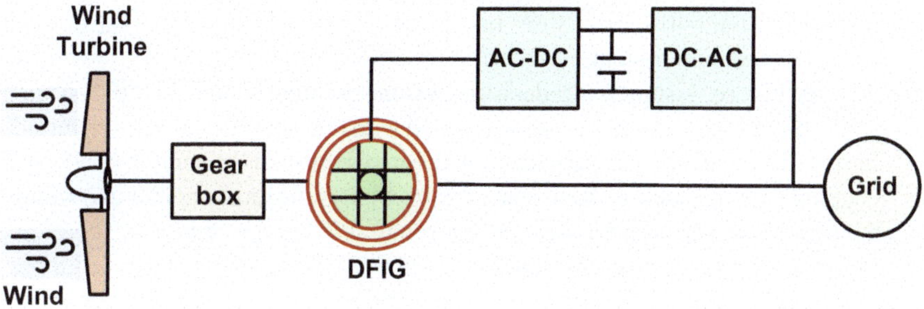

Fig. 3 *Type-3* wind energy conversion system

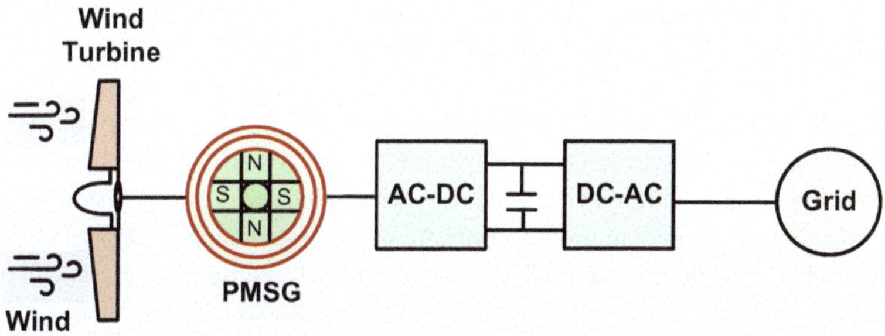

Fig. 4 *Type-4* wind energy conversion system

(1) *Type-1*: Figure 1 shows the detailed schematic of the type-1 system configuration (e.g. known as fixed speed). The squirrel cage induction generator is coupled with the grid. In this configuration [6–8], the soft starter is required to control the current transient during the starting operation. In induction generator, there is no permanent magnet, thereby, the external reactive power source should be provided to build the terminal voltage. In addition, the capacitor bank is also required at the terminal as squirrel cage induction generator requires the reactive power.

(2) *Type-2*: Figure 2 shows the detailed schematic of the type-2 system configuration (e.g. known as limited variable speed). The wound rotor induction generator [6–8] is used herein to develop the wind energy conversion system. The soft starter and capacitor bank are used herein to control the initial current dynamics and voltage profile, respectively. The variable resistance is connected to rotor, which helps to control the speed (which leads to the heat losses); however, it can provide limited speed control from 0 to 10%, unlike type-1 configuration.

(3) *Type-3*: Figure 3 shows the detailed system configuration of type-3 system configuration (e.g. known as variable speed). It is known as doubly fed induction generator [6–8]. The stator is directly connected to the grid and rotor is connected with back-to-back connected converter. It can dynamically control

the speed from –40% to + 30%. In the state of art, it is most popular in wind energy conversion system.

(4) *Type-4*: Figure 4 shows the detailed system configuration of type-4 system configuration (e.g. known as variable speed with full-scale frequency converter). In most of the cases, the permanent magnet synchronous generator (PMSG) is coupled with the back-to-back converter [6–8]. It is the most popular and widely adopted topology configuration in the renewable energy industry. Therefore, this topology has been elaborated further in this chapter, while a detailed three-phase diagram of the system is presented in Fig. 5. It is mainly composed of a wind turbine, a permanent magnet synchronous generator, two VSCs, a Boost converter, a DC link capacitor (C_{dc}), an RC filter (R_f, C_f), a three-phase grid (with grid impedance $R_s + jL_s$), and associated loads. A common DC connection connects these two voltage source converters with DC link voltage (V_{dc}). The converter that is linked to the PMSG is named VSC-1, while the converter that is connected to the grid is named VSC-2. The wind turbine-powered PMSG system is linked to the VSC-1, which assures maximum power point tracking (MPPT). VSC-2 bearing currents (i_{vsc}), is connected to a three-phase grid and a nonlinear load with load currents (i_L), which enhances grid power quality. To reduce the harmonics in the injected grid currents (i_s), a coupling inductor (Ls) is utilized. For the reduction of higher order switching harmonics, an RC ripple filter is utilized.

The PMSG-based wind energy conversion system has certain advantages over other types of configurations. It is explained as follows [6–8].

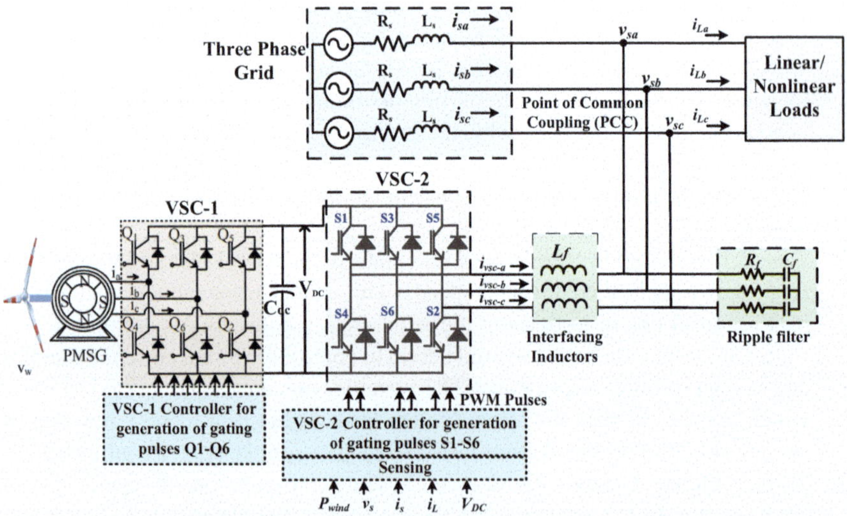

Fig. 5 Block diagram of the grid-connected *type-4* wind energy generation system

- This configuration does not require capacitor bank to supply reactive power as it consists of the permanent magnet in the electrical machine.
- This configuration has full-scale frequency support as back-to-back converters are connected between PMSG and the grid. The generator side converter is used to extract maximum power under variable wind speed. The grid side converter feeds the extracted power into the grid at system frequency.
- The PMSG-based system is direct drive, has slow rotation speed, does not have rotor current, and can be used without gearbox.

Therefore, this system is more popular and widely adopted in the wind energy conversion system. The integration of wind energy conversion system must follow the certain IEEE standards such as IEEE-519.

3 Modelling of Wind Energy Conversion Systems

The modelling plays a very important role for effective control of the permanent magnet synchronous generator [9, 10]. The detailed modelling of the system is divided into two sections, (1) wind turbine modelling, (2) modelling of permanent magnet synchronous generator. Each section is explained as follows.

A. Wind Turbine Modelling

For given wind speed, the kinetic power of wind is expressed as [9, 10],

$$P_{tur} = \frac{1}{2}\rho\pi r^2 v_w^3 \tag{1}$$

where, v_w, r and ρ represent wind speed, wind turbine rotor radius and air density, respectively. Usually, the wind turbine cannot extract all energy from this kinetic energy. The extracted wind turbine can be expressed as [9, 10],

$$P_{tur} = \frac{1}{2}\rho\pi r^2 v_w^3 C_p(\lambda, \beta) \tag{2}$$

where $C_p(\lambda, \beta)$ power coefficient, λ represents tip speed ratio, β represents pitch angle. For given wind turbine speed, λ is defined as

$$\lambda = \frac{\omega_{tur} r}{v_w} \tag{3}$$

where ω_{tur} is the rotational speed of the turbine rotor. The $C_p(\lambda, \beta)$ is defined as [9, 10],

$$C_p(\lambda, \beta) = C_1(C_2/\lambda_i - C_3\beta - C_4)e^{(C_5/\lambda_i)} + C_6\lambda \tag{4}$$

where

$$\frac{1}{\lambda_i} = \frac{1}{\lambda + 0.008\beta} - \frac{0.035}{\beta^3 + 1}. \tag{5}$$

The maximum value of C_p ($\approx$0.48) is determined for $\beta = 0$ and $\lambda = 8.1$. However, the theoretical maximum value of C_p is 0.593, which is known as Betz limit. The utmost wind power is extracted from the wind turbine when C_p and λ reach their optimum values.

B. Modelling of Permanent Magnet Synchronous Generator

The PMSG is modelled in synchronous rotating reference frame (e.g. *dq*-domain). The mathematical formulated as [9, 10],

$$v_{sd} = R_s i_{sd} + p\psi_{sd} - n_p \omega_{rm} \psi_{sq} \tag{6}$$

$$v_{sq} = R_s i_{sq} + p\psi_{sq} + n_p \omega_{rm} \psi_{sd} \tag{7}$$

$$\psi_{sd} = L_{sd} i_{sd} + \psi_{pm} \tag{8}$$

$$\psi_{sq} = L_{sq} i_{sq} \tag{9}$$

where n_p, R_s, L_{sd}, L_{sq}, v_{sd}, v_{sq}, i_{sd} and i_{sq} are the number of pole pairs, stator resistance, d-axis and q-axis inductances, voltages and currents, respectively; p represents derivation of term with respective to time. By considering $L_{sd}= L_{sq}= L_s$, the modelling equation of the PMSG is written as [9, 10],

$$\begin{bmatrix} v_{sd} \\ v_{sq} \end{bmatrix} = \begin{bmatrix} (R_s + pL_s) & -n_p L_s \omega_{rm} \\ n_p L_s \omega_{rm} & (R_s + pL_s) \end{bmatrix} \begin{bmatrix} i_{sd} \\ i_{sq} \end{bmatrix} + \begin{bmatrix} 0 \\ n_p \omega_{rm} \psi_{pm} \end{bmatrix}. \tag{10}$$

Assuming a stiff drive train, the motion equation is written as

$$\frac{d\omega_{rm}}{dt} = \frac{1}{J}(T_{tur} - T_e - T_d) \tag{11}$$

$$n_p \omega_{rm} = \omega_e \tag{12}$$

$$p\theta = \omega_{rm} \tag{13}$$

where θ is the instantaneous rotor angular position, $T_d = K_r \omega_{rm}$, K_r is the viscous friction coefficient of PMSG; T_e and P_e is electrical torque and electrical power, which are defined as [9, 10],

$$T_e = 1.5n_p\left(\psi_{sd}i_{sq} - \psi_{sq}i_{sd}\right) \tag{14}$$

$$P_e = 1.5n_p\left(v_{sd}i_{sd} + v_{sq}i_{sq}\right). \tag{15}$$

4 Controller Design for Wind Energy Conversion Systems

The control strategy for power quality improvement in the WECS is described in this section. The topology considered for this purpose is as shown in Fig. 5; however, the similar control strategy is applicable for the WECS topologies as in Figs. 1, 2 and 3. As seen in Fig. 5, two voltage source converters are being deployed, named as VSC-1 and VSC-2. The VSC-1 is responsible for extracting maximum power from the wind turbine, while the VSC-2 is responsible for the power quality improvement at the PCC. The block diagram for control of VSC-1 is shown in Fig. 6a, which involves feeding a reference angular velocity of the rotor of the PMSM through a maximum power point tracking (MPPT) algorithm. There are various MPPT algorithms available in the literature [11–13], such as perturb and observe, incremental conductance, etc. The reference angular velocity (ω_{ref}) thus obtained is then compared to the actual angular velocity (ω) of the rotor, which is estimated using a sensor-less control. The proportional-integral (PI) regulator makes sure that the actual angular velocity follows the reference signal, thereby, extracting the maximum power from the wind turbine at a given wind speed. The output of the PI regulator is eventually converted to the reference current signals (i_{abcref}) required as an output of the PMSG, at a given instant of wind power 'P'. The reference current signals are then compared to the actual current signals, and the appropriate gating signals are produced through a hysteresis current control. The whole process is illustrated in Fig. 6a [14].

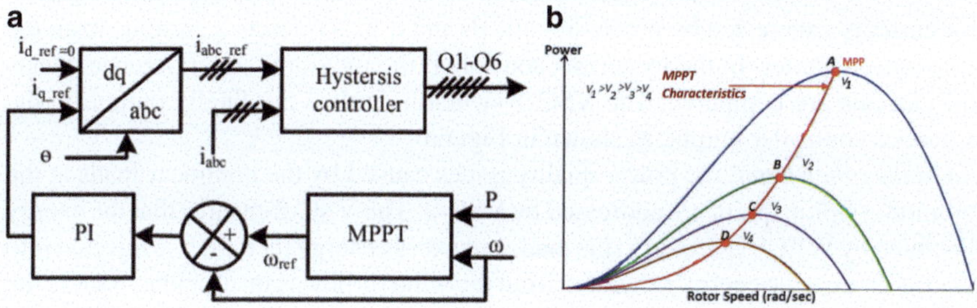

Fig. 6 **a** Block diagram of VSC-1 control, **b** Wind turbine power characteristics

4.1 Maximum Power Point Tracking Control

The characteristics of the turbine are nonlinear in nature. As a result, an MPPT approach is necessary to extract maximum power. The literature [10–13] has a full investigation and description of numerous MPPT algorithms. Because of its simplicity, the P&O (Perturb and Observe) approach is easy to execute. For WEGS, the dynamic responsiveness of the P&O approach is good under shifting wind speed circumstances. The operating point is randomly disrupted in one direction in this MPPT method, and the power is determined at the new operating point. The turbine characteristic, seen in Fig. 6b [15], may be separated into three portions, and the MPPT algorithm operates as follows.

(1) At MPP, $dP/d\omega = 0$; no change in speed of the generator.
(2) Left side of MPP, $dP/d\omega > 0$; the speed is increased by a fixed step ($\Delta\omega$) to track the MPP.
(3) Right side of MPP, $dP/d\omega < 0$; speed is decreased by $\Delta\omega$ to track the MPP.

The rotational speed (ω), rotor position 'θ', torque and power are estimated in this method and are fed to the MPPT controller. When the power generated by the wind generator grows, the operating point is disturbed in one way, and when the power generated by the wind generator falls, the operating point is agitated in the opposite direction. The MPPT algorithm's output calculates the speed of the reference generator. The algorithm's governing equation is as follows:

$$\omega_{ref}(k) = \omega_{ref}(k-1) + \Delta\omega\, if\ \mathrm{dP} > 0\,\&\ \mathrm{d}\omega > 0,\ \text{or}\ \mathrm{dP} < 0\,\&\,\mathrm{d}\omega < 0. \quad (16)$$

The reference generator speed (ω_{ref}) is generated by the MPPT controller. The speed of the reference generator is compared to the estimated speed (ω), and the speed error is sent into a proportional-integral (PI) controller. The output of the PI controller is the q-axis generator reference current (i_{q_ref}). To create a quick dynamic reaction, the current on the d-axis is set to zero. The three-phase reference generator currents are created by processing the d- and q-axis reference currents from the dq to abc converter. In the hysteresis controller, the measured and reference generator currents are compared. The VSC-1 switching pulses are finally output by the hysteresis controller output, as shown in Fig. 6a.

On the other hand, the power quality issues caused by the nonlinear loads at the common coupling point are addressed by VSC-2. The VSC-2 ensures that the energy generated by WECS is served to the load, and excess energy thereafter is injected into the grid. In case the energy generated from the wind turbine is insufficient to serve the load, the utility grid is responsible for fulfilling the load demand. The control strategy for VSC-2 is being elaborately described in the following subsections. There are several control strategies reported for VSC-2 and this chapter illustrates only prominent control schemes, which include mainly—fuzzy logic-based control, adaptive control and stochastic filter-based control. These are described as follows.

4.2 *Fuzzy Logic-Based Control for Power Quality Improvement*

The load connected at PCC are assumed to be highly nonlinear by nature. The initial objective of the power quality improvement is to filter out all the harmonic components in the load current. For this purpose, the fundamental power consuming part of the load current is precisely estimated. For precise estimation, the fuzzy logic tuned generalized integrator is deployed. In general, the transfer function of a generalized integrator can be written as [16],

$$F_1(s) = i_{LQ}(s)/i_L(s) = \alpha\omega_1^2/(s^2 + \alpha\omega_1 s + \omega_1^2) \tag{17}$$

where 'ω_1' is the system frequency (which is obtained from a frequency locked loop as will be described later) and 'α' is parameter deciding the bandwidth of the function, while 'i_L' represents the load current and 'i_{LQ}' represents the quadrature component of the load current. This transfer function can be modified, to provide improved DC-offset rejection capability from the input signal. The transfer functions corresponding to fundamental quadrature component (i_{LaQ}) and error signal (e_{La}) of the algorithm can be identified as

$$Q(s) = \frac{i_{Laq}(s)}{i_{La}(s)} = \frac{\alpha_1\omega^2 s}{s^3 + (\alpha_1 + \alpha_0)\omega s^2 + \omega^2 s + \alpha_0\omega^3} \tag{18}$$

$$E(s) = \frac{e_{La}(s)}{i_{La}(s)} = \frac{s(s^2 + \omega^2)}{s^3 + (\alpha_1 + \alpha_0)\omega s^2 + \omega^2 s + \alpha_0\omega^3} \tag{19}$$

In (18), 'α_1' is the I-SOGI characteristic parameter, $\alpha_1 = 2\xi$, and ξ is the under-damping factor, which is normally 0.707 [17]. Where 'α_0' is the characteristic value that determines DC offset rejection. From [18] it can be depicted that the connection between the characteristic parameters 'α_1' and 'α_0' for a stable system, which is provided as,

$$\alpha_0^3 + 3\alpha_1\alpha_0^2 + (3\alpha_1^2 + 9)\alpha_0 + \alpha_1^2 - 4.5\alpha_1 = 0 \tag{20}$$

As a result, the value of 'α_0' may be determined as 0.27. The frequency response analysis of the enhanced SOGI algorithm (i.e. frequency response corresponding to the transfer function, $i_{LaQ}(s)/i_{La}(s)$) is shown in Fig. 7a. Without a frequency locked loop, the 'central frequency' (equivalent to the dashed vertical line in Fig. 7a) is fixed and cannot be changed even when the system frequency changes, introducing inaccuracy in the fundamental amplitude calculation of load currents. The frequency locked loop (FLL) represented in Fig. 7b establishes an effective mechanism for altering this 'central frequency'.

Conventionally, the FLL uses either single or multiple proportional-integral controllers, however, in the FLL employed here, fuzzy logic-based precise frequency

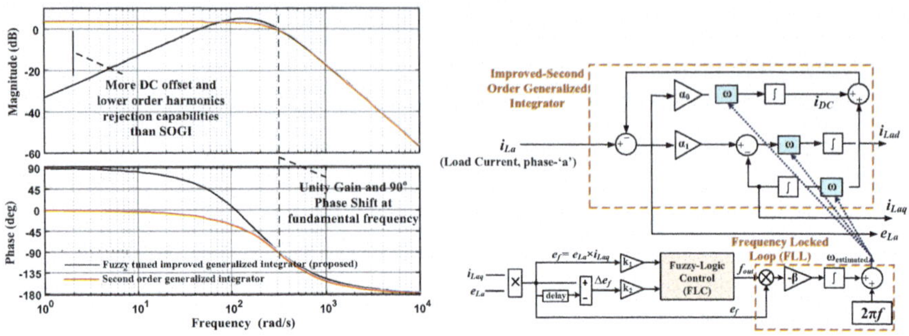

(a) (b)

Fig. 7 **a** Frequency response of the I-SOGI-based algorithm, **b** Fuzzy logic gain tuned ISOGI-based overall control approach

estimation is used in order to ensure good convergence of the algorithm. Thus, the overall schematic of the control to achieve the initial objective of obtaining fundamental power consuming component of load current is depicted in Fig. 7b. In Fig. 7b, 'i_{Lad}' represents the in-phase fundamental component of load current while 'i_{Laq}' represents the quadrature component. The FLL loop is highlighted in Fig. 7b, where the Takagi-type fuzzy logic controller is employed. The fuzzy logic tunes the gain 'β' of FLL for proper frequency estimation. It uses frequency error and its derivative for this purpose.

Equations (18) and (19) show the functions for $i_{LaQ}(s)$ and $e_{La}(s)$, respectively. The bode plot for the transfer functions E(s) and Q(s), with the FLL, tuned via fuzzy logic, is shown in Fig. 8a. The fuzzy logic membership functions are as shown in Fig. 8b. According to the frequency response analysis, the signals i_{LaQ} and e_{La} are in-phase as long as the input frequency is lower than the fundamental frequency. When the input frequency exceeds the fundamental frequency, the phase shift reaches 180 degrees. As a result, the frequency error (e_f) may be defined as the product of i_{LaQ} and e_{La}, as shown in Fig. 7b. Thus, the frequency error is positive when the input frequency is less than the fundamental frequency (as in Fig. 9), negative when the input frequency is larger than the fundamental frequency, and zero when the input frequency is greater than the fundamental frequency.

Therefore, by moving the SOGI central frequency until it matches the input frequency, an integral controller with negative gain could be utilized to provide the zero DC component of the frequency error. As a result, the FLL immediately detects the input frequency as, $\dot{\omega} = -\beta i_{LaQ} e_{La}$. This effectively eliminates frequency inaccuracy. The gain of integrator is appended by adding a fuzzy output gain (f_{out}) as

$$\dot{\omega} = -\beta f_{out} i_{LaQ} e_{La} \text{ or } \dot{\omega} = -\beta f_{out} e_f. \tag{21}$$

The enhanced SOGI with FLL based on fuzzy logic controller [19] guarantees strong convergence capabilities through nonlinear frequency estimation. The fuzzy

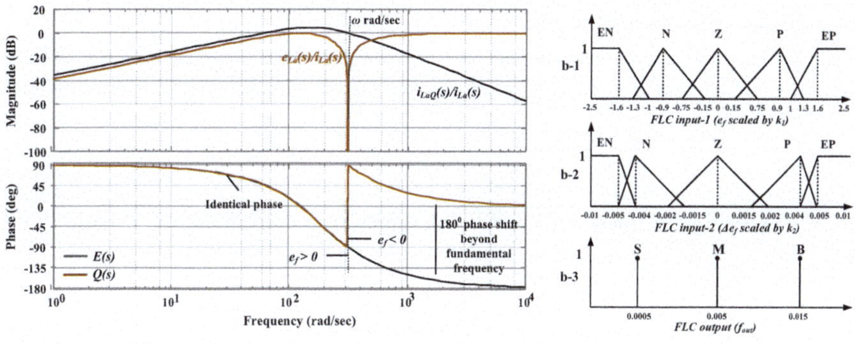

(a) (b)

Fig. 8 **a** Frequency response of the fuzzy logic tuned FLL-based I-SOGI, **b** Membership functions (b-1) Frequency error (b-2) Variation in frequency error (b-3) Internal gain

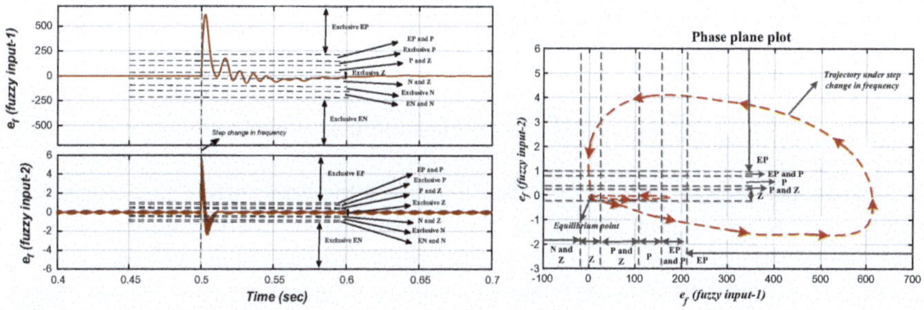

(a) (b)

Fig. 9 **a** Response of FLC inputs to step change in frequency, **b** Corresponding phase-plane plot

logic adjusts the gain 'f_{out}' that is fed into the frequency locked loop for accurate frequency estimation. It accomplishes this by utilizing frequency error and its derivative. The dotted arrows in Fig. 7b indicate that the block of I-SOGI inputs frequency from the estimated frequency ($\omega_{estimated}$) determined via FLC.

Linguistic variables are included in the membership functions for fuzzy logic input–output relations, such as EP (Extreme Positive), P (Positive), Z (Zero), N (Negative) and EN (Extreme Negative), S (Small), M (Medium), B (Big). Figure 8b depicts these functions, while the corresponding plots for step change in frequency are also reported in Fig. 9a along with its phase-plane plot in Fig. 9b. The membership function grade is found from

$$\psi(x) = 1 - |x - m|/0.5w \tag{22}$$

'm' represents the coordinate corresponding to the unity membership function. The variables 'x' and 'w' denote the value of the input variable and the width of the variable, respectively. In this paper, a fuzzy inference technique of the Takagi type is applied. Weights (w_i) are assigned to each rule based on the minimal action specified by

$$w_i = \min\{\psi_{error}(e_f), \psi_{error}(\Delta e_f)\}. \tag{23}$$

As a deflection of integral gain or fuzzy output (f_{out}), the weighted average of all rule outputs is as

$$f_{out} = \sum_{i=1}^{N} w_i z_i / \sum_{i=1}^{N} w_i. \tag{24}$$

In this case, 'z_i' is the coordinate relating to the corresponding output membership function, and 'N' is the number of fuzzy rules. Table 1 contains a list of the rules. The fuzzy inputs, namely frequency error and its derivative, are normalized using appropriate normalizing factors (k_1 and k_2). The fuzzy rule is written [19] as.

$$Rule_k : \text{If } ef(j) \text{ is } P_k \text{ and } \Delta ef(j) \text{ is } Q_k, \text{ then } f_{out} \text{ is } R_k$$

After successful extraction of precise 'i_{Laq}' from the load current 'i_{La}', the subsequent control procedure for generation of VSC-2 gating pulses is illustrated in Fig. 10, and is explained as follows. Line voltages are transformed to phase voltages (v_{sa}, v_{sb} and v_{sc}). The amplitude of the PCC terminal voltage is computed using phase voltages as follows:

$$V_t = \sqrt{(2/3)(v_{sa}^2 + v_{sb}^2 + v_{sc}^2)}. \tag{25}$$

The resulting terminal voltage is utilized to compute in-phase unit templates and quadrature unit templates,

$$\left[x_{pa}\ x_{pb}\ x_{pc}\right] = (1/V_t)\left[v_{sa}\ v_{sb}\ v_{sc}\right] \tag{26}$$

Table 1 Fuzzy logic rules

Δe_f	e_f				
	EN	**N**	**z**	**P**	**EP**
EN	f_{out} = **S**	f_{out} = **S**	f_{out} = **S**	f_{out} = **S**	f_{out} = **S**
N	f_{out} = **S**	f_{out} = **B**	f_{out} = **M**	f_{out} = **B**	f_{out} = **S**
Z	f_{out} = **S**	f_{out} = **M**	f_{out} = **B**	f_{out} = **M**	f_{out} = **S**
P	f_{out} = **S**	f_{out} = **B**	f_{out} = **M**	f_{out} = **B**	f_{out} = **S**
EP	f_{out} = **S**	f_{out} = **S**	f_{out} = **S**	f_{out} = **S**	f_{out} = **S**

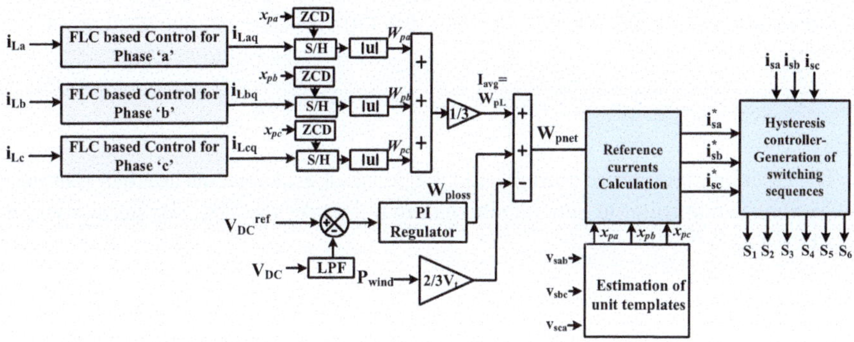

Fig. 10 Fuzzy logic-based control procedure for generation of VSC gating pulses of WECS

$$\begin{bmatrix} x_{qa} \\ x_{qb} \\ x_{qc} \end{bmatrix} = \frac{1}{2\sqrt{3}} \begin{bmatrix} 0 & -2 & 2 \\ 3 & 1 & -1 \\ -3 & 1 & -1 \end{bmatrix} \begin{bmatrix} x_{pa} \\ x_{pb} \\ x_{pc} \end{bmatrix} \tag{27}$$

An adaptive DC link voltage reference is created in proportion to the terminal voltage to decrease VSC converter switching losses and converter tripping under weak grid circumstances (V_t). The PI controller's [16] output represents the loss component (W_{ploss}), which may be recognized as

$$W_{ploss}(k) = k_p \times V_{DCe}(k) + k_i \int V_{DCe}(k)\, dk \tag{28}$$

where $V_{DCe} = V_{DC} - V_{DC}^{ref}$ (reference DC link voltage) and with k_p, k_i being PI controller gains.

Moreover, a realistic system experiences a continual change in wind speed. An extra feed-forward component derived using wind power facilitates the quick dynamic reaction under such continuous disturbances.

$$W_{wff} = (2P_{wind})/(3V_t). \tag{29}$$

Thus, the algorithm's quadrature fundamental component of load current (i_{Laq}) is routed through a sample and hold circuit activated by a zero crossing detector. The amplitude of the fundamental component of load currents (W_{pa}, W_{pb}, and W_{pc}) are determined, using zero crossing detectors. The W_{peq} is calculated by taking the average of the fundamental component amplitudes of three phases. The net active power consuming weight component (W_{pnet}) is calculated as follows:

$$W_{pnet} = W_{Lp} + W_{ploss} - W_{wff} \tag{30}$$

Following that, reference grid currents for three phases are analysed using an in-phase unit template,

$$i_{sa}^* = W_{pnet}.x_{pa},\ i_{sb}^* = W_{pnet}.x_{pb},\ i_{sc}^* = W_{pnet}.x_{pc} \tag{31}$$

The calculated reference currents and the detected grid currents are then sent into the hysteresis controller to generate the gating pulses for the VSC. Hysteresis width may be set as 0.02.

4.3 Adaptive Volterra Filter-Based Control for Power Quality Improvement

This section describes partially decoupled adaptive Volterra filter (PDAVF)-based control technique, which is used to estimate the reference currents, which are used to switch the VSC-2. To ensure that the detected grid currents adjust to the reference currents, the hysteresis current controller is used. This section depicts the major reference current component that is retrieved using PDAVF. Initially, the estimate of the fundamental-frequency component of the load current is made easier by evaluating the in-phase unit templates. For this reason, the in-phase templates are provided by

$$\left[x_{pa}\ x_{pb}\ x_{pc}\right]^T = (1/V_m)\left[v_{sa}\ v_{sb}\ v_{sc}\right]^T \tag{32}$$

where x_{pa}, x_{pb} and x_{pc} are the in-phase voltage templates derived from the PCC line voltages The peak terminal voltage value (V_m) in (32) may be calculated as follows:

$$V_m = \sqrt{(2/3)\left(v_{sa}^2 + v_{sb}^2 + v_{sc}^2\right)} \tag{33}$$

By increasing the unit templates with the respective weights, the active power component of load current is derived (w_{pa}, w_{pb} and w_{pc}). The weights suggested by the LMS method may be calculated for a certain phase (phase 'a') as [20, 21],

$$W_{pa}(k+1) = W_{pa}(k) + \mu_a e_{pa}(k) x_{pa}(k) \tag{34}$$

$$e_{pa}(k) = i_{La}(k) - W_{pa}(k) x_{pa}(k) \tag{35}$$

'μ' in (35) denotes the adaptation constant that governs the LMS algorithm's convergence rate and weight estimation accuracy. The weights for phases 'b' and 'c' are determined in the same way. The PDAVF-based LMS algorithm is produced by substituting the error (e_{pa}) in (34) with the equation [22],

$$e_{pa(j)}(k) = i_{La}(k) - \sum_{i=1}^{j} W^T_{pa(i)}(k)x_{pa(i)}(k) \tag{36}$$

The error ($e_{pa(j)}$) subscript ‘j’ in (36) denotes the error corresponding to the jth Volterra partial filter. It is possible to describe it in the following manner. Taking the order of the Volterra filter as ‘p,’ the set of jth order filter weights is denoted as $W_{pa(j)}$, and $e_{pa(j)}$ reflects the associated error for filter weight updation. The jth order filter weight is then adjusted in the same manner as the LMS, as follows:

$$W_{pa(j)}(k+1) = W_{pa(j)}(k) + \mu_{a(j)}e_{pa(j)}(k)x_{pa(j)}(k) \tag{37}$$

For phase-a, ‘$\mu_{a(j)}$’ denotes the adaptation constant corresponding to the jth order filter weight update. For the jth order PDAVF, $x_{pa(j)}$ for $1 \le j \le p$ is as

$$x_{pa(j)}(k) = \left[x_{pa}(k)\right]^{2j} \tag{38}$$

Equations (35), (36) and (37) may be used to implement the partially decoupled LMS-based Volterra filter. By modifying (36) as follows, the filter may be expanded to an LMF-based algorithm.

$$W_{pa(j)}(k+1) = W_{pa(j)}(k) + \mu_{a(j)}e^3_{pa(j)}(k)x_{pa(j)}(k) \tag{39}$$

Figure 11 depicts the construction of the pth order LMS-based Volterra filter and total weight extraction. The partly decoupled approach makes it easier to determine the boundaries on the adaption constants in order to show the algorithm's convergence. The algorithm's convergence in mean square can be ensured by placing a generic set of upper restrictions on the convergence coefficient. The top boundaries are denoted by [22],

$$0 < \mu_{a(j)} < \frac{2}{\text{Trace}\left[\boldsymbol{C}_{j,j}\right]} \tag{40}$$

$C_{j,j}$ are diagonal elements of the Volterra auto-correlation matrix. This matrix's components are

$$\boldsymbol{C}_{i,k} = E\{x_{pa(i)}x_{pa(k)}\} \tag{41}$$

The diagonal elements for the current purpose may be calculated using (38) and (41), as shown below.

$$\boldsymbol{C}_{j,j} = E\left\{\left[x_{pa}(k)\right]^{2j}\right\} \tag{42}$$

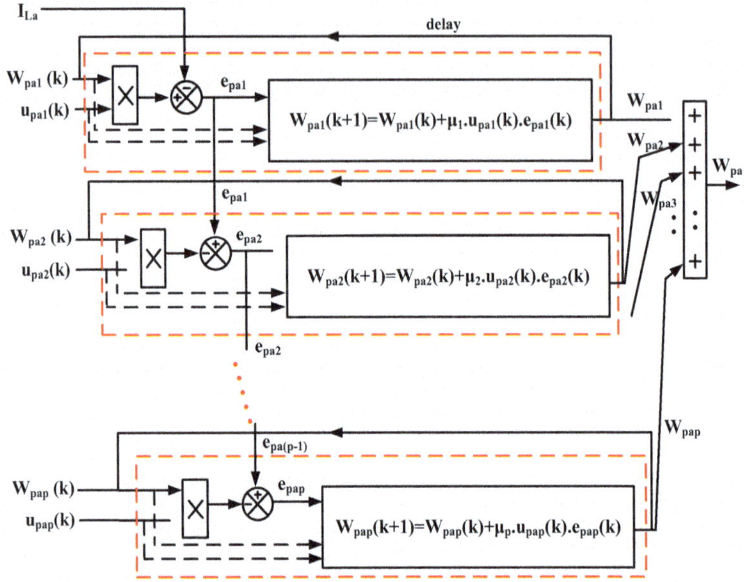

Fig. 11 Structure of pth order PDAVF

Higher order nonlinearities may exist for load current in the system. The higher order PDAVF can achieve low steady-state error while maintaining dynamic behaviour. However, because of the computational complexity of the work at hand, the fifth-order PDAVF is considered. Furthermore, by limited to a fifth-order filter, significant error reduction in weight estimation may be achieved. The upper bound of a(j) for ensuring algorithm convergence for fifth-order PDAVF may be found in (40) and (42) as 2.67, 2.91, 3.05, 3.14 and 3.21, respectively. After the successful extraction of weight component from the load current, the subsequent control procedure for generation of VSC-2 gating pulses is illustrated in Fig. 12, and is explained as follows.

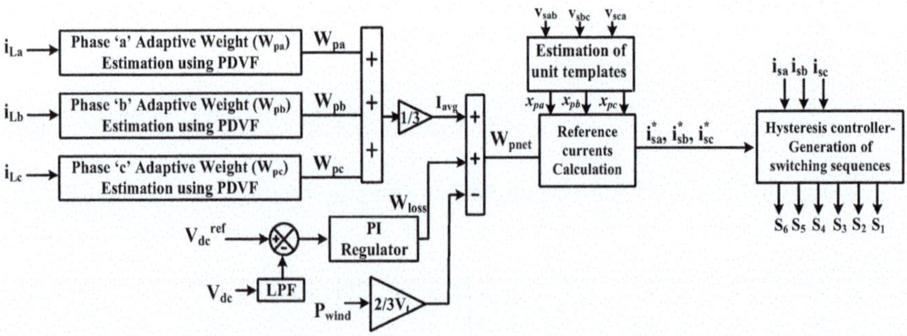

Fig. 12 Adaptive Volterra-based control overview for generation of VSC gating pulses of WECS

The error signal between the measured DC bus voltage and the reference voltage (V_{dce}) is then processed by the PI controller [23] to produce the loss component, as shown below.

$$W_{ploss}(k) = k_p \times V_{dce}(k) + k_i \int V_{dce}(k)dk \quad (43)$$

W_{ploss} reflects the loss component in this case. The controller's integral and proportional gains are denoted by k_i and k_p, respectively. In a realistic system, the fluctuation of wind speed is a continual disturbance. A faster dynamic reaction under such continuous disturbance is always desirable, which is reduced by the addition of an extra feed-forward component. This may be calculated for terminal voltage (V_t) as follows:

$$W_{wff} = (2P_w)/(3V_t) \quad (44)$$

This component guarantees that the instantaneous effect of wind power change is smoothly reflected in the active power component, and therefore the grid currents are smoothly adjusted to feed the necessary wind power.

The reference grid currents are then calculated using the net active power weight consuming component. The net active power component (W_{pnet}) comprises the total of previously analysed weight components; however, the *wff* component must be removed to get the desired impact. As a result, W_{pnet} is rated as follows:

$$W_{\text{pnet}} = W_{Lp} + W_{\text{ploss}} - W_{wff} \quad (45)$$

with $W_{Lp} = I_{\text{avg}} = (W_{pa} + W_{pb} + W_{pc})/3$. The grid reference currents are then computed as follows:

$$i^*_{sa} = W_{\text{pnet}}.x_{pa}, i^*_{sb} = W_{\text{pnet}}.x_{pb},\ i^*_{sc} = W_{\text{pnet}}.x_{pc} \quad (46)$$

The deviations between the calculated reference currents and the detected grid currents are sent into the hysteresis current controller to generate the switching pulses for the VSC.

4.4 Stochastic Kalman Filter-Based Control for Power Quality Improvement

This section describes the Kalman filter (KF)-based control for the VSC-2. Grid currents (i_{sa}, i_{sb}, i_{sc}), load currents (i_{La}, i_{Lb}, i_{Lc}), PCC voltages (v_{sa}, v_{sb}, v_{sc}) and the DC link voltage (V_{DC}) are required inputs to the VSC-2. The control method is used to assess the reference grid currents, and the hysteresis controller is utilized to

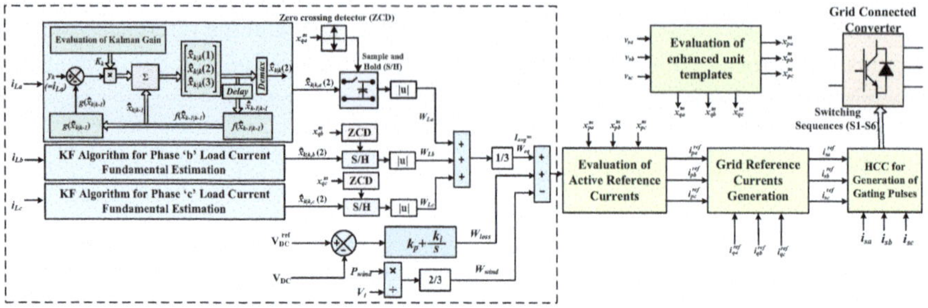

Fig. 13 Stochastic Kalman Filter-based control overview for generation of VSC gating pulses of WECS

guarantee that the detected grid currents follow the reference grid currents. Figure 13 depicts the overall control scheme for generating gating pulses for VSC-2, to obtain the desired power quality improvement features.

The KF is a mathematical process that works via a prediction and correction mechanism. Because it is a recursive structure, it may be executed in real time without saving observations or previous estimates. It offers information about the estimation's quality by including the variance of the estimation error in addition to the best estimate. The KF has the capacity to correct itself and update itself by taking advantage of adjustments between unknown dynamics. In essence, the KF algorithm predicts a new state based on its prior estimation by introducing a correction term proportionate to the expected inaccuracy. This mistake is statistically decreased in this manner.

Initially, the load current is denoted by [24],

$$i_{La} = \sum_{n} I_{Lan} \cos(\omega_n t + \phi_n) \tag{47}$$

where I_{pn}, ϕ_n, ω_n are the magnitude, phase and frequency of the nth harmonic. The load current at the kth sampling moment (y_k), is written as

$$y_k = I_{La1} \cos(k\omega_1 T_s + \phi_1) + \varepsilon_k \tag{48}$$

where 'T_s' is the sampling period and ε_k is Gaussian noise with a mean of zero and a variance of σ_v^2. Equation (48) is written in terms of anticipated fundamental load current ($\hat{y}_k$) as

$$y_k = \hat{y}_k + \varepsilon_k \tag{49}$$

The nonlinear recursive filter is used to estimate the size and frequency of the signal even in the presence of noise as [24],

$$\hat{x}_{k|k-1} = f(\hat{x}_{k-1|k-1}) \tag{50}$$

$$\hat{x}_{k|k} = \hat{x}_{k|k-1} + K_k(y_k - g(\hat{x}_{k|k-1})) \tag{51}$$

Thus, the predicted state vector $\hat{x}_{k|k}$ from (51) is applied to the system for the kth instant. In (51), the Kalman gain (K_k) is calculated as

$$K_k = \hat{P}_{k|k-1} \frac{\partial g^T}{\partial \hat{x}}\bigg|_{(k|k-1)} \left[\frac{\partial g}{\partial \hat{x}}\bigg|_{(k|k-1)} \hat{P}_{k|k-1} \frac{\partial g^T}{\partial \hat{x}}\bigg|_{(k|k-1)} + R_o \right]^{-1} \tag{52}$$

As shown below, the error covariance vector ($\hat{P}_{k|k-1}$) is assessed and updated.

$$\hat{P}_{k|k-1} = \frac{\partial f}{\partial \hat{x}}\bigg|_{k|k-1} \hat{P}_{k-1|k-1} \frac{\partial f^T}{\partial \hat{x}}\bigg|_{k|k-1} + Q_o \tag{53}$$

$$\hat{P}_{k|k} = \hat{P}_{k|k-1} - K_k \frac{\partial g}{\partial \hat{x}}\bigg|_{k|k-1} \hat{P}_{k|k-1} \tag{54}$$

$$\frac{\partial f}{\partial \hat{x}}\bigg|_{x_k=\hat{x}_{k|k-1}} = \begin{bmatrix} 1 & 0 & 0 \\ \hat{x}_{k|k-1}(2) & \hat{x}_{k|k-1}(1) & -1 \\ 0 & 1 & 0 \end{bmatrix}, \quad \frac{\partial g}{\partial \hat{x}}\bigg|_{x_k=\hat{x}_{k|k-1}} = \begin{bmatrix} \hat{x}_{k|k-1}(2) & \hat{x}_{k|k-1}(1) & -1 \end{bmatrix} \tag{55}$$

The covariance matrices associated with the process noise (w_k) and the measurement noise are denoted by Q_o and R_o, respectively. EKF's R_o and Q_o are commonly started at 10^{-4} and $10^{-4}I_{3\times3}$, respectively [25]. However, in (52), $R_k = R_o e^{\|y_k - g(\hat{x}_{k|k-1})\|^2}$ is evaluated instead of R_o, by giving the error covariance term the exponential term of error between measured load current and calculated basic load component. If the error between detected and estimated components is large, the exponential factor grows quicker, increasing the error covariance term and achieving error mitigation. Thus, at significant disturbances in the load current, the EKF achieves robust performance by including an exponential term.

Based on the estimated value at time point (k–1), the state prediction is utilized to compute the value at time point k. If the present dynamics are significantly different at time point k–1, the estimator is unable to track the situation and offer suitable weighting. As a result, it cannot forecast a value near to the true value at time point k. The estimated three-phase fundamental load currents are thus collected and subsequently processed to create the VSI switching pulses. As seen previously, the EKF control output is the estimated state vector $\hat{x}_{k|k}$ from (51), which is calculated for the kth instant. Thus, whenever the error at the (k–1)th instant is confined by the bounded limits and the solution is exponentially stable with finite constrained limits [26].

A sample and hold (S/H) control logic is then used to assess the active power component of load currents. S/H logics are used to pass the basic component calculated from (50)–(52). Because the fundamental component is in phase with the grid voltages, S/H logic triggering pulses are generated using quadrature unit templates and a zero crossing detector (ZCD). The in-phase (x_{pa}, x_{pb}, x_{pc}) and quadrature (x_{qa}, x_{qb}, x_{qc}) unit templates are generated for this purpose. A new approach of extracting unit templates is provided here to ensure good performance under grid voltage imbalances and harmonic distortions. The mathematical framework for the derivation of improved unit templates is as follows [27].

(1) The positive sequence phase voltages are first analysed. In the αβ reference frame, the instantaneous positive sequence voltage components are assessed as

$$v_{\alpha\beta}^{+} = [T_{\alpha\beta}][T_p][T_{\alpha\beta}]^{-1} v_{\alpha\beta} = \frac{1}{2}\begin{bmatrix} 1 & -e^{-j\pi/2} \\ e^{-j\pi/2} & 1 \end{bmatrix} v_{\alpha\beta} \quad (56)$$

$$[T_{\alpha\beta}] = \sqrt{\frac{2}{3}}\begin{bmatrix} 1 & -0.5 & -0.5 \\ 0 & 0.866 & -0.866 \end{bmatrix}, [T_p] = \frac{1}{3}\begin{bmatrix} 1 & a & a^2 \\ a^2 & 1 & a \\ a & a^2 & 1 \end{bmatrix}; a = e^{j2\pi/3} \quad (57)$$

Thus, using inverse Clarke's transformation, the positive sequence voltages '$v_{\alpha\beta}^{+}$' and associated values in the abc-domain are calculated from (56). Clarke's transformation is used to create the αβ-components of phase voltages (v_α, v_β) at first. These are passed through second-order generalized integrators to get the equivalent in-phase and quadrature components.

(2) The voltage amplitude is then calculated as

$$V_t = \sqrt{(2/3)\left(v_{sa}^{+2} + v_{sb}^{+2} + v_{sc}^{+2}\right)} \quad (58)$$

(3) The improved in-phase (x_{pa}^{m}, x_{pb}^{m}, x_{pc}^{m}) and quadrature (x_{qa}^{m}, x_{qb}^{m}, x_{qc}^{m}) unit templates are then evaluated as

$$\boldsymbol{x_p^m} = \left[x_{pa}^{m}\ x_{pb}^{m}\ x_{pc}^{m}\right]^T = (1/V_t)\left[v_{sa}^{+}\ v_{sb}^{+}\ v_{sc}^{+}\right]^T \quad (59)$$

$$\boldsymbol{x_q^m} = \begin{bmatrix} x_{qa}^{m} \\ x_{qb}^{m} \\ x_{qc}^{m} \end{bmatrix} = \frac{1}{2\sqrt{3}}\begin{bmatrix} 0 & -2 & 2 \\ 3 & 1 & -1 \\ -3 & 1 & -1 \end{bmatrix}\begin{bmatrix} x_{pa}^{m} \\ x_{pb}^{m} \\ x_{pc}^{m} \end{bmatrix} \quad (60)$$

Lastly, the outcome of the S/H logics is used to calculate the load current weights (W_{La}, W_{Lb}, W_{Lc}). The comparable active power element of load current is the load current weight average (W_{Leq}).

Active power is required by the VSC-2 to control the DC link voltage (V_{DC}) to its reference (V_{DC}^{ref}). During the change of the DC link voltage, the VSC consumes some power from the common coupling point, which is referred to as the loss component of active power (W_{loss}). It is used to adjust the DC link voltage in relation to the reference DC link voltage and is computed as [28],

$$W_{loss} = k_p\left(V_{DC}^{ref} - V_{DC}\right) + k_i \int \left(V_{DC}^{ref} - V_{DC}\right)dt \quad (61)$$

where 'k_p' and 'k_i' are the controller's proportional and integral gains, respectively. The DC link capacitor's reference voltage is fixed at 1.2 times the peak line voltage. Under such continual variation in the wind speed to the system, a quick dynamic reaction is necessary. The feed-forward term (W_{wind}) makes this possible by reducing transients in grid currents. The evaluation of the wind feed-forward term is as depicted in Sect. 4.3. Then after, the reference currents generation procedure as shown in Fig. 13, which goes with similar lines as depicted in Sects. 4.2 and 4.3. Finally, the difference in the sensed and the reference currents is fed to the hysteresis controller, which outputs the VSC gating pulses.

5 System Simulation Results and Discussion

The system simulation with aforementioned control strategies is performed in this section and the results with the controllers with different test cases are reported. The PMSG-based wind energy conversion system (as shown in Fig. 5) is modelled in MATLAB/Simulink platform using Simpower system library. The WECS performance results are put forward to give an idea to the reader, about the operation of the system, under manifold cases such as load unbalancing scenario, load balancing scenario, distribution static compensator (DSTATCOM) mode, wind energy conversion system mode, simultaneous change in load unbalancing and DSTATCOM mode, etc. The results illustrated in this section depict the fuzzy logic-based control, while the similar results could be observed for other control methods illustrated in Sect. 1.4. The system parameters are shown in the Appendix, while an illustrative example simulation file of the system with these parameters and complete control design in the MATLAB/Simulink environment (version: 2020b) is provided in this book.

(A) Case-I Performance of WECS at Load Balancing to Unbalancing Scenario

Figure 14 depicts the system behaviour under load unbalancing scenario. The system dynamics are presented using significant signals such as grid currents (i_s), load current (i_L), VSC currents (i_{vsc}), DC link voltage (V_{DC}), wind power (P_{wind}), reference grid currents (i_{ref}) generated by the controller and the average load current (i_{avg}) generated by the controller, along with grid active power (P_g) and reactive power (Q_g). The grid voltages are balanced and sinusoidal as illustrated in Fig. 14. The grid currents are balanced even with the presence of nonlinear loads at the coupling point. The converter provides compensating currents at the coupling point to make balanced grid currents. The converter also provides reactive power in case of reactive load loads at the coupling point to compel the in-phase operation of grid voltage and phase current. The nonlinear load is connected at the coupling point, which can be demonstrated with help of load current dynamics in Fig. 14. The system is stable, which can be illustrated using DC bus voltage dynamics in Fig. The DC bus voltage is regulated using proportional-integral controller. It is fed to the local loads, and surplus power is fed to the grid side network. The dynamics of grid power is shown in Fig. 14. The reactive power flow in the grid is null, which depicts the unity power factor operation

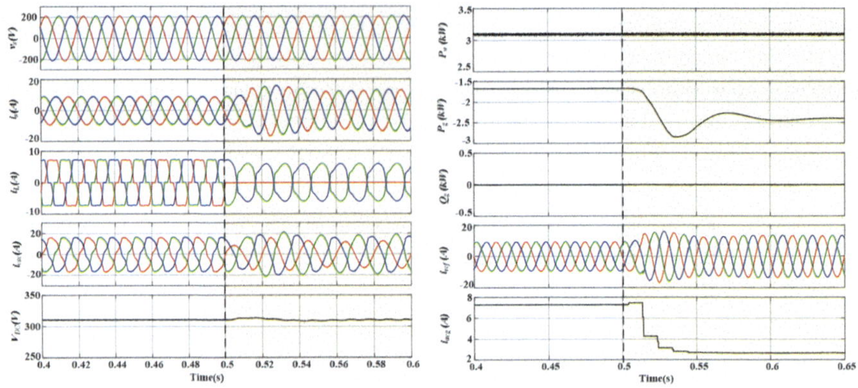

Fig. 14 System performance under load unbalancing

of the distribution grid. Furthermore, the dynamics of reference grid currents are depicted in Fig. 14. The corresponding dynamics of actual grid currents are shown in Fig. Henceforth, the grid currents adhere to the IEEE-519–2014 standard.

(B) Case-II Performance of WECS at Load Unbalancing to Balancing Scenario

Figure 15 shows the waveforms of v_s, i_s, i_L, i_{vsc}, V_{DC}, P_{wind}, P_g, Q_g i_{ref}, i_{avg}. As load balancing occurs at second, the typical value of the grid currents is decreased as power consumption in the load side network is increased. The grid currents follow the revised IEEE-519 standard. The dynamics of the load side network is depicted in Fig. The VSC provides compensating currents as illustrated in Fig. 15. As load balancing occurs, all three phases are connected to the network and power consumption is relatively increased, which impacts the power feed to the grid side network. However, the system maintained stable during this period and the corresponding dynamics are illustrated using the DC bus voltage in Fig. The active power dynamics of wind power and grid power are depicted in Fig. The change in grid power is easily observed in Fig. 15. One can easily observe that there is no reactive power flow in the grid side network, which illustrates the unity power factor operation of the grid. The performance of the system is satisfactory and follows the IEEE standard.

(C) Case-III Performance of WECS at Variation of Wind Speed from 11 m/s to 8 m/s

Figure 16 shows the system performance under variation of wind speed from 11 m/s to 8 m/s. As wind speed is decreased, the wind power generation is decreased, and it impacts the grid side dynamics. The grid currents are decreased as illustrated in Fig. 16. The typical magnitude of grid current is decreased as active power generation from the wind energy conversion system is decreased. The VSC compensate the load side network as depicted in Fig. 16. The load currents are not affected by the variation in the wind speed. The DC bus voltage is regulated as per reference command as shown in Fig. 16. The proportional-integral controller is used herein to regulate the DC bus voltage. The active power variation in wind energy conversion system is

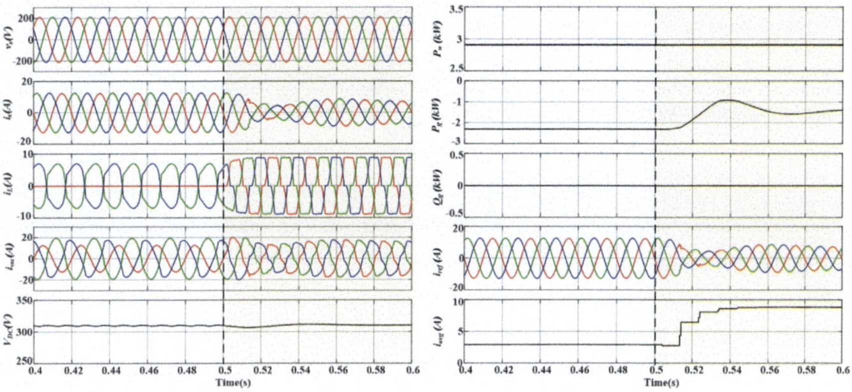

Fig. 15 System performance under load balancing

shown in Fig. The active and reactive power flow in the grid side network is depicted in Fig. 16. It validates the unity power factor operation of the distribution grid. The dynamics of the reference grid currents are also shown in Fig. 16. The tracking of the grid currents can be observed in Fig. 16.

(D) Case-IV Performance of WECS at Variation of Wind Speed from 8 m/s to 11 m/s

Figure 17 shows the system performance under variation of wind speed from 8 m/s to 11 m/s. The magnitude of grid currents is increased as the active power injection from the wind energy conversion system is relatively increased. The grid currents are balanced and strictly follow the IEEE-519 standard. The compensating currents are provided by the local VSC as illustrated in Fig. 17. There is no impact on the load side network as validated using the load current dynamics. The DC bus voltage is maintained as per reference value, and it is found stable as shown in Fig. 17. Figure 17 shows the waveforms of v_s, $i_{s,}$ i_L, $i_{vsc,}$ $V_{DC,}$ $P_{wind,}$ P_g, Q_g $i_{ref,}$ i_{avg}. The

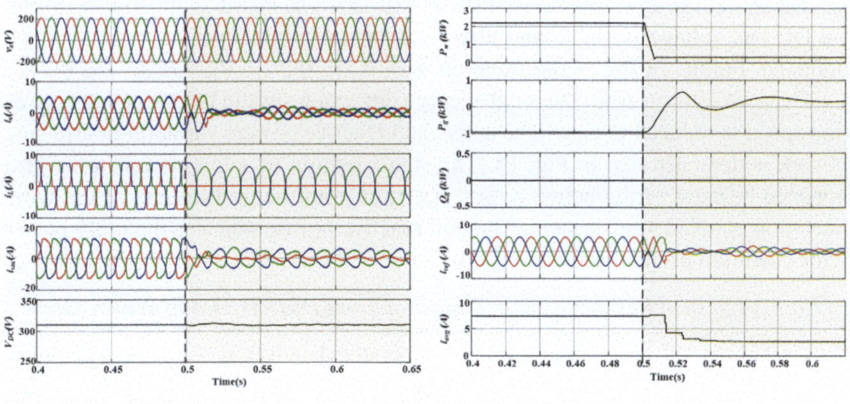

Fig. 16 System performance variation of wind speed from 11 m/s to 8 m/s

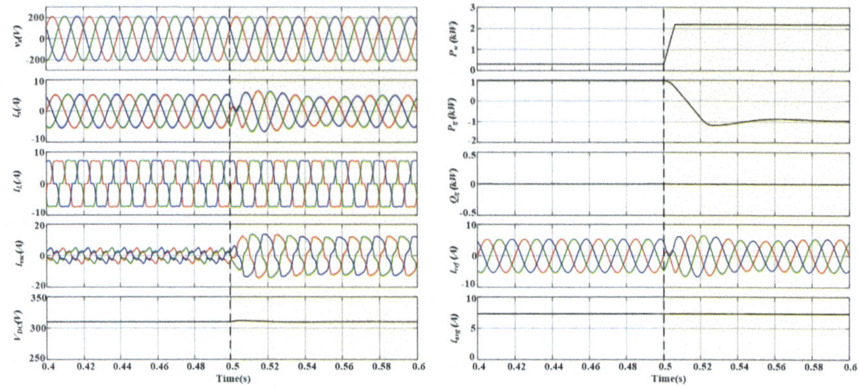

Fig. 17 System performance variation of wind speed from 8 m/s to 11 m/s

active power variation in wind energy system and grid side network is demonstrated in Fig. 17. As wind speed increased, the mechanical power input is increased, and it reflects the noticeable increment in electrical output power of PMSG-based wind energy conversion system. The reactive power flow in the grid side network is null, which illustrates the unity power factor operation.

(E) Case-V System Response while Transition from WECS Mode to DSTATCOM Mode

Figure 18 shows the waveforms of v_s, $i_{s,}$ i_L, $i_{vsc,}$ $V_{DC,}$ $P_{wind,}$ P_g, Q_g $i_{ref,}$ i_{avg}, enabling the capabilities of distribution static compensator (DSTATCOM) mode. As wind speed is quite low, then, the system behaves as distribution static compensator. The active power is provided to the load from the grid side network. The reactive power is provided from the local VSC to make unity power factor operation. The VSC only provides compensating current, not active power injection. The load side network is not affected as depicted in Fig. the phase reversal of the grid currents is observed in Fig. The system follows the revised IEEE-519–2014 standard as illustrated in Fig. The DC bus voltage is stable and effectively maintained as per reference voltage command. Figure 18 shows the waveforms of v_s, $i_{s,}$ i_L, $i_{vsc,}$ $V_{DC,}$ $P_{wind,}$ P_g, Q_g $i_{ref,}$ i_{avg}. The active power from the wind energy conversion system is lessened to zero as wind speed is reduced to zero. The active and reactive powers variation in grid side network is demonstrated in Fig. 18. The sign variation in active power in the grid happened because of the active power injection from the grid instead of the wind generation coupled VSC system. The null reactive power indicates the unity power factor operation of the distribution grid.

(F) Case-VI System Response while Transition from DSTATCOM to WECS Mode

As the wind speed is recovered to 10 m/s, the system behaves as wind energy conversion system. Figure 19 shows the waveforms of v_s, $i_{s,}$ i_L, $i_{vsc,}$ $V_{DC,}$ $P_{wind,}$ P_g, Q_g $i_{ref,}$ i_{avg}. As the wind energy conversion system is recovered, the system starts supplying

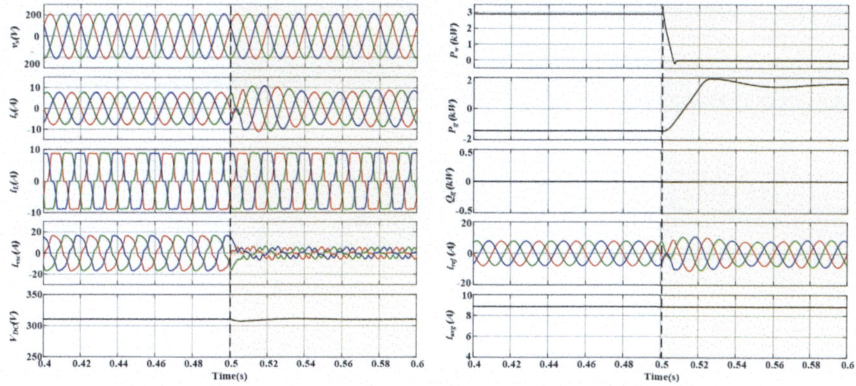

Fig. 18 System performance under transition from WECS mode to DSTATCOM mode

the active power to the coupling point. The phase reversal in the grid currents is observed as the system recovers to the WECS mode. The typical dynamics of VSC and load currents are illustrated in Fig. 19. The presented work follows the IEEE-519 standard. The DC bus voltage is found stable as illustrated in Fig. 19. The active and reactive power variation in grid and wind side network is also depicted. The power reversal in the grid side network is observed in Fig. 19.

(G) Case-VII System Response with DSTATCOM to WECS with Load Unbalancing Scenario

Figure 20 shows the system performance for WECS mode with load unbalancing scenario. As the simultaneous change of load unbalancing and wind speed occurs, the dynamics of system parameters are observed in Fig. 20. As load unbalancing occurs, the change of power in load side network is reflected in the grid side network. Additionally, the wind power is recovered and is also reflected in the grid side network.

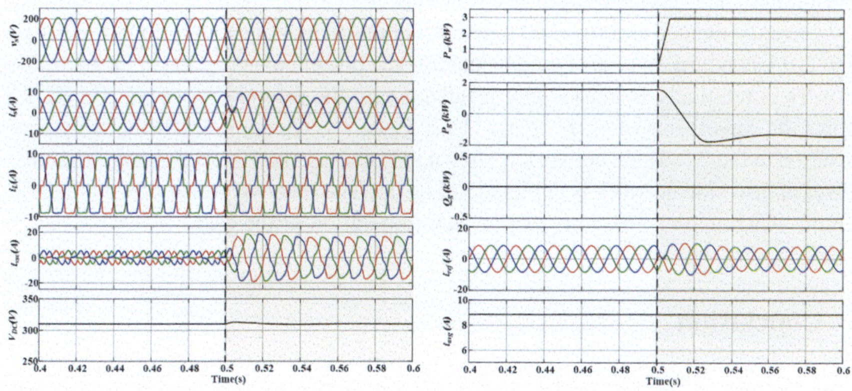

Fig. 19 System performance under transition from DSTATCOM mode to WECS mode

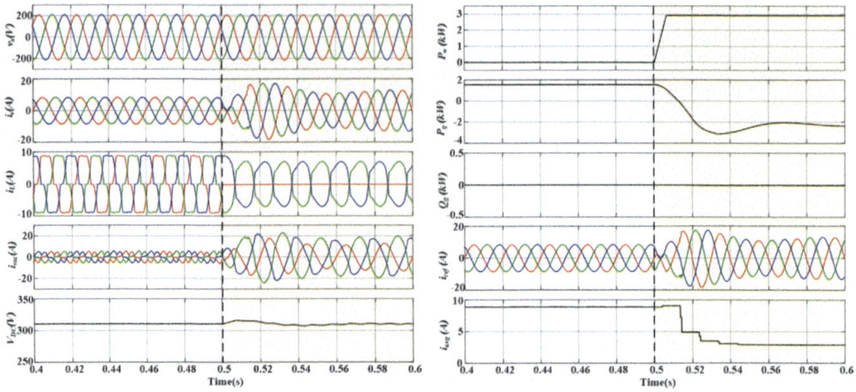

Fig. 20 System performance under mode transition and load unbalancing (1 phase load disconnected)

From Fig. 20, one can observe that the load currents are nonlinear and its THD is quite high, thereby, it violates the IEEE-519–2014 standard. The VSC currents provide compensating currents and make grid currents balanced. The grid currents are balanced and sinusoidal, which adheres to the IEEE-519–2014 standard. Fig. shows the waveforms of v_s, $i_{s,}$ i_L, $i_{vsc,}$ $V_{DC,}$ $P_{wind,}$ P_g, Q_g $i_{ref,}$ i_{avg}.. The change in active power of wind energy conversion system is shown in Fig. The active and reactive power flows in the grid side network are shown in Fig. 20. The reactive power is null in the system, which indicates the unity power factor operation of distribution grid.

(H) Case-VIII System Response with WECS to DSTATCOM with Load Balancing Scenario

Figure 21 shows the system performance for DSTATCOM mode with load balancing scenario. As wind speed is reduced beyond the threshold, the system behaves as distribution static compensator. The VSC provides compensation current to make grid current balanced and sinusoidal, which depicts that the system follows the IEEE-519–2014 standard. In case of load balancing, the power consumption in the load side network is relatively increased and it reflects into the grid side network. Figure 21 shows the waveforms of v_s, $i_{s,}$ i_L, $i_{vsc,}$ $V_{DC,}$ $P_{wind,}$ P_g, Q_g $i_{ref,}$ i_{avg}. The active power variation in the wind side network is shown in Fig. 21. Likewise, the change in active power of grid side network is depicted in Fig. 21. The reactive power flow in the grid is null, which depicts the unity power factor operation of the grid.

6 Conclusion

In this chapter, the detailed implementation of the wind energy conversion system is analysed. The detailed modelling of wind turbine and converters are explained for

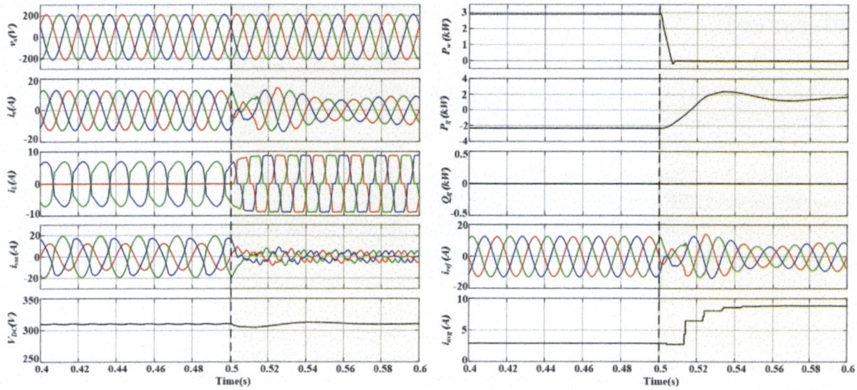

Fig. 21 System performance under mode transition and load unbalancing (1 phase extra load connected)

PMSG-based wind energy conversion system. Various control strategies for optimal operation under various operating scenarios are explicitly discussed in this chapter. The detailed block realization of MATLAB/Simulink model is also demonstrated. Simulation results demonstrate the satisfactory operation of the PMSG system with various operating scenarios such as load unbalancing, load balancing, DSTATCOM mode, WECS mode, variation of wind speed, etc. The grid currents are balanced and sinusoidal, which indicates that the system follows the revised IEEE-519–2014 standard.

Acknowledgements Vedantham Lakshmi Srinivas is with the department of electrical engineering at the Indian Institute of Technology Dhanbad and Nishant Kumar is with the department of electrical engineering at the Indian Institute of Technology Jodhpur (Corresponding author: Vedantham Lakshmi Srinivas). The authors express their deep gratitude to Professor Bhim Singh and Professor Sukumar Mishra from Indian Institute of Technology Delhi for their support. This work is done in partly funded by the SLE-Institute Fund of the Indian Institute of Technology Dhanbad under the project no. FRS(168)/2021-2022/EE.

Appendix

Wind power generation = 3 kW, Stator resistance = 0.4578 Ω; Stator inductance = 3.334 mH; $\lambda_m = 0.171$ Wb; Rated torque = 14 Nm; $V_{DC} = 700$ V, $C_{DC} = 6.5$ mF; $\xi = 0.7$; $k_{pv} = 0.01$, $k_{iv} = 0.01$, $k_{p\omega} = 0.1$, $k_{i\omega} = 0.01$; Interfacing inductor = 4 mH, Filter resistance = 10 Ω, Filter capacitance = 10 μF; Sampling time = 35 μs. An illustrative example simulation file of the wind energy conversion system with these parameters and complete control design in the MATLAB/Simulink environment (version: 2020b) is provided with this book.

References

1. Murray L (2021) Abuoyant future for offshore wind: Floating wind systems are on the cusp of a huge transformation as they transition from pilot projects to commercial-sized wind farms. Eng Technol 16(10):1–4
2. Rezkallah M, Chandra A, Singh B, Singh S (2019) Microgrid: configurations, control and applications. IEEE Trans Smart Grid 10(2):1290–1302
3. Blaabjerg F, Chen Z, Kjaer SB (2004) Power electronics as efficient interface in dispersed power generation systems. IEEE Trans Power Electron 19(5):1184–1194
4. Chen Z, Guerrero JM, Blaabjerg F (2009) A review of the state of the art of power electronics for wind turbines. IEEE Trans Power Electron 24(8):1859–1875
5. Key Highlights, 'Global Wind Energy Council'. https://gwec.net/global-wind-report-2021/
6. Ackermann T (2005) Wind power in power systems. John Wiley & Sons, Ltd., Chichester, West Sussex, England
7. Abad G, Lopez J, Rodrıguez M, Marroyo L, Iwanski G (2011) Doubly fed induction machine. John Wiley & Sons Inc., Hoboken, New Jersey
8. Xu D, Blaabjerg F, Chen W, Zhu N (2018) Advanced control of doubly fed induction generator for wind power systems. John Wiley & Sons Inc., Hoboken, New Jersey
9. Pradhan S, Singh B, Panigrahi BK, Murshid S (2019) A composite sliding mode controller for wind power extraction in remotely located solar PV–wind hybrid system. IEEE Trans Ind Electron 66(7):5321–5331
10. Joo YH, Antonysamy R, Ramasamy T, Lee SR (2022) Stable maximum power extraction and DC link voltage regulation for PMVG-based WECS. IEEE Trans Ind Electron Early Access. https://doi.org/10.1109/TIE.2022.3153813
11. Dalala ZM, Zahid ZU, Yu W, Cho Y, Lai J-S (2013) Design and analysis of an MPPT technique for small-scale wind energy conversion systems. IEEE Trans Energy Convers 28(3):756–767
12. Srinivas VL, Singh B, Mishra S (2020) Self-synchronizing VSM with seamless operation during unintentional islanding events. IEEE Trans Ind Inf 16(9):5680–5690
13. Esram T, Chapman PL (2007) Comparison of photovoltaic array maximum power point tracking techniques. IEEE Trans Energy Convers 22(2):439–449
14. Gupta TN, Murshid S, Singh B (2018) Single-phase grid interfaced WEGS using frequency adaptive notch filter for power quality improvement. In: 2018 2nd IEEE international conference on power electronics, intelligent control and energy systems (ICPEICES), Delhi, India, pp 630–635
15. Kumari S, Kushwaha V, Gupta TN (2018) A maximum power point tracking for a PMSG based variable speed wind energy conversion system. In: 2018 International conference on power energy, environment and intelligent control (PEEIC), Greater Noida, India, pp 789–794
16. Jain C, Singh B (2017) Solar energy used for grid connection: a detailed assessment including frequency response and algorithm comparisons for an energy conversion system. IEEE Ind Appl Mag 23(2):37–50
17. Xie M, Wen H, Zhu C, Yang Y (2017) DC offset rejection improvement in single-phase SOGI-PLL algorithms: methods review and experimental evaluation. IEEE Access 5:12810–12819
18. Karimi-Ghartemani M, Khajehoddin SA, Jain PK, Bakhshai A, Mojiri M (2012) Addressing DC component in PLL and notch filter algorithms. IEEE Trans Power Electron 27(1):78–86
19. Vedantham S, Kumar S, Singh B, Mishra S (2018) Fuzzy logic gain-tuned adaptive second-order GI-based multi-objective control for reliable operation of grid-interfaced photovoltaic system. IET Gener Transm Distrib 12(5):1153–1163
20. Singh B, Solanki J (2009) An implementation of an adaptive control algorithm for a three-phase shunt active filter. IEEE Trans Ind Elec 56(8):2811–2820
21. Srinivas VL, Kumar S, Singh B, Mishra S (2018) Partially decoupled adaptive filter based multifunctional three-phase GPV system. IEEE Trans Sustain Energy 9(1):311–320
22. Griffith D, Arce G (1997) Partially decoupled Volterra filters: formulation and LMS adaptation. IEEE Trans Signal Process 1485–1494

23. Vedantham S, Bhim S, Sukumar M (2022) Enhanced power quality PV inverter with leakage current suppression for three-phase SECS. IEEE Trans Ind Electron 69(6):5756–5767
24. Srinivas VL, Singh B, Mishra S (2019) Fault ride-through strategy for two-stage grid-connected photovoltaic system enabling load compensation capabilities. IEEE Trans Ind Electron 66(11):8913–8924
25. Anagnostou G, Pal BC (2018) Derivative-free Kalman filtering based approaches to dynamic state estimation for power systems with unknown inputs. IEEE Trans Power Syst 33(1):116–130
26. Rapp K (2004) Nonlinear estimation and control in the iron ore pelletizing process: an application and analysis of the extended Kalman filter. PhD thesis, Norwegian University of Technology, NTNU
27. Srinivas VL, Singh B, Mishra S (2020) Seamless mode transition technique for virtual synchronous generators and method thereof. IEEE Trans Ind Inf 16(8):5254–5266
28. Vedantham S, Bhim S, Kumar S, Sukumar M (2021) A normalized adaptive filter for enhanced optimal operation of grid-interfaced PV system. IEEE Trans Ind Appl 57(2):1715–1724. (Mar–Apr 2021).

Chapter 2
Design of Wind Farm Infrastructure for Reduction of Power Loss in Radial Distribution System

Aliva Routray, Khyati D. Mistry, and Sabha Raj Arya

Abstract Wind energy as DG from the wind farm is applied to the distribution system for loss minimization. In the wind farm, the downstream wind turbines experience a reduction in wind speed due to the upstream turbines ahead of them. Wind wake has a significant impact on wind turbine efficiency. Single and multi-wake effects are considered during the effective wind power calculation. A new irregular infrastructure of the wind farm is designed to have maximum wind power extraction with minimum wind wake loss and wind farm area. The hourly data distribution pattern of wind speed is non-parametric. This issue proceeds to the implementation of Kernel Density Estimation technique to the thesis work. Load Impedance Matrix method is implemented to the distribution system for load flow analysis. Daily load variation and air density are applied to the distribution system for an approach towards a practical system. IEEE-28 and IEEE-69 bus systems are studied for the active power loss reduction.

Keywords Air density · Distributed generation · Grey Wolf Optimization · Kernel Density Estimation · Radial distribution system · Wake effect · Wind farm layout

Nomenclature

RDS	Radial Distribution System
DG	Distribution Generation
LIM	Load Impedance Matrix
KDE	Kernel Density Estimation
MCS	Monte Carlo Simulation
PDF	Probability Density Function

A. Routray · K. D. Mistry · S. R. Arya (✉)
Department of Electrical Engineering, Sardar Vallabhbhai National Institute of Technology Surat, Gujarat 395007, India
e-mail: sabharaj79@gmail.com

A. K. Giri et al. (eds.), *Power Quality: Infrastructures and Control*,
Studies in Infrastructure and Control,
https://doi.org/10.1007/978-981-19-7956-9_2

RTS	Reliability Test System
GWO	Grey Wolf Optimization
WF	Wind Farm

1 Introduction

In the last few decades, the efficiency of the distribution system has increased because of the increasing research studies of electric distribution companies. The limitation on energy sources and the cost effective way of delivering power to the consumer end, motivate the system planners, engineers and researchers to bring greater ideas for better implementation in the distribution system. In other hand, it causes higher electric losses which is a matter of concern as 40% of the total power is consumed in the distribution lines. Hence, the loss reduction in the distribution system is important for electric distribution network companies. Several strategies such as network reconfiguration, capacitor bank allocation, optimum Distributed Generation placement etc. are adopted to improve the power loss reduction effect. However, the application of renewable energy as DG is appreciated for its several socio-economic benefits.

Wind energy is the most available and exploitable renewable source to generate electricity. Issues such as penetration of the DG in the network, environmental aspects and wind energy economy are some of the major challenges facing the power society. Few techniques for this green technology have already been adopted and introduced, and some new other applications are still in a demanding process [1]. Adaptation of energy distribution assets lowers greenhouse gas emissions by reducing the devastating effects of conventional power plants. The electrical power generated is fed to the grid or the customers in a small-scale application as distributed wind turbine. The optimum size and capacity of wind-based DG can improve system efficiency, support the grid, provide additional power during the peak load condition, power loss minimization, bus voltage improvement, etc. [2]. These advantages only take place when the wind turbines are properly sized and located at the optimum position in the system. On the other side, the wind turbine generator also implements few challenges and the new objective function changes the optimization problem to be more complex. The application of DG provides primary exposure of energy utilization to the public and the small-scale application at the distribution level gives benefits to reduce utility loads including the reduced energy losses. A lot of research studies have been materialized on the application of wind energy as DG. Besides optimum placement of DG, uncertainty in wind power generation, demand-side management using wind power, the effect of atmospheric parameters on wind, etc. are some of the challenging areas in wind energy application [3]. However, this work is primarily based on the optimum placement of wind energy as DG in the distribution system using optimization technique.

The uncertainty of wind data and the micro sitting are explained in [4]. The wind speed assessment and its respective power generation for several locations are mentioned in many works of literature. In [5], the potential of wind energy is addressed with wind speed data analysis. Wind energy also depends on many parameters like wind speed at a certain height, ambient temperature, air density, wind farm location, onshore or offshore wind farm layouts, type of wind turbine etc. Effective energy extraction faces few challenges in the fields of wind speed prediction or wind energy generation, system stability and reliability with integrated protection and measuring equipment.

The downstream turbines experience a reduction in wind speed at the wind turbine rotor since the incoming wind is scattered by the wind turbines ahead of the downstream turbines. This effect is also referred to as wind turbulence generated in turbines. Therefore, wind farm layout design is very essential for maximum power extraction with low wake loss. It has been observed from previous work that wind farm power generation is reduced by 10–15% due to the wake effect [6]. In addition to lower power production, the maintenance cost also gets higher in downstream wind turbines. This gives rise to the optimized placement of wind turbines to mitigate the wind wake effects. Another criterion for selecting wind farm layout is farm location that is based on the respective wind speed and wind direction. There are few assumptions that are needed to be considered before any wind farm layout design problem formulation. The assumptions can be the number of fixed turbines with turbine availability. The wind farm terrain should have a small variation in roughness. The wind turbines that are placed in the wind farm should have the same power curve function with the same power output at a fixed turbine height and wind speed.

In Ref. [7], a regular infrastructure of wind farm is used as the Distributed Generation in which the wind turbines are placed in a grid structured array. The capacity of the wind farm is 1000 MW (northeast China) and the number of rows and columns for the wind farm array is 6 and 11, respectively. The distance between the turbines is 400 m horizontally and 600 m vertically. In this arrangement, the wind turbines are equally distant from each other forming symmetric wind turbine placements. This is a conventional way of placing wind turbines. This grid structure design layout provides minimum lifetime cost and lowers the losses. However, this arrangement faces a significant power reduction when it comes to distribution level application. The wake losses are significant and need to be taken care of. Therefore, wind turbines are positioned as per the low wake loss in the wind farm. The wind farm design can be irregular and spatial with crossed wind turbine placements. In this arrangement, the downstream turbine is placed in between the upstream turbines in a cross manner, so that it does not stay immediately after the upstream turbine and the efficiency gets improved. Wind Farm developers should leave some space between turbines in a cross manner as per the National Institute of Wind Energy (NIWE) guidelines [8]. There are two types of wind farms as per location which is a concern and they are onshore and offshore wind farms. Offshore wind turbines are the latter concepts or advanced concepts of wind power generation from the wind above sea level, where the wind farms are those built on land serving as onshore wind farms. Unlike onshore wind farm, offshore wind farm considers some design parameters like depth of water, distance from the shore, tower size, pitch actuator, wind resource dependability etc. This makes the installation cost as well as the maintenance cost to be expensive. However, the potential from offshore wind farms is much higher due to faster and more constant winds which provide wind energy at a very stable rate [9].

Wind turbines are the medium to extract wind energy from the wind. Down the stream of wind flow, there is a decrease in the speed of the wind because of the wake produced by the wind turbine. This effect is known as the wake effect and it is mostly seen in downstream wind turbines [10]. The process is continued till the wake speed achieves the free wind speed. The wake effect significantly lowers wind power production and increases the cost of electricity. Many controlling schemes on the wake effect are designed to improve the power production of the wind farm by deflecting the wake from the downstream wind turbines. The wake loss also indirectly affects the wind speed prediction issue by making it anonymous. The key issue for a wind far operation is its variability in wind speed. The aerodynamic losses of several wind turbines in a large wind farm can be a matter of effective power generation concern due to the wake effects. There are many optimized wind farms that are designed to keep the wake effect below the threshold limit of wind wake loss. However, the wake properties also depend on the type of terrain or the surface roughness of the wind farm apart from the wind turbine position. Complex terrain shows different wake characteristics as compared to the plain and homogeneous terrain. This happens because of the local wind flow turbulence which depends on the wind farm location. Hence, the wind turbines that are installed in the complex terrain face more complexity due to the additional wake effect than flatlands [7]. In this work, the terrain is considered to be homogenous for the further analysis. Hence, the heights of all the wind turbines are same all over the wind farm and make the work analysis easier.

The Onshore and offshore layout designs of wind farm infrastructure are designed to minimize wake losses, thus maximizing wind turbine power generation. Depending on the maximum power generation, the wind farm design can be grid structured, irregular, or optimized [7]. In Refs. [7, 10], the regular wind farm is investigated with constant inter-turbine distance. The conventional wind turbine arrangements suffer maximum wake effect as compared to unusual spatial arrangements. In Refs. [11, 12], there are wind farms in which the wind turbines follow irregular placements.

The turbines that interact among themselves, produce turbine shadowing in the farm. In this mutual interaction, the downstream turbines can be fully shaded or partially shaded forming the common shadow area [12]. In many works of literature, the minimum distance that is kept between the turbines is five times the turbine rotor diameter to minimize or neglect the turbulence pressure exertion [7]. In Ref. [13], the minimum distance between two turbines is kept to four times the rotor diameter for the study purpose. For small scale application, wind farms are of relatively small land area and they can be used as Distributed Generation in a distribution system. Wind farms used in distribution systems are relatively small with a lower rating of energy.

Wake loss can be because of single or multiple wind turbines. The wake effect caused by a single wind turbine is known as the single wake effect and similarly, the wake effect caused by multiple wind turbines is known as the multiple wake effect [13]. The wake model is the frequently used tool for determining the wake loss in any wind farm infrastructure. Again, this wake model is analyzed based on kinematics, field and length of roughness classification. The kinematic model of analysis is widely used for its precision, simple mathematical formulation and faster calculation [13]. The arrangement of wind turbines in the wind farm is primarily responsible for the type of wake effect caused to any downstream turbine. The upstream turbine can shadow the downstream turbine completely if it is just ahead of the free speed wind turbine or the upstream turbine can partially shadow the downstream turbine causing a common overlapping area between the two turbines. This common area is known as the shadowing area. There is a mathematical expression to calculate the area to find the decrease in wind speed because of the wake effect [12]. Out of several wake models, an analytical or semi-empirical Jensen wake Model is implemented to find the wake speed. This model is simple and relatively provides a good degree of accuracy. This Jensen wake model is generally used for far wake analysis and uses momentum equation to model the deficit in downstream wind velocity [14]. This model assumes the linearly expanding wake behind the wind turbine and it carries a decrease in wind velocity as compared to the velocity of free wind speed. This model gives good results for offshore wind farms and far wake regions with efficient prediction [15].

The next assumption is the wind speed distribution at the wind farm site. Conventionally, wind speed follows Weibull distribution throughout the day. However, in the current days, the wind speed data pattern adopts different non-parametric distribution methods and artificial intelligence techniques to get implemented in many research works. To minimize the wake effect and improve the power extraction, the distance between two consecutive wind turbines should be at least four times the rotor diameter of the wind turbine.

To minimize the losses between the wind turbines under certain conditions, the distance between the wind turbines is increased. Wind turbines in farms are spaced with minimum three times of the rotor diameter to avoid too much wind turbulence around the wind turbine [16]. The distance between the turbines is maintained to nine times the rotor diameter [17], and ten times in [18]. In the same way, the distance is kept to more than five times the diameter of the turbine rotor in [19]. This ultimately ends up in providing reliable, cost-effective and efficient energy loads. The potential to improve the efficiency and prediction of power generation can be achieved through wake loss reduction [20].

In this work, a new infrastructure of the wind farm is designed to minimize the wake loss and increase the wind farm efficiency. Single wake and multi-wake effects are considered during wind power output calculation. The wake loss is calculated using graphical approach and a new formula is derived to determine this wake loss at any horizontal distance from the wind turbine. Wind speed is generated using Kernel Density Estimation followed by Monte Carlo Simulation because of its high uncertainty. The power loss is minimized with the wind power which is penetrated to the distribution system. The optimal location and capacity of DG are obtained with the help of the Grey Wolf Optimization technique. The hourly load profile and air density are considered for the power flow analysis. The IEEE-28 bus and IEEE-69 bus are studied for the loss minimization objective. The bus voltages and the power losses are discussed in the result section.

2 Parameters Affecting Wind Power Generation

This chapter deals with the loss minimization in the distribution system with available wind power as distributed generation placement and wake effects caused by the wind turbines. The basic model for wake effect analysis is considered for the study purpose. The wake analysis is carried out with the different spatial wind farm configuration for different wind farm layouts. A new wind farm layout is considered for optimum power generation. The power is fed to the distribution system and line loss is minimized from the hourly wind power generation. It also discusses the research gap and the identified research areas of the thesis organization.

2.1 Wind Farm Location

The wind farm location is a vital parameter on wind power productivity as it is wind speed and wind direction specific. In this work, wind speed is considered for the wind farm located in Surat, India. The direction of wind speed and its speed are extracted for over 9 years from 2010 to 2017 from a weather website [21]. From the wind speed data observation for the wind farm located in Surat, it can be found that maximum wind is coming from the west direction and the minimum wind is from the north direction.

2.2 Wind Speed Data Prediction

Wind speed is input data for wind power generation. The wind power as DG is penetrated to the distribution system for loss minimization. The new set of wind speeds is generated from wind speed prediction systems. The old historical wind speed data are used as training data being input to the predictive system. The hourly wind speed is fetched for an entire day to analyze the system at hourly intervals.

2.2.1 Kernel Density Estimation and Monte Carlo Simulation

The wind power prediction system is highly uncertain because of the conditional parameters of the environment. The wind speed conventionally follows the Weibull distribution pattern for a given day where the hourly wind speed doesn't follow any standard distribution function. By observing the statistical non-parametric properties of wind speed, a Kernel Density Estimation (KDE) method of probabilistic technique is used [22]. In this technique, wind speed is treated as the random variable and the smoothing parameter is helpful to get the Probability Density Function of the random function. The wind speeds are available at an interval of six hours in the meteorological website. Hence, the new data set of wind speed is generated in terms of four time intervals of 24 h. Eight years of previous wind speed data are collected for the wind farm location from 2010 to 2017. The histogram plot and new PDF for the wind speed data distribution during 7.00 AM to 12.00 PM are given in Fig. 1. In the similar way, the wind speeds can be generated for the other hours of the day. The PDFs are further used for obtaining the Cumulative Distribution Functions (CDF) which are mainly very helpful for Monte Carlo Simulation.

2.3 *Layout Designs of Wind Farm*

Small-scale wind energy is a micro-generation technique to supply power to the consumers at the distribution level. A number of wind turbines in a land form a wind farm. To meet their growing power demand, large-scale wind farms are installed at onshore and offshore. These wind farms make a significant contribution towards the power grids with consumer demand satisfaction. Wind farm infrastructures can be formed by having several options of wind turbine positions. The position of a

Fig. 1 PDF of wind speed

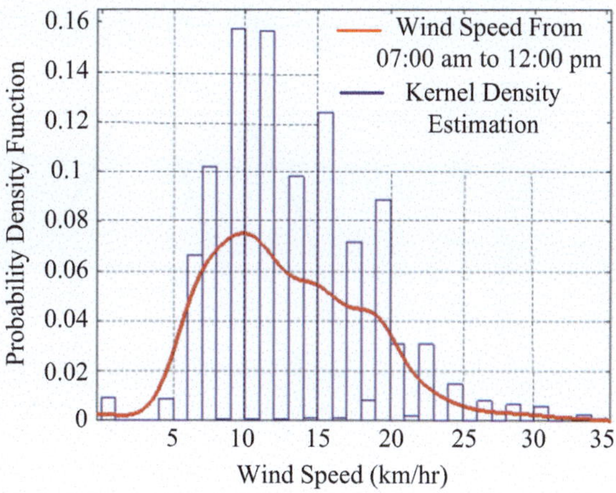

wind turbine decides the relative amount of wind power that can be extracted from the wind. Traditionally, grid structured layouts from wind farms are used as a DG. In this work, crossed structured wind farm layout is considered for the study. This layout is surveyed because of low wind energy loss with less wind wake effect which is described in the subsection below [23].

The wind farm under study consists of ten turbines. The wind turbine is having 100 kW of rated power and the specifications are considered from Ref. [24]. The power curve for the 100 kW wind turbine is shown in Fig. 2. The 100 kW wind turbine is used for the IEEE-28 and IEEE-69 bus systems. The wind power generated, thrust coefficient, Tip Speed Ratio (TSR_i) are mathematically expressed in Eqs. (1), (2), (3) and (4).

$$P_{wk} = \begin{Bmatrix} 0\ S_i < S_{ci}\ or\ S_i > S_{co} \\ P_{wki}\ S_{ci} \leq S_i < S_r \\ P_{wkr} S_r \leq S_i \leq S_{co} \end{Bmatrix} \tag{1}$$

$$P_{wki} = \frac{1}{2} \times rou \times F_0 \times C_p \times S_i^3 \tag{2}$$

$$TSR_i = \frac{2 \times \pi \times b \times N_r}{S_i} \tag{3}$$

$$C_{ti} = \frac{C_{pi}}{TSR_i} \tag{4}$$

where,

P_{wk} is wind power, P_{wki} is the wind power generated when the wind speed lies between the cut-in and rated wind speed, P_{wkr} is the rated wind power, S_i is the wind speed, S_{ci} is the cut-in wind speed, S_{co} is the cutout wind speed and S_r is the rated

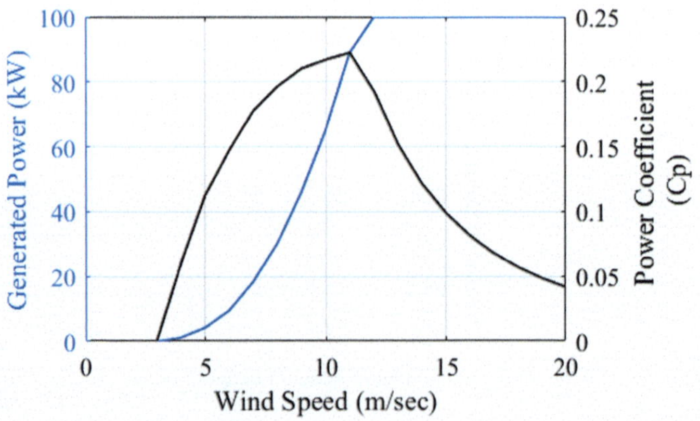

Fig. 2 Power curve for 100 kW wind turbine

wind speed of the specified wind turbine, C_{pi} is the power coefficient of the wind turbine, C_{ti} is the thrust coefficient, 'rou' is the air density, F_0 is the swept area by the wind turbine blade, b is the wind turbine rotor diameter and N_r is the frequency of revolutions.

This irregular wind farm layout has similar wind turbines placed spatially. To draw maximum wind power from the available wind, the consecutive horizontal distance between the wind turbines must be taken care of. The more the inter-turbine distance, the more is the expansion in the land area of the wind farm. Hence, minding the economy, the varying horizontal distance between the turbines is limited to four times the rotor diameter length in meter. In this work, the wind speed of downstream turbines is studied for the changing distance between consecutive turbines in the wind farm. The changing distance is in terms of rotor diameter of the wind turbine (b) and the wind farm is examined with a horizontal gap of b, $2 \times b$, $3 \times b$ and $4 \times b$, respectively. When the interval is b, the wind farm is named as WF-I. Similarly, for $2 \times b$, $3 \times b$ and $4 \times b$ intervals, the corresponding wind farms are represented as WF-II, WF-III and WF-IV. The graphical representation of WF-I is given in Fig. 3a. Similarly, the graphical representation can be given for WF-II, WF-III and WF-IV, respectively.

These four wind farms are experimented based on wind wake analysis and hence the wind speed is determined for the best case. The upstream wind turbines get the incoming wind speed where the downstream wind turbines don't get the free wind speed as the wind is scattered by the upstream wind turbines. Hence, the wake losses are more when the distance between the wind turbines is less. After analyzing the wind speeds of individual wind turbines for the respective infrastructures of wind

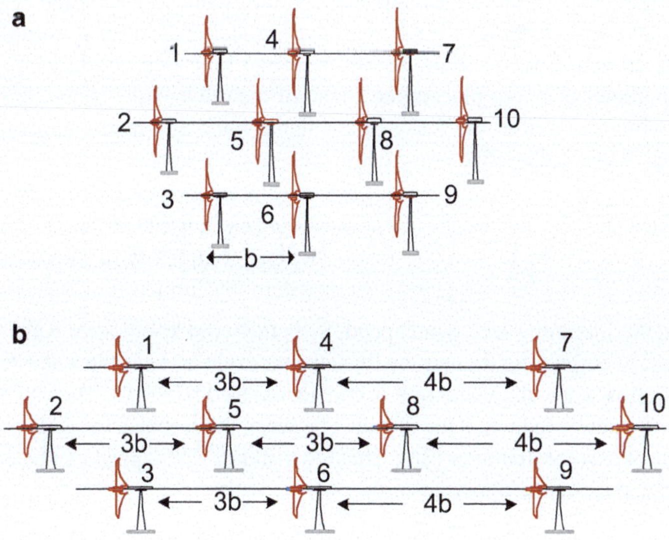

Fig. 3 **a** Wind Farm-I infrastructure with horizontal distance of b. **b** Wind Farm-V infrastructure with horizontal distance of 3b and 4b

farms, a new wind farm is designed. This new farm layout is termed as WF-V with the corresponding consecutive turbine gaps of 3 × b and 4 × b, respectively, and it is shown in Fig. 3b.

2.4 *Wind Wake Analysis*

Wind energy is extracted from wind speed in a wind farm. The wind turbine generator helps in converting wind energy into electricity. The wind energy that leaves the wind turbine always has a lower content of energy than the upstream wind energy. Therefore, the wind turbines behind the upstream turbines face a reduction in wind speed and turbulence. This is known as the wake effect. In this work, an irregular wind farm is considered because of its low wake loss. To analyze more about wake loss, different wind farms are considered for study purposes. When the wake effect happens to wind turbine due to one upstream turbine, it is called a single wake effect. Similarly, when the wind turbine is getting affected by multiple upstream wind turbines in the wind farm, it is known as multi-wake effect. In this work, both the effects are examined and the results are obtained for all the wind turbines as given in this section.

2.4.1 Single Wake Model

In this section, one downstream turbine is analyzed with a single wake model. A simple Jensen Wake model is implemented in this work to obtain the wind speed of downstream wind turbines [25]. This Jensen model is used for its simplicity and less computational time. To demonstrate the wake loss, two wind turbines having the rotor diameter of b, are placed one ahead of the other with a separable distance of h. The wind speed after the wake loss is determined and mathematically can be expressed as given in Eq. (5).

$$S_0\left[\left(1-\left(1-\sqrt{1-C_{ti}}\right)\left(\frac{b}{b_h}\right)^2\right)\right]=S_1 \tag{5}$$

Here, S_0 is the incoming free wind speed, S_1 is the wind speed after wake loss, h is the distance, b_h is the rotor diameter with a distance of h and ε is the wake decreasing constant. In this work, the value of ε is considered as 0.075 [12]. The wake effect is assumed to be neglected when the distance between the turbines exceeds eight times the wind turbine rotor diameter [26]. The downstream wind speed can be calculated at any distance using the above formula.

2.4.2 Multiple Wake Model

In a wind farm, as the wind wake spreads gradually from one wind turbine to another wind turbines, it interacts with the swept area of each turbine. This common area between two turbines is known as the shadowed area. It can be completely shadowed, partially shadowed or not shadowed at all. This wake analysis involves the wake effect of all the downstream wind turbines and collectively presents a multi-wake analysis. A mathematical formula is expressed by combining all the single and multi-wake effects of one wind turbine into a single effect [18]. The wind speed at wind turbine 'i' (WTi) is assumed to be undisturbed free wind speed as S_0. The wind turbine 'j' (WTj) is affected by the upstream wind turbines. The upstream wind turbines can be directly in front of it with a single wake effect or at some angular position causing shadowed area. The vertical distance between these two turbines (WTi and WTj) is 'h' as mentioned before. As this vertical distance increases the wake radius also gets bigger. The shadowed area between two turbines is calculated and is given in Eq. (6). The relation between the shadowed area and vertical distance between the turbines is obtained and is given in Table 1. It shows the changing shadow area with a variation of distance between the turbines from 0.5 × b to 8 × b.

$$F_{\text{shadow, i}} = (b_0)^2 \times COS^{-1}\left(\frac{C_{ij}^2 + b_0^2 - a_i(h_{ij})^2}{2 \times C_{ij} \times b_0}\right) + (a_i(h_{ij}))^2 \times COS^{-1}\left(\frac{C_{ij}^2 + a_i(h_{ij})^2 - b_0^2}{2 \times C_{ij} \times b_0}\right)$$
$$-\frac{1}{2}\sqrt{(-C_{ij} + b_0 + a_i(h_{ij})) \times (C_{ij} + b_0 - a_i(h_{ij})) \times (C_{ij} - b_0 + a_i(h_{ij})) \times (C_{ij} + b_0 + a_i(h_{ij}))} \quad (6)$$

Here, b_0 is the radius of the wind turbine i,e turbine diameter (b)/2, C_{ij} is the vertical distance between the rows of the wind turbines and considered to be a constant as b meters, $a_i(h_{ij})$ is the wake radius which is a function of the horizontal distance between the turbines. The linear as well as the nonlinear relationship between the wake radius and shadowed area with horizontal distance can be observed from Fig. 4.

In this work, the wake analysis is analyzed for different wind farm layouts. Hence, the shadow area and the wake radius are found to be unique for individual cases. The wind speed is determined after finding the shadowed area and wake radius using Eq. (7) [13].

$$S_j(h_{ij}) = S_i\left[1 - \sum_{i=1}^{n}\left(1 - \sqrt{1 - C_{ti}}\right)\left(\frac{b}{b_{h_i}(h_{ij})}\right)^2\left(\frac{F_{shadow,i}}{F_0}\right)\right] \quad (7)$$

Here, n is the total number of downstream wind turbines present in the wind farm, 'i' is any wind turbine being affected by wakes and b_{hi} is the wake rotor diameter affecting the ith wind turbine. The wind turbines (WT1,WT2 and WT3) have same incoming wind speed for being the upstream wind turbines in the wind farm. The wind speeds for the above turbines are generated and is mentioned in Table 2.

Table 1 Wake effect parameter calculations

Sl.No	Horizontal distance (h_{ij})	Single wake vertical distance (b_h)	Wake radius (a_i)	Shadowed area ($F_{shadow,i}$)
1	0.5b	1.075b	0.5625b	$0.0150b^2$
2	1b	1.15b	0.625b	$0.0431b^2$
3	1.5b	1.225b	0.6875b	$0.0802b^2$
4	2b	1.3b	0.75b	$0.1244b^2$
5	2.5b	1.375b	0.8125b	$0.1745b^2$
6	3b	1.45b	0.875b	$0.2297b^2$
7	3.5b	1.525b	0.9375b	$0.2887b^2$
8	4b	1.6b	1b	$0.3508b^2$
9	4.5b	1.675b	1.0625b	$0.4147b^2$
10	5b	1.75b	1.125b	$0.4795b^2$
11	5.5b	1.825b	1.1875b	$0.5438b^2$
12	6b	1.9b	1.25b	$0.6062b^2$
13	6.5b	1.975b	1.3125b	$0.6649b^2$
14	7b	2.05b	1.375b	$0.7176b^2$
15	7.5b	2.125b	1.4375b	$0.7607b^2$
16	8b	2.25b	1.5b	$0.7854b^2$

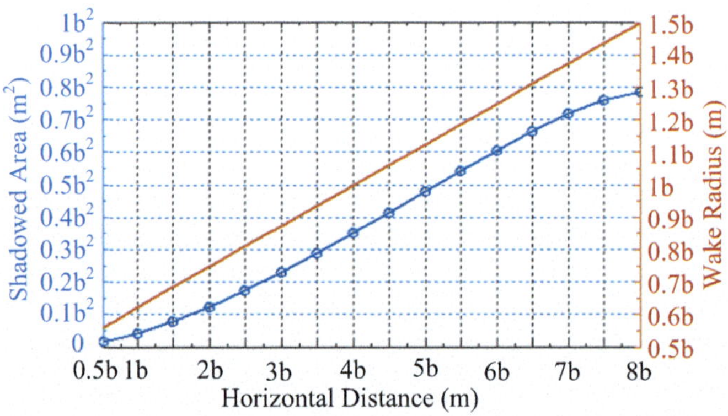

Fig. 4 Variation of wake radius and shadow area with horizontal distance

The wind speeds of all the wind turbines are determined with single and multi-wake effect for separate wind farms and the results are given in Tables 3, 4, 5, 6 and 7, respectively. Table 3 gives the wind speeds for all the wind turbines present in the WF-I layout. Table 4 represents the wind speeds at individual wind turbines for WF-II layout. Similarly, Tables 5 and 6 represent the wind speeds of turbines for WF-III and WF-IV layouts, respectively. The wind speeds of different wind turbines of the WF-V layout are mentioned in Table 7.

Table 2 Wind speed for upstream wind turbines (WT1,WT2 and WT3) (m/s)

Time (Hours)	*0100*	*0200*	*0300*	*0400*	*0500*	*0600*
Wind speed (m/s)	5.3	4.8	3.6	3.3	2.7	4.2
Time (Hours)	*0700*	*0800*	*0900*	*1000*	*1100*	*1200*
Wind speed (m/s)	3	2.5	4.5	5	6.4	5.5
Time (Hours)	*1300*	*1400*	*1500*	*1600*	*1700*	*1800*
Wind speed (m/s)	4.7	3.8	4.2	5.3	4.7	3.6
Time (Hours)	*19 × 00*	*2000*	*2100*	*2200*	*2300*	*0000*
Wind speed (m/s)	2.3	3.3	4.1	5	5.8	6.4

Table 3 Wind turbine speed for layout WF-I (m/s)

Time (Hours)	WT4 WT6	WT5	WT7 WT9	WT8	WT10
0100	5.2341	5.2382	5.1785	5.1811	5.1254
0200	4.746	4.7493	4.7005	4.7026	4.6569
0300	3.5699	3.5717	3.5444	3.5456	3.5202
0400	3.2747	3.2762	3.2533	3.2543	3.2329
0500	2.6803	2.6815	2.6636	2.6644	2.6477
0600	4.1597	4.1621	4.1257	4.1272	4.0931
0700	2.9756	2.9771	2.9551	2.956	2.9354
0800	2.4831	2.4841	2.4688	2.4695	2.4552
0900	4.4526	4.4555	4.4125	4.4144	4.3743
1000	4.9414	4.945	4.892	4.8943	4.8447
1100	6.3025	6.3085	6.2203	6.2242	6.1418
1200	5.4281	5.4325	5.3675	5.3703	5.3095
1300	4.6482	4.6514	4.6046	4.6066	4.5628
1400	3.767	3.769	3.7392	3.7405	3.7126
1500	4.1597	4.1621	4.1257	4.1272	4.0931
1600	5.2341	5.2382	5.1785	5.1811	5.1254
1700	4.6482	4.6514	4.6046	4.6066	4.5628
1800	3.5699	3.5717	3.5444	3.5456	3.5202
1900	2.2857	2.2866	2.2736	2.2742	2.2621
2000	3.2747	3.2762	3.2533	3.2543	3.2329
2100	4.0616	4.0639	4.0292	4.0307	3.9982
2200	4.9414	4.945	4.892	4.8943	4.8447
2300	5.72	5.7249	5.6526	5.6557	5.5881
0000	6.3025	6.3085	6.2203	6.2242	6.1418

Table 4 Wind turbine speed for layout WF-II (m/s)

Time (Hours)	WT4 WT6	WT5	WT7 WT9	WT8	WT10
0100	5.2395	5.2471	5.1933	5.1947	5.1418
0200	4.7504	4.7567	4.7125	4.7137	4.6703
0300	3.5723	3.5758	3.5512	3.5518	3.5276
0400	3.2767	3.2797	3.259	3.2596	3.2392
0500	2.6819	2.6842	2.668	2.6685	2.6526
0600	4.163	4.1676	4.1347	4.1356	4.1031
0700	2.9776	2.9804	2.9605	2.9611	2.9415
0800	2.4845	2.4864	2.4726	2.473	2.4594
0900	4.4564	4.4619	4.4231	4.4242	4.3861
1000	4.9462	4.953	4.905	4.9063	4.8592
1100	6.3105	6.3218	6.2421	6.2443	6.1659
1200	5.434	5.4423	5.3835	5.3851	5.3274
1300	4.6524	4.6585	4.6161	4.6173	4.5757
1400	3.7697	3.7735	3.7465	3.7473	3.7208
1500	4.163	4.1676	4.1347	4.1356	4.1031
1600	5.2395	5.2471	5.1933	5.1947	5.1418
1700	4.6524	4.6585	4.6161	4.6173	4.5757
1800	3.5723	3.5758	3.5512	3.5518	3.5276
1900	2.2869	2.2885	2.2768	2.2771	2.2656
2000	3.2767	3.2797	3.259	3.2596	3.2392
2100	4.0647	4.0692	4.0377	4.0386	4.0077
2200	4.9462	4.953	4.905	4.9063	4.8592
2300	5.7265	5.7358	5.6704	5.6722	5.6079
0000	6.3105	6.3218	6.2421	6.2443	6.1659

The wind speeds are determined for a complete day. One upstream wind speed of 5.3 m/s at 1.00 AM is considered for the study purpose. As per the farm layout configuration, WT-1, WT-2 and WT-3 are the three upstream wind turbines without any wake losses. From the results, it can be noticed that the downstream turbine that involves more wakes from other turbines ahead of it, has the lowest wind speed to produce wind power.

During the analysis of the wind speed of all the wind turbines for WF-I, WF-II, WF-III and WF-IV layouts, the wind speeds of downstream wind turbines are examined. The wind speeds after the wake effect are almost same or the difference is very low for WT-4, WT-5,WT-6 and WT-8 turbines considering four farm layouts and can be observed from Tables 3, 4, 5 and 6. But, the wind turbines WT-7, WT-9 and WT-10 face more wake loss than other turbines. The wind speed difference is more

Table 5 Wind turbine speed for layout WF-III (m/s)

Time (Hours)	WT4 WT6	WT5	WT7 WT9	WT8	WT10
0100	5.2426	5.2521	5.2041	5.2009	5.1672
0200	4.7529	4.7607	4.7214	4.7187	4.6912
0300	3.5737	3.5781	3.5561	3.5547	3.5393
0400	3.2779	3.2816	3.2632	3.2619	3.249
0500	2.6828	2.6856	2.6713	2.6703	2.6602
0600	4.1648	4.1707	4.1413	4.1393	4.1187
0700	2.9788	2.9823	2.9645	2.9633	2.9509
0800	2.4853	2.4877	2.4754	2.4746	2.4659
0900	4.4586	4.4655	4.431	4.4286	4.4044
1000	4.9489	4.9574	4.9147	4.9118	4.8819
1100	6.315	6.3291	6.2582	6.2533	6.2036
1200	5.4373	5.4477	5.3954	5.3918	5.3552
1300	4.6549	4.6623	4.6247	4.6221	4.5957
1400	3.7712	3.776	3.752	3.7504	3.7335
1500	4.1648	4.1707	4.1413	4.1393	4.1187
1600	5.2426	5.2521	5.2041	5.2009	5.1672
1700	4.6549	4.6623	4.6247	4.6221	4.5957
1800	3.5737	3.5781	3.5561	3.5547	3.5393
1900	2.2875	2.2896	2.2792	2.2785	2.2712
2000	3.2779	3.2816	3.2632	3.2619	3.249
2100	4.0665	4.072	4.0441	4.0422	4.0226
2200	4.9489	4.9574	4.9147	4.9118	4.8819
2300	5.7303	5.7418	5.6836	5.6797	5.6389
0000	6.315	6.3291	6.2582	6.2533	6.2036

and significant for WF-III layout when compared with WF-IV layout as observed from the results.

A new infrastructure of wind farm as WF-V is designed to reduce the wake loss with a smaller land area. From the results, it can be observed that the relative difference of wind speed has got a significant decrement for WT-7, WT-9 and WT-10 in the case of WF-V configuration. The wind speed difference is changed from 0.0257 m/s to 0.0046 m/s for WT-7 and WF-9. Similarly, the speed is improved by 0.0018 m/s for WT-10.

The wind wake losses are calculated to observe the pattern when the turbines are added gradually. For WF-III, WF-IV and WF-V wind farm layouts, the wake loss is found to be rising when the interaction of the number of turbines is more. The relative gap between the WF-III and WF-IV case scenario is quite large but the gap is minimized with the WF-V case scenario.

Table 6 Wind turbine speed for layout WF-IV (m/s)

Time	WT4 WT6	WT5	WT7 WT9	WT8	WT10
0100	5.2453	5.2547	5.2298	5.2057	5.2057
0200	4.7552	4.7629	4.7425	4.7227	4.7227
0300	3.575	3.5793	3.5679	3.5569	3.5569
0400	3.279	3.2826	3.273	3.2638	3.2638
0500	2.6836	2.6864	2.679	2.6718	2.6718
0600	4.1665	4.1723	4.157	4.1423	4.1423
0700	2.9798	2.9832	2.974	2.9651	2.9651
0800	2.486	2.4884	2.482	2.4758	2.4758
0900	4.4606	4.4674	4.4495	4.4321	4.4321
1000	4.9513	4.9597	4.9376	4.9161	4.9161
1100	6.3191	6.333	6.2962	6.2605	6.2605
1200	5.4403	5.4506	5.4234	5.3971	5.3971
1300	4.657	4.6644	4.6449	4.6259	4.6259
1400	3.7726	3.7773	3.7648	3.7528	3.7528
1500	4.1665	4.1723	4.157	4.1423	4.1423
1600	5.2453	5.2547	5.2298	5.2057	5.2057
1700	4.657	4.6644	4.6449	4.6259	4.6259
1800	3.575	3.5793	3.5679	3.5569	3.5569
1900	2.2881	2.2902	2.2848	2.2795	2.2795
2000	3.279	3.2826	3.273	3.2638	3.2638
2100	4.0681	4.0736	4.0591	4.045	4.045
2200	4.9513	4.9597	4.9376	4.9161	4.9161
2300	5.7336	5.745	5.7148	5.6855	5.6855
0000	6.3191	6.333	6.2962	6.2605	6.2605

3 Load Flow Implementation

In this work, the load flow formulation is based on the formation of the LIM matrix. A topological structure of the radial distribution system is considered for an instance in this section. To formulate the matrix, there is an avoidance of matrix inversion and its upper or lower factorization which makes the computation faster. A simple idea of set theory is implemented to get the LIM matrix.

For the execution of the load flow in any distributed system, the LIM matrix construction is the one requirement for this method. To demonstrate the matrix formulation, a radial distribution system with 'nn' nodes is considered in this theory section as shown in Fig. 5. The number of branches for the 'nn' nodal system are 'nn-1'. The Branch Path Impedance (BPI) is defined as the impedance of the respective nnth

Table 7 Wind turbine speed for layout WF-V (m/s)

Time	WT4 WT6	WT5	WT7 WT9	WT8	WT10
0100	5.2426	5.2521	5.2252	5.2009	5.2075
0200	4.7529	4.7607	4.7387	4.7187	4.7242
0300	3.5737	3.5781	3.5658	3.5547	3.5577
0400	3.2779	3.2816	3.2713	3.2619	3.2645
0500	2.6828	2.6856	2.6776	2.6703	2.6723
0600	4.1648	4.1707	4.1542	4.1393	4.1434
0700	2.9788	2.9823	2.9723	2.9633	2.9658
0800	2.4853	2.4877	2.4808	2.4746	2.4763
0900	4.4586	4.4655	4.4462	4.4286	4.4334
1000	4.9489	4.9574	4.9335	4.9118	4.9177
1100	6.315	6.3291	6.2894	6.2533	6.2632
1200	5.4373	5.4477	5.4184	5.3918	5.3991
1300	4.6549	4.6623	4.6413	4.6221	4.6273
1400	3.7712	3.776	3.7626	3.7504	3.7537
1500	4.1648	4.1707	4.1542	4.1393	4.1434
1600	5.2426	5.2521	5.2252	5.2009	5.2075
1700	4.6549	4.6623	4.6413	4.6221	4.6273
1800	3.5737	3.5781	3.5658	3.5547	3.5577
1900	2.2875	2.2896	2.2838	2.2785	2.2799
2000	3.2779	3.2816	3.2713	3.2619	3.2645
2100	4.0665	4.072	4.0564	4.0422	4.046
2200	4.9489	4.9574	4.9335	4.9118	4.9177
2300	5.7303	5.7418	5.7093	5.6797	5.6877
0000	6.315	6.3291	6.2894	6.2533	6.2632

node. The BPI set is the set of impedances from the substation bus to the nnth node. The BPI set and impedance are denoted by 'Π_{Bn}' and Z_{zn}, respectively. The BPI set for the radial system is given in Eqs. (8)–(13).

$$\prod_{B2} = \{Z_{z1}\} \tag{8}$$

$$\prod_{B3} = \{Z_{z1}, Z_{z2}\} \tag{9}$$

$$\prod_{B4} = \{Z_{z1}, Z_{z2}, Z_{z3}\} \tag{10}$$

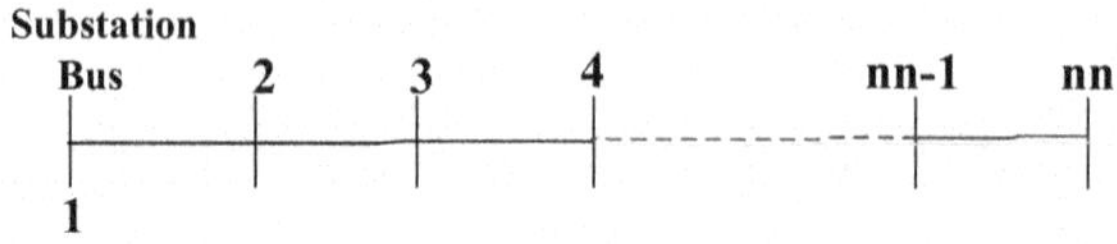

Fig. 5 Radial distribution system of 'nn' nodes

$$\prod_{B(nn-1)} = \{Z_{z1}, Z_{z2}, Z_{z3}, \ldots.., Z_{z(nn-1)}\} \tag{11}$$

$$\prod_{Bnn} = \{Z_{z1}, Z_{z2}, Z_{z3}, Z_{z4} \ldots\ldots, Z_{z(nn-1)}, Z_{znn}\} \tag{12}$$

Similarly, the branch path impedances can be found for any nodal system. The BPI matrix is always a 1 × 1 matrix as observed from the above equations. The load value at each node is referred as '∂_{nn}' and it is extracted from the consumed load by the customers connected at that respective load bus. The LIM matrix is a combination of load matrix and impedance matrix.

The BPI matrix and the load matrix are 1 × 1 matrices in case of single-phase systems. The voltage drop between the customers nodes which is caused by the load connected at that system bus 'i', is denoted by 'dP_{D1i}'. The load bus is connected at jth node and denoted as '∂_j'. The voltage drop mathematically can be expressed as in Eq. (13).

$$dP_{D1i} = \sum_{j=1}^{nn} \frac{\partial_j}{P_{Dj}}^{*} \sum\left(\prod_{Bj} \cap \prod_{Bi}\right) \tag{13}$$

Now, the total voltage drop in the system along the branch path of ith bus is the subtraction of ith bus voltage from the primary substation voltage. Hence, the mathematical equation for total voltage drop can be expressed as,

$$P_{Di} = P_{D1} - \sum_{j=1}^{nn} \left(\frac{\partial_j}{P_{Dj}}\right)^{*} \sum\left(\prod_{Bj} \cap \prod_{Bi}\right) \tag{14}$$

Here, P_{D1} is the substation voltage. From this Eq. (14), it is quite observed that the bus voltages can be expressed as a function of substation voltage, network impedance and the previous iteration value of bus voltage. Now, the LIM matrix can be formed as per the nn-bus system with reference to Eq. (14). For the radial system given in Fig. 5, the bus voltages can be formed as given in Eq. (15).

$$\begin{bmatrix} P_{D1} \\ P_{D2} \\ P_{D3} \\ \vdots \\ \vdots \\ P_{D(nn-1)} \\ P_{Dnn} \end{bmatrix} = \begin{bmatrix} P_{D1} \\ P_{D1} \\ P_{D1} \\ \vdots \\ \vdots \\ P_{D1} \\ P_{D1} \end{bmatrix} - \begin{bmatrix} \partial_1^{\bullet}\sum(\Pi_{B1}\cap\Pi_{B1}) & \partial_2^{\bullet}\sum(\Pi_{B2}\cap\Pi_{B1}) & \cdots\ \cdots & \partial_{nn-1}^{\bullet}\sum(\Pi_{B(nn-1)}\cap\Pi_{B1}) & \partial_{nn}^{\bullet}\sum(\Pi_{Bnn}\cap\Pi_{B1}) \\ \partial_1^{\bullet}\sum(\Pi_{B1}\cap\Pi_{B2}) & \partial_2^{\bullet}\sum(\Pi_{B2}\cap\Pi_{B2}) & \cdots\ \cdots & \partial_{nn-1}^{\bullet}\sum(\Pi_{B(nn-1)}\cap\Pi_{B2}) & \partial_{nn}^{\bullet}\sum(\Pi_{Bnn}\cap\Pi_{B2}) \\ \partial_1^{\bullet}\sum(\Pi_{B1}\cap\Pi_{B3}) & \partial_2^{\bullet}\sum(\Pi_{B2}\cap\Pi_{B3}) & \vdots\ \vdots & \partial_{nn-1}^{\bullet}\sum(\Pi_{B(nn-1)}\cap\Pi_{B3}) & \partial_{nn}^{\bullet}\sum(\Pi_{Bnn}\cap\Pi_{B3}) \\ \vdots & \vdots & \vdots\ \vdots & \vdots & \vdots \\ \partial_1^{\bullet}\sum(\Pi_{B1}\cap\Pi_{B(nn-1)}) & \partial_2^{\bullet}\sum(\Pi_{B2}\cap\Pi_{B(nn-1)}) & \cdots\ \cdots & \partial_{nn-1}^{\bullet}\sum(\Pi_{B(nn-1)}\cap\Pi_{B(nn-1)}) & \partial_{nn}^{\bullet}\sum(\Pi_{Bnn}\cap\Pi_{B(nn-1)}) \\ \partial_1^{\bullet}\sum(\Pi_{B1}\cap\Pi_{Bnn}) & \partial_2^{\bullet}\sum(\Pi_{B2}\cap\Pi_{Bnn}) & \cdots\ \cdots & \partial_{nn-1}^{\bullet}\sum(\Pi_{B(nn-1)}\cap\Pi_{Bnn}) & \partial_{nn}^{\bullet}\sum(\Pi_{Bnn}\cap\Pi_{Bnn}) \end{bmatrix} \begin{bmatrix} \frac{1}{P_{D1}^{\bullet}} \\ \frac{1}{P_{D2}^{\bullet}} \\ \frac{1}{P_{D3}^{\bullet}} \\ \vdots \\ \vdots \\ \frac{1}{P_{D(nn-1)}^{\bullet}} \\ \frac{1}{P_{Dnn}^{\bullet}} \end{bmatrix} \tag{15}$$

$$\begin{bmatrix} P_{D1} \\ P_{D2} \\ P_{D3} \\ \vdots \\ \vdots \\ P_{D27} \\ P_{D28} \end{bmatrix} = \begin{bmatrix} P_{D1} \\ P_{D1} \\ P_{D1} \\ \vdots \\ \vdots \\ P_{D1} \\ P_{D1} \end{bmatrix} - \begin{bmatrix} \partial_1^{\bullet}\sum(\Pi_{B1}\cap\Pi_{B1}) & \partial_2^{\bullet}\sum(\Pi_{B2}\cap\Pi_{B1}) & \cdots\ \cdots & \partial_{27}^{\bullet}\sum(\Pi_{B27}\cap\Pi_{B1}) & \partial_{28}^{\bullet}\sum(\Pi_{B28}\cap\Pi_{B1}) \\ \partial_1^{\bullet}\sum(\Pi_{B1}\cap\Pi_{B2}) & \partial_2^{\bullet}\sum(\Pi_{B2}\cap\Pi_{B2}) & \cdots\ \cdots & \partial_{27}^{\bullet}\sum(\Pi_{B27}\cap\Pi_{B2}) & \partial_{28}^{\bullet}\sum(\Pi_{B28}\cap\Pi_{B2}) \\ \partial_1^{\bullet}\sum(\Pi_{B1}\cap\Pi_{B3}) & \partial_2^{\bullet}\sum(\Pi_{B2}\cap\Pi_{B3}) & \vdots\ \vdots & \partial_{27}^{\bullet}\sum(\Pi_{B27}\cap\Pi_{B3}) & \partial_{28}^{\bullet}\sum(\Pi_{B28}\cap\Pi_{B3}) \\ \vdots & \vdots & \vdots\ \vdots & \vdots & \vdots \\ \partial_1^{\bullet}\sum(\Pi_{B1}\cap\Pi_{B27}) & \partial_2^{\bullet}\sum(\Pi_{B2}\cap\Pi_{B27}) & \cdots\ \cdots & \partial_{27}^{\bullet}\sum(\Pi_{B27}\cap\Pi_{B27}) & \partial_{28}^{\bullet}\sum(\Pi_{B28}\cap\Pi_{B27}) \\ \partial_1^{\bullet}\sum(\Pi_{B1}\cap\Pi_{B28}) & \partial_2^{\bullet}\sum(\Pi_{B2}\cap\Pi_{B28}) & \cdots\ \cdots & \partial_{27}^{\bullet}\sum(\Pi_{B27}\cap\Pi_{B28}) & \partial_{28}^{\bullet}\sum(\Pi_{B28}\cap\Pi_{B28}) \end{bmatrix} \begin{bmatrix} \frac{1}{P_{D1}^{\bullet}} \\ \frac{1}{P_{D2}^{\bullet}} \\ \frac{1}{P_{D3}^{\bullet}} \\ \vdots \\ \vdots \\ \frac{1}{P_{D27}^{\bullet}} \\ \frac{1}{P_{D28}^{\bullet}} \end{bmatrix} \tag{16}$$

From Eq. (14), it is obvious that,

$$\sum\left(\prod_{Bj}\cap\prod_{Bi}\right)=\sum\left(\prod_{Bi}\cap\prod_{Bj}\right) \tag{17}$$

The intersection of BPI sets between two different nodes ‘i’ and ‘j’ is always the same as the impedance path is common to both the branch paths. The LIM matrix is formed for a 28-bus radial distribution system which is used as a single-phase balanced test bus system in the work. Equation (15) can be written as Eq. (16) for the IEEE-28 bus system. The LIM matrix is a square matrix. The order of the LIM matrix is 28 × 28 and the order of bus voltage is 28 × 1 in the case of a 28-bus RDS radial distribution system. Equation (16) can be generalized as,

$$[P_D] = [P_{D1}] - [LIM] \times \left[\frac{1}{P_D^*}\right] \tag{18}$$

The flow chart for the above load flow analysis is depicted in Fig. 6. This LIM power flow technique of finding power loss, line currents and bus voltages is simple and faster with a single-step approach. This technique has been implemented in many radial distribution systems (balanced, unbalanced and weak mesh systems) and a proper description of the method can be found in literature [27].

Fig. 6 Flow chart of proposed load flow method [27]

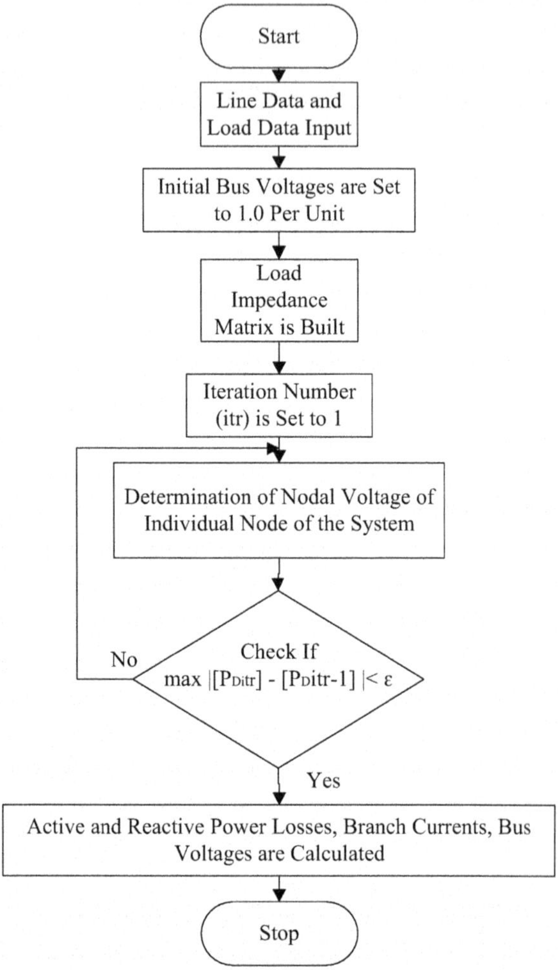

4 Effect of Air Density on Wind

The wind energy is affected by wind speed, rotor swept area, thrust coefficient of turbine and the air density of the atmosphere. The degree of air density depends on the temperature (T_{amb}) and pressure (P_{res}) for an absolute location and it is universally known as the Gas law. The mathematical expression is given in Eq. (19). Here, the value of R_g is 287.085 J/Kg/K [28].

$$rou = \frac{P_{res}}{R_g \times T_{amb}} \tag{19}$$

The hourly temperature and pressure data are available on the weather websites. Hence, the air density is calculated from these available data and implemented in the work. Air density is calculated for all hours of the day from the above formula given in Eq. (19).The temperature, pressure and air density values [21] are given in Table 8 for that day on an hourly basis and the parameter variations with time can be observed.

Out of several aerodynamic factors, air density and wind speed variables are considered for wind power generation. The swept area and the thrust coefficient of a wind turbine are constant and independent of weather conditions. Upstream wind speeds are generated on six-hourly bases using the Kernel Density Estimation and Monte Carlo Simulation. After wake loss, the wind speed is determined and then the effect air density factor is added to the wind power calculation.

Table 8 Hourly temperature, pressure and air density [21]

Sl.No	Time (hr)	Temperature (°C)	Pressure (mbar)	Air density (kg/m^3)
1	0100	34	1005	1.1398
2	0200	33	1005	1.1436
3	0300	32	1005	1.1473
4	0400	31	1005	1.1511
5	0500	30	1005	1.1549
6	0600	29	1006	1.1599
7	0700	30	1006	1.156
8	0800	31	1007	1.1534
9	0900	33	1007	1.1458
10	1000	34	1007	1.1421
11	1100	35	1007	1.1384
12	1200	37	1006	1.1299
13	1300	38	1005	1.1252
14	1400	39	1004	1.1205
15	1500	40	1003	1.1158
16	1600	41	1002	1.1111
17	1700	41	1001	1.11
18	1800	41	1001	1.11
19	1900	40	1001	1.1136
20	2000	39	1002	1.1182
21	2100	36	1002	1.1291
22	2200	35	1003	1.1339
23	2300	35	1004	1.135
24	0000	35	1006	1.1373

5 Load Modelling Profile

The test systems in this work are studied hourly. The parameters like hourly changing loads with hourly varying atmospheric measurable factors are incorporated in this proposed work. The hourly load demands are changing as per the deterministic type of constant power modelling [29]. The DG-generated power and the generated power from the substation are assumed to be constant with the only variable load demand. The load demands are dependent on the peak load and hourly load demands are given in terms of the percentage of peak demand. The values are different for weekdays and weekends. The detailed data for the consumer demand modelling is considered from the IEEE-RTS system [30].

6 Problem Formulation

The objective function of the work is to determine the active power losses in the distribution system using DG. Here, bus number varies from i to nn. Few operational constraints are to be satisfied to ensure system integrity. The power balance equation at bus 'i' after DG is connected to the grid is given in Eq. (20).

$$P_{\mathrm{xDGi}} = P_{\mathrm{xi}} - P_{\mathrm{xDi}} \tag{20}$$

Here, P_{xDGi} is the power generation from DG, P_{xi} is the bus power generation and P_{Di} is the power supplied to the loads connected at the nodes at bus 'i', respectively. Mathematically, the objective function in this work can be expressed as,

$$\text{Total Power Loss } = \sum_{i=1}^{nn} P_{xi_{\mathrm{Loss}}} \tag{21}$$

The constraints to the objective function are described below.

Constraints of Equality

The power losses in the lines and the power consumed by the consumers as loads must be equal to the sum of power generation from the grid as well as from the DG. The power balance equation is given by,

$$P_{\mathrm{xG}} + P_{\mathrm{xDG}} = P_{\mathrm{xD}} + P_{\mathrm{xLOSS}} \tag{22}$$

Here, P_{xG} is the power generated from the grid, P_{xD} is the consumed power by the consumers, P_{xDG} is the power generation from the DG source and P_{xLOSS} is the losses in the distribution lines.

Constraints of Inequality

In every distribution line, there is some limit to the current flow in the line due to the current-carrying capacity. If I_{xmax} is the maximum current limit for the respective line and Ix is the flowing current, then the equation is given as in Eq. (23).

$$I_x \leq I_{xmax} \tag{23}$$

Similarly, the nodal voltage of the system must be in its limitation or tolerance range. If V_x is the actual voltage at any node and V_{xmin} is the minimum and V_{xmax} is the maximum voltage limits, then the equation can be given as in (23).

$$|V_{xmin}| \leq |V_x| \leq |V_{xmax}| \tag{24}$$

Constarints of DG

DG is connected as an additional part of the existing system. Hence it also follows some limitations as, the maximum capacity of DG should be less than its maximum limits. If P_{xDG} is the DG power generation at any time and P_{xDGmax} is the maximum power that can be generated using the DG, then the constraint is given in Eq. (25).

$$|P_{xDG}| \leq P_{xDGmax} \tag{25}$$

This helps the system to stay stable and maximum benefits can be obtained for the system with consideration of DG constraints.

7 Optimization Technique

Wind power is considered as DG and used in the standard IEEE test systems. The location and capacity of DG are found using the Grey Wolf Optimization (GWO) technique [31]. Power loss minimization is considered an objective function for DG placement. The mathematical model is well described with the exploitation and exploration of search space. Few algorithm specific parameters are tuned which makes the results promising and effective. There are some algorithm specific parameters that need to be initialized during the implementation of the algorithm.

8 Distribution System Under Study

Two radial distribution systems based on the benchmark 28-node [32] and 69-node test system [33] are studied in this work. The system base voltage for the 28-bus system is considered as 11 kV and the base power is 100 MVA. The system base voltage and base MVA of the system 69-bus system are 12.66 kV and 100 MVA, respectively.

9 Result and Discussion

The wind speed that is generated from Kernel Density Estimation is considered as the free upstream wind speed. A new irregular wind farm layout WF-V is designed after a wake loss investigation. The wind speed for all the wind turbines of the farm is calculated after the single and multi-wake effect. The effect of air density is also taken into consideration. The power generation from the individual wind turbines is found for every hourly wind speed. Total wind power produced from the farm is penetrated to the test systems under study and it is considered as DG. To analyze the system performance, the LIM matrix method of load flow technique is implemented. The hourly analysis also involves hourly load modelling. The power losses in the distribution lines are then calculated and bus voltages are determined.

9.1 Hourly Wind Power Generation

The wind power is generated from the ten wind turbines in the wind farm. The hourly generated power is calculated with the swept area, thrust coefficient, air density and wind speed available at that particular hour. All the turbines except WT-1, WT-2 and WT-3 produce different wind powers at a specific time interval because of the wake effect on them. The total wind power that is generated from the wind farm is the sum of the individual power generated by each wind turbine in the farm and it is given in Table 9.

The total power generated by all the turbines is fed to the optimum bus number. The power demand from the grid is decreased for the DG power injection through wind power. The power injection has been considered to be the same power generated by the turbine.

9.2 Active and Reactive Power Loss

The active and reactive power loss in the distribution lines is determined for both the test systems. The power losses with hourly DG capacity and hourly time are measured. All the generated wind power is fed to the test system. The amount of power loss that is minimized is calculated in terms of percentage. The load flow is carried out for the test system IEEE-28 radial distribution system with wind power as DG and the power losses are given in Table 10. Similarly, the power losses are determined for the IEEE-69 bus radial distribution system with no DG and with DG penetration.

The DG power improves the power loss minimization with additional wind power supply and the results can be observed from Tables 10 and 11. During the peak load demand that is at 12.00 PM, the line losses are minimized from 68.81 kW to

Table 9 Hourly generated DG power from 100 kW wind turbine

Time	WT1 (W)	WT2 (W)	WT3 (W)	WT4 (W)	WT5 (W)	WT6 (W)	WT7 (W)	WT8 (W)	WT9 (W)	WT10 (W)	Total (W)
1:00 AM	49,499	49,499	49,499	48,189	48,116	48,189	46,988	46,666	46,988	46,714	480,347
2:00 AM	38,262	38,262	38,262	37,250	37,193	37,250	36,322	36,073	36,322	36,110	371,306
3:00 AM	18,119	18,119	18,119	17,640	17,613	17,640	17,200	17,082	17,200	17,100	175,832
4:00 AM	14,453	14,453	14,453	14,070	14,049	14,070	13,720	13,626	13,720	13,640	140,254
5:00 AM	8580.2	8580.2	8580.2	8353.1	8340.4	8353.1	8145	8089.2	8145	8097.5	83,263.9
6:00 AM	27,045	27,045	27,045	26,329	26,289	26,329	25,673	25,497	25,673	25,523	262,448
7:00 AM	11,282	11,282	11,282	10,984	10,967	10,984	10,710	10,637	10,710	10,648	109,486
8:00 AM	7024.8	7024.8	7024.8	6838.9	6828.5	6838.9	6668.5	6622.8	6668.5	6629.6	68,170.1
9:00 AM	32,355	32,355	32,355	31,499	31,451	31,499	30,714	30,504	30,714	30,535	313,981
10:00AM	42,544	42,544	42,544	41,418	41,355	41,418	40,386	40,109	40,386	40,151	412,855
11:00AM	80,800	80,800	80,800	78,662	78,542	78,662	76,702	76,176	76,702	76,255	784,101
12:00PM	54,499	54,499	54,499	53,057	52,976	53,057	51,735	51,381	51,735	51,434	528,872
1:00 PM	36,226	36,226	36,226	35,267	35,214	35,267	34,389	34,153	34,389	34,188	351,545
2:00 PM	20,852	20,852	20,852	20,300	20,270	20,300	19,795	19,659	19,795	19,679	202,354
3:00 PM	27,045	27,045	27,045	26,329	26,289	26,329	25,673	25,497	25,673	25,523	262,448
4:00 PM	49,499	49,499	49,499	48,189	48,116	48,189	46,988	46,666	46,988	46,714	480,347
5:00 PM	36,226	36,226	36,226	35,267	35,214	35,267	34,389	34,153	34,389	34,188	351,545
6:00 PM	18,119	18,119	18,119	17,640	17,613	17,640	17,200	17,082	17,200	17,100	175,832
7:00 PM	5656.5	5656.5	5656.5	5506.8	5498.5	5506.8	5369.6	5332.8	5369.6	5338.3	54,891.9
8:00 PM	14,453	14,453	14,453	14,070	14,049	14,070	13,720	13,626	13,720	13,640	140,254
9:00 PM	25,403	25,403	25,403	24,731	24,694	24,731	24,115	23,950	24,115	23,974	246,519

(continued)

Table 9 (continued)

Time	WT1 (W)	WT2 (W)	WT3 (W)	WT4 (W)	WT5 (W)	WT6 (W)	WT7 (W)	WT8 (W)	WT9 (W)	WT10 (W)	Total (W)
10:00PM	42,544	42,544	42,544	41,418	41,355	41,418	40,386	40,109	40,386	40,151	412,855
11:00PM	62,564	62,564	62,564	60,908	60,816	60,908	59,391	58,984	59,391	59,045	607,135
12:00AM	80,800	80,800	80,800	78,662	78,542	78,662	76,702	76,176	76,702	76,255	784,101

Table 10 Hourly active and reactive power loss for IEEE-28 Bus

Civil time (Hr)	Without DG		DG capacity (kW)	With DG		Minimized active power loss (%)	Minimized reactive power loss (%)
	Ploss (kW)	Qloss (kVAR)		Ploss (kW)	Qloss (kVAR)		
1.00 AM	26.54	17.75	26.38	22.89	15.32	13.75	13.69
2.00 AM	23.18	15.50	19.72	20.62	13.80	11.04	10.96
3.00 AM	21.59	14.44	8.34	20.52	13.73	4.95	4.91
4.00AM	20.06	13.42	6.46	19.27	12.89	3.93	3.94
5.00 AM	20.06	13.42	4.13	19.55	13.08	2.54	2.53
6.00 AM	21.59	14.44	13.13	19.93	13.33	7.68	7.68
7.00 AM	26.54	17.75	4.87	25.83	17.28	2.67	2.64
8.00 AM	38.16	25.53	3.28	37.57	25.14	1.54	1.52
9.00 AM	50.93	34.07	16.30	47.57	31.83	6.59	6.57
10.00 AM	61.57	41.19	22.23	56.48	37.80	8.26	8.23
11.00 AM	67.33	45.04	46.69	56.46	37.78	16.14	16.11
12.00 PM	68.81	46.04	29.56	61.66	41.27	10.39	10.36
1.00 PM	67.33	45.04	18.22	62.90	42.09	6.57	6.54
2.00 PM	68.81	46.04	9.41	66.47	44.47	3.40	3.41
3.00 PM	68.81	46.04	12.63	65.68	43.95	4.54	4.53
4.00 PM	64.41	43.09	25.72	58.40	39.08	9.33	9.30
5.00 PM	62.98	42.13	17.97	58.78	39.34	6.66	6.62
6.00 PM	62.98	42.13	8.07	61.06	40.86	3.04	3.01
7.00 PM	58.80	39.33	2.47	58.23	38.96	0.96	0.94
8.00 PM	57.44	38.43	6.27	56.03	37.49	2.45	2.44
9.00 PM	57.44	38.43	11.9	54.79	36.66	4.61	4.60
10.00 PM	58.80	39.33	22.07	53.89	36.06	8.35	8.31
11.00 PM	50.93	34.07	34.76	43.98	29.43	13.64	13.61
12.00 AM	34.02	22.76	46.65	26.86	17.97	21.04	21.04

61.66 kW with 29.56 kW of hourly generated wind power and the reactive power loss is minimized from 46.04kVAR to 41.27kVAR, respectively. The load flow is carried out for test system IEEE-69 radial distribution system with wind power as DG and the power losses are given in Table 11. Grey Wolf Optimization method is implemented to find the best location and capacity of DG in the distribution system. The active power loss is minimized from 224.92 kW to 119.37 kW with a DG capacity of 528.87 kW at 12.00PM. Similarly, the reactive power loss for the IEEE-69 bus system is minimized from 102.12 kVAR to 57.17kVAR with a DG capacity of 375.75kVAR at peak hours of the day.

It can be observed that, the active and reactive power losses are minimized with DG penetration at the optimum node of the test bus system. Apart from improving the

Table 11 Hourly active and reactive power loss for IEEE-69 Bus

Time	Wind Power	%age Power Injection	PDG (kW)	QDG (kVAR)	Load percentage	Without DG		With DG	
						Ploss (kW)	Qloss (kVAR)	PlossDG (kW)	QlossDG (kVAR)
1:00 AM	480,347	12.66775	480.347	341.2817	64	86.4844	39.4146	39.4146	17.3627
2:00 AM	371,306	9.792109	371.306	263.8092	60	75.5103	34.4267	36.5577	17.8098
3:00 AM	175,832	4.637054	175.832	124.9269	58	70.329	32.0708	32.0708	23.6305
4:00 AM	140,254	3.698789	140.254	99.64907	56	65.3488	49.8664	29.8055	23.2242
5:00 AM	83,263.9	2.195842	83.2639	59.15817	56	65.3488	55.8386	29.8055	25.7664
6:00 AM	262,448	6.921298	262.448	186.4667	58	70.329	32.0708	32.0708	20.0922
7:00 AM	109,486	2.887373	109.486	77.78871	64	86.4844	72.0073	39.4146	33.2649
8:00 AM	68,170.1	1.797787	68.1701	48.43418	76	124.4597	113.085	56.6536	51.8259
9:00 AM	313,981	8.28033	313.981	223.0804	87	166.2679	110.1288	75.5995	51.7198
10:00 AM	412,855	10.88784	412.855	293.3294	95	201.1326	121.0217	91.3758	57.2785
11:00 AM	784,101	20.67837	784.101	557.0959	99	220.038	81.0177	99.9225	40.5274
12:00 PM	528,872	13.94745	528.872	375.7583	100	224.9219	119.3729	102.1295	57.1655
1:00 PM	351,545	9.27097	351.545	249.7692	99	220.038	145.9498	99.9225	68.4138
2:00 PM	202,354	5.336494	202.354	143.7705	100	224.9219	179.154	102.1295	82.6918
3:00 PM	262,448	6.921298	262.448	186.4667	100	224.9219	166.8877	102.1295	77.4694
4:00 PM	480,347	12.66775	480.347	341.2817	97	210.4601	116.988	95.5932	55.7862
5:00 PM	351,545	9.27097	351.545	249.7692	96	205.7652	167.7685	93.4706	77.3352
6:00 PM	175,832	4.637054	175.832	124.9269	96	205.7652	167.7685	93.4706	77.3352
7:00 PM	54,891.9	1.447613	54.8919	39.00015	93	192.0526	180.1384	87.269	82.2146
8:00 PM	140,254	3.698789	140.254	99.64907	92	187.6044	158.577	85.2566	72.9328

(continued)

Table 11 (continued)

Time	Wind Power	%age Power Injection	PDG (kW)	QDG (kVAR)	Load percentage	Without DG		With DG	
						Ploss (kW)	Qloss (kVAR)	PlossDG (kW)	QlossDG (kVAR)
9:00 PM	246,519	6.501217	246.519	175.1493	92	187.6044	138.6948	85.2566	64.4709
10:00 PM	412,855	10.88784	412.855	293.3294	93	192.0526	114.3256	87.269	54.1826
11:00 PM	607,135	16.01141	607.135	431.3634	87	166.2679	70.8086	75.5995	34.8538
12:00 AM	784,101	20.67837	784.101	557.0959	72	110.9408	26.822	50.5201	14.4278

Table 12 Bus voltages for IEEE-28 bus system

Bus.no	Node voltage without DG (p.u)	Node voltage with DG (p.u)	Bus.no	Node Voltage without DG (p.u)	Node voltage with DG (p.u)
1	1	1	15	0.943438	0.946846
2	0.986369	0.987311	16	0.937748	0.942557
3	0.966818	0.969169	17	0.926684	0.932668
4	0.952873	0.956248	18	0.925718	0.931708
5	0.938865	0.943668	19	0.924067	0.930068
6	0.92845	0.934422	20	0.923217	0.929224
7	0.919379	0.926787	21	0.92259	0.9286
8	0.91695	0.924377	22	0.91653	0.924888
9	0.916673	0.924102	23	0.91501	0.924033
10	0.91643	0.923861	24	0.913846	0.923424
11	0.946749	0.950146	25	0.91359	0.92317
12	0.944998	0.948401	26	0.913426	0.923008
13	0.943953	0.947359	27	0.91646	0.923892
14	0.943676	0.947083	28	0.916338	0.92377

loss minimization, the DG power naturally improves the bus voltages of the system which is described in the next subsection.

9.3 *Bus Voltage Improvement*

The bus voltages are computed for the test bus systems and the results are given for the IEEE-28 bus and IEEE-69 bus systems in Tables 12 and 13, respectively. After DG connection, the bus voltages are improved significantly as observed from the results.

The bus voltages are determined for the IEEE-69 bus system before and after DG placement. The analysis takes place during the peak hour of the day i.e., noon when the load demand is 100% of the total peak demand.

The bus voltages for other hours of the day can also be determined. The bus voltages for all the nodes of the distribution test system can be evaluated in terms of hourly nodal voltages with hourly DG penetration and hourly load variation.

9.4 *Optimal DG Location and Capacity*

The power flow analysis is carried out with hourly load demand and hourly DG power generation. The test systems are studied at optimum DG location and optimum DG

Table 13 Bus voltages for IEEE-69 bus system

Node N	Node Voltage without DG (p.u)	Node Voltage With DG (p.u)	Node No	Node Voltage without DG (p.u)	Node Voltage With DG (p.u)	Node No	Node Voltage without DG (p.u)	Node Voltage With DG (p.u)	Node Number	Node Voltage without DG (p.u)	Node Voltage With DG (p.u)
1	1	1	18	0.9581	0.9631	35	0.9992	0.9992	52	0.9785	0.9832
2	1	1	19	0.9576	0.9626	36	0.9999	0.9999	53	0.9747	0.9805
3	0.9999	0.9999	20	0.9573	0.9623	37	0.9997	0.9998	54	0.9714	0.9783
4	0.9998	0.9999	21	0.9568	0.9618	38	0.9996	0.9996	55	0.9669	0.9753
5	0.999	0.9992	22	0.9568	0.9618	39	0.9995	0.9996	56	0.9626	0.9724
6	0.9901	0.9922	23	0.9568	0.9618	40	0.9995	0.9996	57	0.9401	0.9576
7	0.9808	0.985	24	0.9566	0.9616	41	0.9988	0.9989	58	0.929	0.9502
8	0.9786	0.9832	25	0.9564	0.9614	42	0.9986	0.9986	59	0.9248	0.9474
9	0.9774	0.9824	26	0.9564	0.9614	43	0.9985	0.9985	60	0.9197	0.9442
10	0.9724	0.9774	27	0.9563	0.9614	44	0.9985	0.9985	61	0.9123	0.9394
11	0.9713	0.9763	28	0.9999	0.9999	45	0.9984	0.9984	62	0.9121	0.9391
12	0.9682	0.9731	29	0.9999	0.9999	46	0.9984	0.9984	63	0.9117	0.9388
13	0.9653	0.9702	30	0.9998	0.9998	47	0.9998	0.9998	64	0.9098	0.9369
14	0.9624	0.9674	31	0.9997	0.9998	48	0.9985	0.9986	65	0.9092	0.9363
15	0.9595	0.9645	32	0.9997	0.9997	49	0.9947	0.9947	66	0.9713	0.9762
16	0.959	0.964	33	0.9995	0.9995	50	0.9942	0.9942	67	0.9713	0.9762
17	0.9581	0.9631	34	0.9992	0.9992	51	0.9785	0.9832	68	0.9679	0.9728
									69	0.9679	0.9728

capacity to minimize the line losses in the distribution systems. Using the GWO optimization technique, the best DG capacities and locations are found for IEEE-28 and IEEE-69 bus radial distribution systems. The capacities of DG are assumed to be 100% of wind power generation from the wind farm which is different for both the test systems. The optimal location for the IEEE-28 bus system and IEEE-69 bus system is found to be bus number 24 and bus number 61, respectively.

10 Conclusion

In this chapter, an irregular infrastructure of a wind farm is adopted for its low wind power losses due to the wake effect. A new set of wind speeds is generated using the non-parametric data distribution method of Kernel Density Estimation (KDE) and Monte Carlo Simulation (MCS) for the wind farm located at Surat, Gujarat, India. The obtained wind speed is implemented with single as well as multiple wake effect caused by downstream wind turbines ahead of the upstream wake-free wind turbines. The wind power is calculated for all the wind turbines in the farm and the total wind power generated is fed to the distribution system under study. The analysis is carried out with the IEEE-28 bus and IEEE-69 bus radial distribution systems. To observe the system characteristics after DG penetration, the LIM matrix method of the load flow technique is implemented to the test systems for being simpler and faster. The analysis also involves hourly varying load demands. Moreover, this work explains the loss minimization in the distribution system using wind power source as DG. To get the practical system output in terms of power loss that happened in the lines and the power produced by the DG, the wind energy undergoes wind turbulence and wind wake effect. The resultant power from the generated wind speed through the Kernel Density Estimation method is used for the DG power penetration to the distribution system. The optimum location and capacity of DG are determined by the GWO optimization technique. From the results of this work, the power losses in the lines are minimized when the bus voltages of the test system are increased.

11 Numerical

Q.1. A wind turbine (WT-d) has rotor diameter of 14 m (b) with a rated power of 25 kW. The shadowed area for the above wind turbine is $0.7857b^2$ and thrust coefficient is 0.103. Consider the wind turbine WT-a is ahead of 3 numbers of wind turbines (WT-c, WT-b, WT-a) with a inter horizontal distance of 2b between consecutive wind turbines. Find the wind power produced by the last wind turbine WT-a. (Assume free wind speed = 3 m/s, rou = 1.23 kg/m^3).

Answer:

b = 14 m,

Total horizontal distance between WT-a and other turbines can be given as,

(From Table 1).

Wind Turbine	Horizontal distance from WT-a (in terms of b)	b_h	F_{shadow}
WT-d	6b	1.9b	$0.6062b^2$
WT-c	4b	1.6b	$0.3508b^2$
WT-b	2b	1.3b	$0.1244b^2$

$S_0 = 3$ m/s.

Now,

The wind speed at WT-a (using Eq. 7),

$$S_{WT-a} = 3 \times \left[1 - \sqrt{1 - 0.103}\left(\sum_{j=1}^{3}\left(\frac{b}{b_{hj}}\right)^2\left(\frac{F_{shadow,i}}{F_0}\right)\right)\right]$$

$$= 3 \times \left[1 - \sqrt{1 - 0.103}\left\{\left(\left(\frac{b}{1.9b}\right)^2\left(\frac{0.6062b^2}{0.7857b^2}\right) + \left(\frac{b}{1.6b}\right)^2\left(\frac{0.3508b^2}{0.7857b^2}\right) + \left(\frac{b}{1.3b}\right)^2\left(\frac{0.1244b^2}{0.7857b^2}\right)\right)\right\}\right]$$

= 1.6316 m/s.

Now,

The power generated by the WT-a is,

$$P = \frac{1}{2} \times rou \times F_0 \times S_{WT-a}^3$$
$$= 411.38 \text{ W}$$

References

1. Goswami DY, Kreith F (eds) (2007) Handbook of energy efficiency and renewable energy. CRC Press
2. Ackermann T (2005) Wind power in power systems, vol 140. John Wiley, Chichester, UK
3. Haider W, Hassan S, Mehdi A, Hussain A, Adjayeng GOM, Kim CH (2021) Voltage profile enhancement and loss minimization using optimal placement and sizing of distributed generation in reconfigured network. Machines 9(1):20
4. Topaloğlu F, Pehlivan H (2018) Analysis of wind data, calculation of energy yield potential, and micrositing application with WAsP. In: Advances in meteorology
5. El Yazidi M, Redouane A, Benzirar M, Zazoui M (2019) Analysis of wind data and assessment of wind energy potential in Lamhiriz Village, Morocco. Appl Solar Energy 55(6):429–437

6. Barthelmie RJ, Larsen GC, Frandsen ST (2006) Comparison of wake model simulations with offshore wind turbine wake profiles measured by sodar. J Atmos Oceanic Tech 23(7):888–901
7. Zhuonan L, Xin A (2013) Flatlands wind farm power generation and wake effect analysis based on Jensen model. In: 2013 IEEE PES Asia-Pacific power and energy engineering conference (APPEEC), Kowloon, pp 1–5
8. https://niwe.res.in/
9. https://www.theenergyfix.com/onshore-vs-offshore-wind-power/
10. Kuenzel S, Kunjumuhammed LP, Pal BC, Erlich I (2014) Impact of Wakes on wind farm inertial response. IEEE Trans Sustain Energy 5(1):237–245
11. Sethi JK, Deb D, Malakar M (2011) Modeling of a wind turbine farm in presence of wake interactions. 2011 International conference on energy, automation and signal, Bhubaneswar, Odisha, pp 1–6
12. Zhang X, Wang W (2009) Wind farm and wake effect modeling for simulation of a studied power system. In: 2009 IEEE/PES power systems conference and exposition, vol 30. Seattle, WA, pp 1–6
13. Shao Z, Wu Y, Li L, Han S, Liu Y (2019) Multiple wind turbine wakes modelling considering the faster wake recovery in overlapped wakes. Energies 12(4):680
14. Tian L, Zhu W, Shen W, Song Y, Zhao N (2017) Prediction of multi-wake problems using an improved Jensen wake model. Renew Energy 1(102):457–469
15. Kusiak A, Song Z (2010) Design of wind farm layout for maximum wind energy capture. Renew Energy 35(3):685–694
16. Charhouni N, Mohammed S, Khalifa M (2019) Realistic wind farm design layout optimization with different wind turbines types. Int J Energy Environ Eng 10(3):307–318
17. Jacobsen A, Godvik M (2020) Influence of wakes and atmospheric stability on the floater responses of the Hywind Scotland wind turbines. Wind Energy
18. Rodrigues RV, Lengsfeld C (2019) "Development of a computational system to improve wind farm layout, part I: model validation and near wake analysis. Energies 12(5):940
19. Hansen KS et al (2016) Wind turbine wake measurement in complex terrain. J Phys Conf Ser 753(3). (IOP Publishing)
20. Shakoor R, Hassan MY, Raheem A, Wu YK (2016) Wake effect modeling: a review of wind farm layout optimization using Jensen's model. Renewab Sustain Energy Rev 58:1048–1059
21. TIMENDATE: Homepage. https://www.timeanddate.com/weather/india/bagalkot/historic?month=6&year=2019
22. Miozzo M, Zordan D, Dini P, Rossi M (2014) SolarStat: modeling photovoltaic sources through stochastic Markov processes. In: 2014 IEEE international energy conference (ENERGYCON). IEEE, pp 688–695
23. Koch F, Gresch M, Shewarega F, Erlich I, Bachmann U (2005) Consideration of wind farm wake effect in power system dynamic simulation. In: Proceedings IEEE Power Technical. Russia, pp 1–7
24. http://www.urbanwind.net/pdf/CATALOUGE_V2.pdf
25. Højstrup KJ, Jensen NO (1986) A simple model for cluster efficiency. In: European wind energy conference and exhibition, pp 407e10
26. Natalia M, Rudion K, Orths A (2010) Study of wake effects for offshore wind farm planning. In: Modern electric power systems (MEPS), 2010 proceedings of the international symposium, vol.7, Wroclaw, MEPS
27. Ghatak U, Mukherjee V (2017) A fast and efficient load flow technique for unbalanced distribution system. Int J Electric Power Energy Syst 84:99–110
28. Essa KS, Etman SM, Embaby M (2004) The effect of air density variation on wind power flux. Wind Eng 28(3):305–310
29. Zhuonan L, Xin A (2004) Prediction of wind turbine power using deterministic and nondeterministic methods. In: IEEE power engineering society general meeting, pp 2017–2024
30. IEEE Reliability Test System (1979) IEEE Trans Power Appar Syst 98(6):2047–2054
31. Mirjalili SM, Lewis A (2014) Grey wolf optimizer. J Adv Eng Softw 69:46–61

32. Das D, Negi HS, Kothari DP (1994) Novel method for solving radial distribution networks. In: Proceedings of IEE generation transmission and distribution, vol 141, no 4. (July 1994)
33. Chakravorty M, Das D (2001) Voltage stability analysis of radial distribution networks. Int J Electric Power Energy Syst 23(2):129–135

Chapter 3
Electric Power Management and Control in DC Buildings—State-Of-The-Art and Emerging Technologies

Andrei Blinov, Indrek Roasto, Andrii Chub, Pietro Emiliani, and Dmitri Vinnikov

1 Introduction

The problem of global climate change has strenghened efforts for minimizing fossil fuel use to reduce the associated carbon footprint. As a result, over the last decade, the installed capacity of alternative energy sources like wind and solar has experienced large and sustained growth [1]. Moreover, significant investments have been made to intensify the transition to electric transportation. At the same time, according to [2] the construction and use of various buildings account for around 40% of EU energy consumption and 36% of CO2 emissions. To address this issue, a number of initiatives have been launched to increase the overall energy efficiency of existing and future buildings. Evidently, the problem of efficiency can be addressed from multiple aspects—use of advanced material and construction technologies, advanced heating and air conditioning systems, installation of local renewable sources and energy storage, use of efficient appliances, etc. Combined adoption of these approaches is revealed in the concept of nearly zero energy building (nZEB), which generates enough energy locally to cope with consumption over the course of a year. In the EU, all new buildings from 2021 should be nZEBs [3]. The recommendations for the nZEB provided by the European Commission are considering both the energy

A. Blinov (✉) · I. Roasto · A. Chub · P. Emiliani · D. Vinnikov
Tallinn University of Technology, Tallinn, Estonia
e-mail: andrei.blinov@taltech.ee

I. Roasto
e-mail: indrek.roasto@taltech.ee

A. Chub
e-mail: andrii.chub@taltech.ee

D. Vinnikov
e-mail: dmitri.vinnikov@taltech.ee

A. K. Giri et al. (eds.), *Power Quality: Infrastructures and Control*,
Studies in Infrastructure and Control,
https://doi.org/10.1007/978-981-19-7956-9_3

efficiency of the building as well as the local renewable regeneration and depend on one of the four climate zones.

One step ahead and logical evolution of the nZEB is a concept of a positive energy building (PEB). PEB can not only be viewed as a net-producer of energy, but it also has a number of extra properties and distinguished features [4–7]:

- capability of generating enough energy not only to fully cover its own demands, but also supply consumer devices or even an electric vehicle;
- energy support to other buildings in the vicinity, with the possible aim of achieving energy neutrality or even positivity at the neighborhood level;
- active management of consumption, production and storage of energy, aiming at maximization of performance in a system-based approach;
- involvment of a complex set of energy partnerships, offering new services to grid operators and utilities;
- aiming towards full-time energy positivity, possibly using biomass as a fuel during low renewable generation periods, instead of focusing on the year-by-year energy balance.

At the same time, the evolution of an already efficient nZEB into PEB will inevitably hit diminishing returns in terms of investment cost and obtained performance improvement. In this regard, the transition of a building that has already realized much of its savings potential to the DC power could be one of the largest remaining sources of energy-efficiency increase.

The advances in power electronics have made the transition to DC distribution in a building easier than ever before. The fact that PEB is most likely to feature local PV and battery storage, which are natively DC, makes the DC power inherently available in such buildings. Moreover, the widespread adoption of energy-efficient technologies, like LED lamps and inverter-driven motors together with the extensive use of DC-supplied consumer electronic devices (personal computers, laptops, cell phones, robotic vacuum cleaners, etc.) has made the DC loads to contribute appreciably, or even dominate in the residential applications. Therefore, the obvious and remarkable advantage of DC distribution is the elimination of power losses due to extra DC-AC-DC conversions. According to some studies, even by adapting existing devices to DC by removing the power factor correction (PFC) stage, savings from a few to over 20 percent can be obtained in certain use cases [8]. In the case of PEB, additional potential benefits of DC are in simplified communication and control, power quality, reliability and resilience [9–12].

Despite these potential advantages, the adoption of DC buildings is facing major barriers, mainly related to immaturity of the technology, lack of standards and DC-ready appliances as well as general awareness among the potential users. As a result, necessary economics of scale has not been established yet, which in general, makes DC distribution more costly than that of AC, diminishing the feasibility of its adoption.

In the last years, studies related to DC buildings have intensified; the barriers of implementation mentioned above have been addressed from various directions. The present overview consolidates existing knowledge regarding the various aspects of residential DC buildings. The main focus is on electric power management and control that are essential enablers of the PEB concept. Although some studies assume the DC building as a part of the DC microgrid, which would be the case for some time in the future, in our analysis, the focus is on a closer perspective, when the PEB is connected to an existing low voltage AC distribution grid. In this case, the grid interface converter (GIC) that interfaces a DC building with the AC grid is the key element of the DC PEB, responsible for the realization of most of the distinguished PEB features listed earlier.

To get a clearer view on how these PEB functions can be realized in practice, it is necessary to analyze the state-of-the-art developments in the area of DC components, systems and grids as well as advanced functions and interactions with the AC distribution grid. To address these problems, this chapter focuses on the following topics: DC grid types, standards, protection and realization aspects (Section 2); control and communication (Section 3); the role of a DC building and relevant AC services (Section 4); possible topological realizations of the GIC (Section 5); perspectives of DC buildings and conclusions (Section 6).

2 DC Infrastructure

2.1 DC Microgrid Voltage Standards and Architecture

There are several typical architectures of DC microgrids, like radial or ring. Among them, radial microgrids, where all devices are connected to one DC bus, are typical for residential applications due to their simplicity and low implementation cost. Therefore, this section addresses only the architecture of radial microgrids. An example of a DC house based on a radial DC microgrid is shown in Fig. 1.

Despite obvious benefits of the DC distribution technology, its proliferation was historically limited by lack of timely standards and overall concerns about protection and safety of DC grids. Low voltage DC (LVDC) also requires a proper definition. For example, the European Union has defined the voltage limits between 75 and 1500 V for LVDC systems in the Low Voltage Directive LVD 2014/35/EU [13]. However, when it comes to residential LVDC systems, there is no proper standardization. One attempt to standardize commercial and residential LVDC systems is open standard REbus with a rated voltage of 380 V, which has gained some popularity in commercial products, but typically, it requires communication between DC microgrid devices, which limits its widespread applicability in a residential area. Other attempts of standardization, like EMerge Alliance coming from data centers or IEEE DC@Home committee initiative, have not attracted particular interest in the residential sector.

Currently, only the Holland has provided a national practice guideline NPR 9090 for LVDC systems [14]. It defines three types of DC voltages: 350, 700, and 1400 V for distribution dc grids [15]. Moreover, it requires the implementation of a galvanic isolation between the AC grid and the DC grid and covers both unipolar (two-wire)

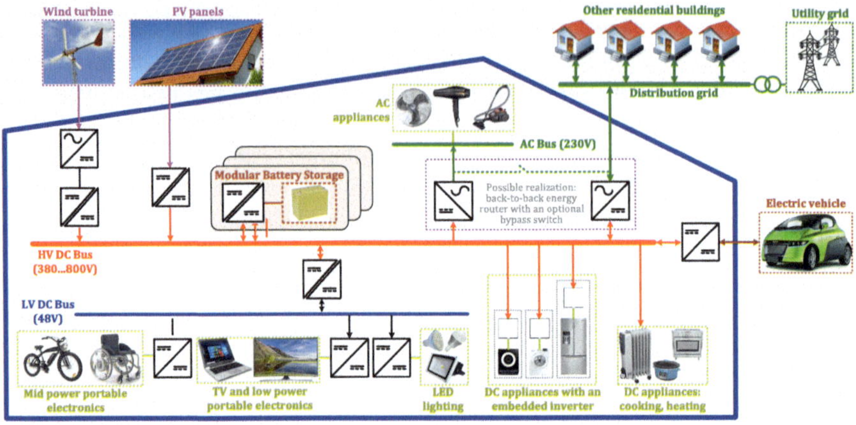

Fig. 1 Sketch of a residential DC house based on a radial DC microgrid

and bipolar (three-wire) DC systems. Unipolar DC grids are useful in systems where low-voltage transmission lines of short distance are needed. Even though they provide no redundancy and only one voltage level, their attractive cost and no problems with load asymmetry make them a superior solutions for small residential houses. The bipolar DC grids can find their use in more powerful applications, like DC street or district due to their higher power handling capability, inherent redundancy and flexibility. However, load asymmetries can require an additional converter to balance pole currents.

The NRP 9090 guidelines establish cross-reference between typical AC and DC systems considering maximum current of 16 A, i.e., power rating when reusing the same wiring for AC and DC systems:

- 350 V DC system can supply 5.6 kW, which is the closet analog of 1-phase 230 V AC system supplying only 3.7 kW.
- 700 V or ±350 V DC systems can supply 11.2 kW, similar to 3-phase 400 V AC systems supplying 11.1 kW.
- 1400 V or ±700 V DC systems can supply 22.4 kW, which is close to 3-phase 690 V AC systems supplying only 19.1 kW.

2.2 *Earthing and Protection of DC Systems*

Today, only NPR 9090 guidelines provide a clear picture of protection zones in LVDC systems. Depending on the maximum current and short-circuit current characteristics, there are five levels of protection: from Zone 0 to Zone 4, as defined in Fig. 2, to distinguish risks in several classes. Each zone has its own requirements regarding disconnection mechanism and time. Zone 0 contains unprotected power sources with voltages up to 1500 V and current of over 500 A. Starting from Zone 1,

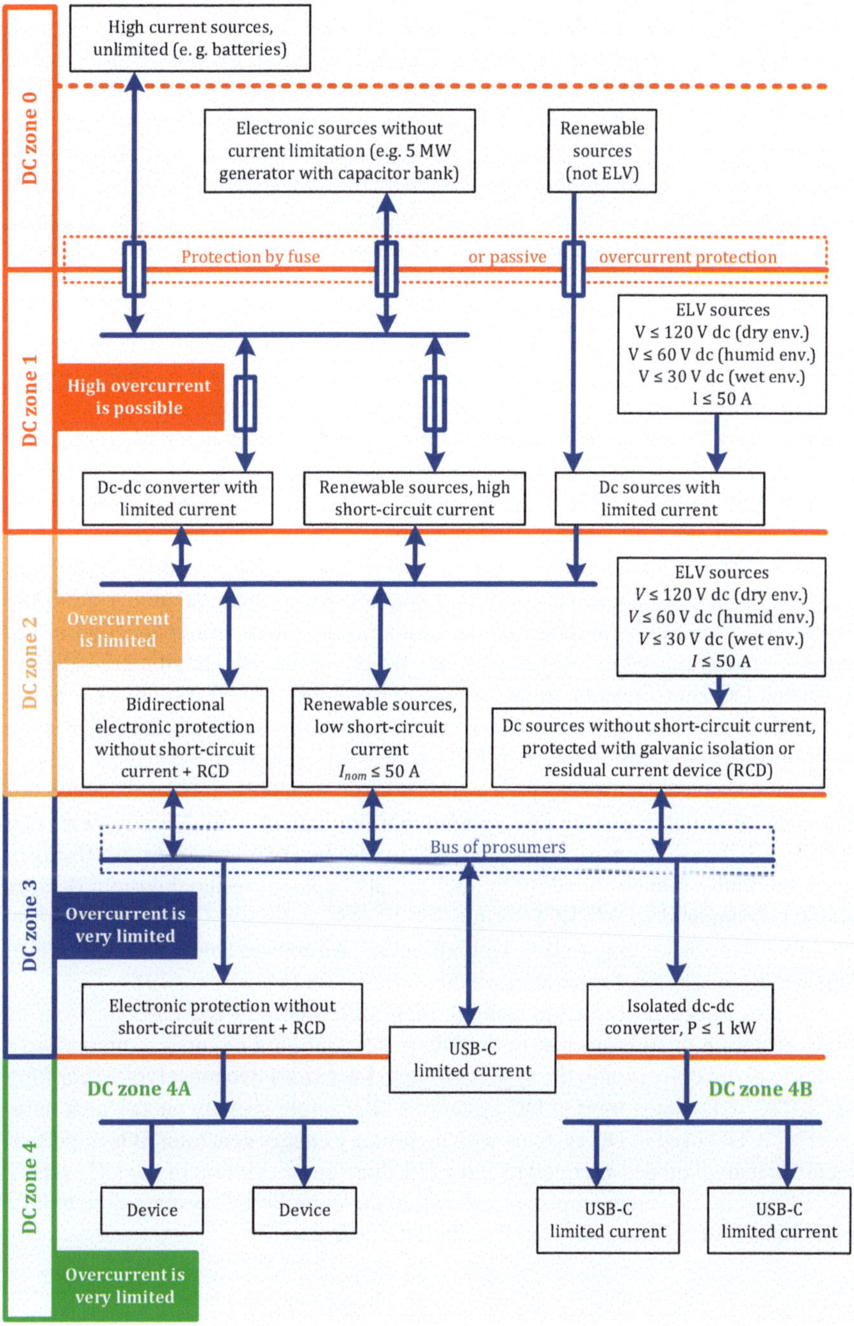

Fig. 2 LVDC protection zones defined in NPR 9090. Zones 2 and 3 are not always present, and there could be a direct link between Zones 1 and 4

extra low voltage (ELV) energy sources could be used, but their maximum allowed voltage depends on the environment (dry, humid, or wet.). In Zone 1 the nominal current is limited to 500 A, and, consequently, the protection is done using fuses or circuit breakers rated for high currents are used. Zone 2 contains devices with maximum current of 50 A. Zones 3 and 4, at limited currents, can benefit from modern semiconductor-based protection, like solid-state circuit breakers. Zone 3 contains prosumers, i.e., devices that can both generate and consume energy. Also, in this zone, voltage is limited to 400 V for unipolar and to 800 V for bipolar DC systems, both operating with current below 50 A. It allows for several energy sources and bidirectional power flow. Zone 4 deals only with energy consumers powered from a single power source, i.e., unidirectional power flow only. Among other recommendations, NEN also recommends not to distribute capacitive loads to avoid tripping protection.

Earthing of DC systems is different from that in AC systems, which is mostly caused by corrosion that appears due to metal contacts. As a result, residential DC systems utilize only TN type of earthing when one pole of the energy generator is connected with the earth, which diminishes most of the issues with corrosion. In the AC grids, TN-S type of earthing is most common as it has a separate Neutral and PE wires, which are connected only at a single point at the generator. Due to the presence of capacitive couplings to the ground, IT systems (no connection to the earth) are not preferred in DC installations. Based on the risk classification and the associated DC zone division, in DC zone 1 to 3 and 4A earthed systems (TN-S) are preferred. In practice, systems in DC zone 4B are usually designed as an IT system, due to the USB-C connection. Therefore, Zone 4 was split into two groups.

In some systems, extra measures could be taken to achieve improved safety. For example, the Dutch company DC Systems BV implements a 48 V safety wire that defines system operation—it is de-energized in case of a fault, which causes the turn-off of the faulty branch. In addition, DC can also contain residual current devices (RCDs). Compared to AC RCDs rated for 30 mA, domestic DC RCDs could be rated for 80 mA according to IEC TS 63,053 Ed1. Moreover, Dutch regulations also consider 20 mA RCDs for home appliances (based on IEC TS 60,479).

It is also important to distinguish the main differences of DC earthing from its implementation in AC systems. In the DC system earthing requires semiconductor diodes to avoid corrosion in the system. Figure 3 provides two most typical earthing types: TN-SCD that is used in DC systems with a single primary energy generator, and TN-SCDD used in DC systems with no primary energy generator. It is important to ensure that all diodes are rated to carry full short-circuit current of the DC system. Even though the given examples are shown in the unipolar DC system, they will be implemented in the same way in a bipolar DC system.

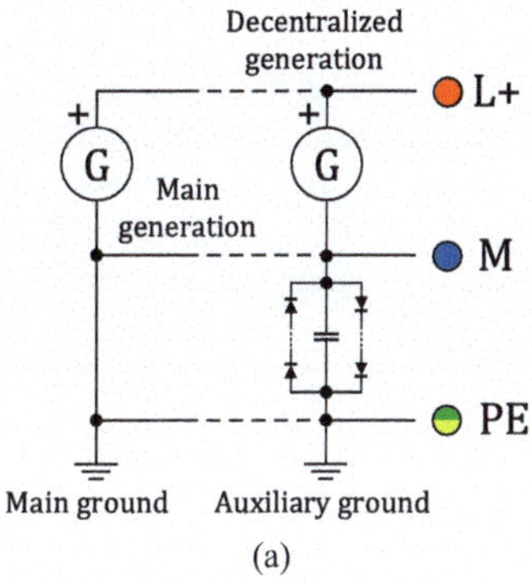

Fig. 3 TN-S earthing implementation in DC systems: single-grounded TN-SCD **a**, multiple-grounded TN-SCDD **b**

2.3 *Wiring Standards*

As in the AC systems, the DC systems require proper standard of wire colors. The existing recommendations take into account retrofitting of AC buildings into those based on DC using the same wire colors. There are two main wiring schemes: one defined by the IEC and the other by the NEN (Dutch standardization authority). Both color schemes are presented in Fig. 4. As the figure shows, the IEC standard is more confusing in practice; the M wire carrying the earth potential can be blue, white, or red, while the coloring is consistent in the NEB standard. Classical 3- and 5-wire household cables could be easily retrofitted to operate with DC as shown in Fig. 5.

Fig. 4 Wire color coding in DC systems according to: IEC **a**, and NEN **b**

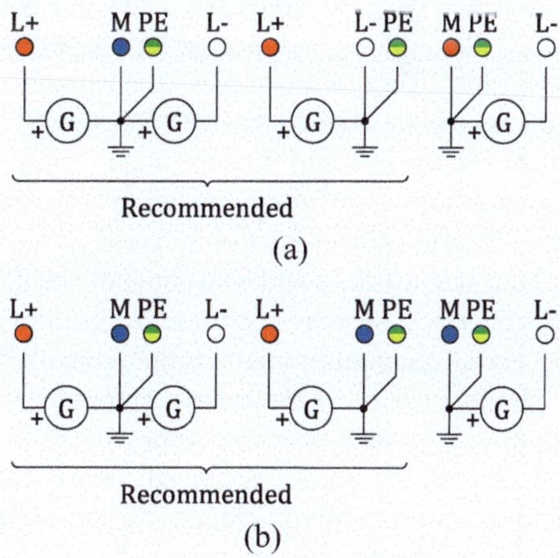

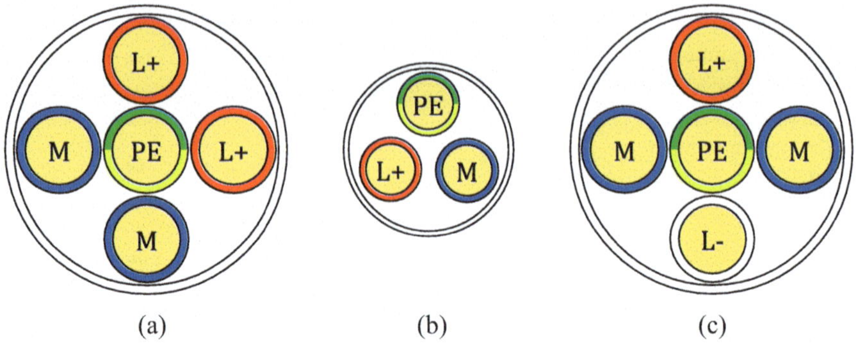

Fig. 5 Use of 5-wire cable in a unipolar DC microgrid **a**, 3-wire cable in a unipolar DC microgrid **b**, 5-wire cable in a bipolar DC microgrids **c**

3 Control and Communication

3.1 Residential Dc Grid Control

Today, the excess of on-site renewable energy is mostly fed back to the grid. However, the grid capacity is a limited resource and in many countries it is approaching its limits. Thus, in the near future, residential buildings may be expected to have also local storages. As a result, such buildings possess independent controls and can operate either in grid connected or islanded mode. A DC building should be able to switch smoothly between island and grid connected modes. In all operating modes, the DC building must maintain regulated voltage. In the island mode, the needed power will be provided by the renewable sources and storage. In the grid connected mode, the building can decide how much energy is taken from the utility grid and how much from local sources. Such small-scale DC microgrids (typically less than 20 kW) can be classified as DC nanogrids [16–19].

A microgrid could be controlled centrally by one master device [19, 20] but it requires high-speed communication between the master and other devices. Since the microgrid is a power electronic-based grid, it is highly dynamic covering a wide range of transient times. It will be technically challenging to design a suitable communication interface. Moreover, economic feasibility will be another issue to face. Therefore, decentralized control methods are typically implemented in microgrids.

Studies in [21–2425] propose hierarchical control for AC and DC microgrids. This multilevel control structure combines advantages of the classical control methods, i.e., centralized and decentralized control. Hierarchical control consists of three main control levels: primary, secondary, and tertiary. The primary control is a decentralized local control strategy responsible for active power sharing and voltage control. Another goal is to avoid circulating currents. Due to reliability reasons, no communication will be implemented. The secondary control level is based on the centralized control structure. It deploys slow control loops (typically 5–15 min) and low bandwidth communication. The secondary control is improving the control result by restoring amplitude and/or frequency deviations caused by the primary control. The tertiary control is managing energy flow between the microgrids and between the utility and the microgrid.

The goal of the dc nanogrid control is to provide power balance and stability in fluctuating loads, sources and storage. This can be achieved either via load shedding or proper storage control [16]. Typically, droop control with DC bus signaling is used for that purpose [1926]. Droop control belongs to the primary control level. In [20–2728], the two-level control structure for DC nanogrid was proposed. The first level of the control utilizes droop control according to the storage state of charge. As a result, the steady state DC bus voltage will be variable and could have deviations from the nominal value. If needed, the secondary control can restore nominal dc bus voltage levels by utilizing low bandwidth communication. Studies in [23] propose the secondary control for improving the current sharing accuracy in DC microgrids. Since DC droop control shares power based on the local voltage, its performance is susceptible to line impedance. Long lines will have noticeable voltage drop that will reduce the accuracy of the droop control. However, in the case of a residential nanogrid, the line impedance is negligible and this problem is non-existent.

In [26], a three level control for DC buildings was discussed. Those principles are also adaptable to DC nanogrids, however usually not implemented. It generally requires a central control unit that could also communicate with a distribution system operator. The economic feasibility of this solution is questionable. Thus, only the primary control level is typically used for DC nanogrids.

The control of DC nanogrids is considered simpler than that of AC microgrids. It does not require frequency or reactive power control. The main control parameter is the DC bus voltage that is typically regulated via the droop control. The droop control could be divided into two types [28]: voltage-current (*V-I*) and current–voltage (*I-V)* droop. Algorithm (1) calculates the output voltage reference based on the measured output current as follows:

$$v_o^*(t) = v_{dc} - i_o(t) \cdot r_d, \tag{1}$$

where $v_o{}^*$ is the reference output voltage of the converter, v_{dc} is the output voltage of the converter in no load condition, i_o is the output (load) current of the converter and r_d is the droop coefficient (virtual resistance).

The *I-V* droop algorithm is used to calculate the output current reference of the converter as follows:

$$i_o^*(t) = \frac{v_{dc-}v_o(t)}{r_d} \tag{2}$$

The droop coefficient must be calculated for each distributed energy source (DES) separately [29]. The main goal of the droop control is to share power and avoid circulating currents between DES. All the sources must be loaded proportionally, and overload must be avoided. This can be done by selecting the correct droop coefficients. The droop coefficient can be calculated as follows [27, 29]:

$$r_d = \frac{\Delta v_{dc}}{I_{o\_max}} = \frac{\Delta v_{dc} \cdot v_{dc\_min}}{P_n} \tag{3}$$

where Δv_{dc} is the DC bus voltage variation range, I_{o_max} is the maximum output current of the corresponding DES, v_{dc_min} is the minimum DC bus voltage value and P_n is the rated power of the DES.

In general, the *V-I* droop is a double loop and *I-V* is a single loop control. The *V-I* control loop consists of an outer voltage and an internal current loop, as shown in Fig. 6a. The droop algorithm (1) derives output voltage reference that will be used by the voltage regulator $G_V(s)$ to program reference for the internal current loop. In the *I-V* droop method, only the current regulator will be used. The droop algorithm (2) directly calculates the current reference, as shown in *b*. Usually the current regulator $G_I(s)$ and the voltage regulator $G_V(s)$ are the PI type regulators.

The *V-I* droop algorithm is the most widely used. However, in the case of the *I-V* droop algorithm, the converter output impedance will be always less than the virtual resistance r_d, which means that the converter regulates its output voltage without any over- or undershoot [28].

The droop control can effectively handle similar types of DES e.g., wind and solar power. But problems occur if there are more sources with different nature e.g., storage or active front-end converter for AC grid connection. Those are bidirectional converters with different control strategies. To handle different loads, sources and storage effectively, DC bus signaling has been applied on top of the droop control. It is a distributed control method for power management in a decentralized grid without a direct communication link [19]. The information will be extracted from the DC

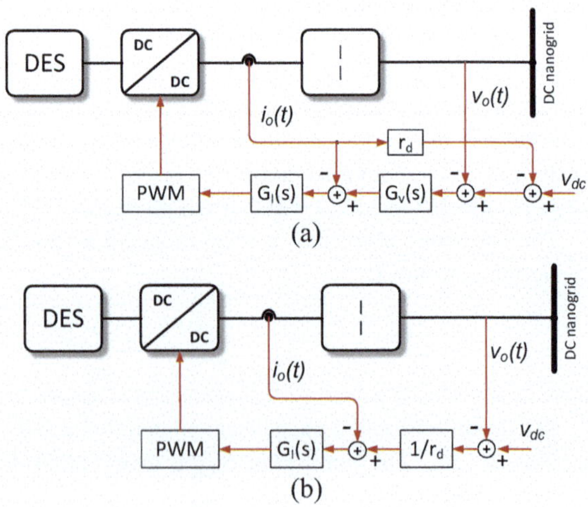

Fig. 6 Main types of droop control: V-I droop control **a**, I-V droop control **b**

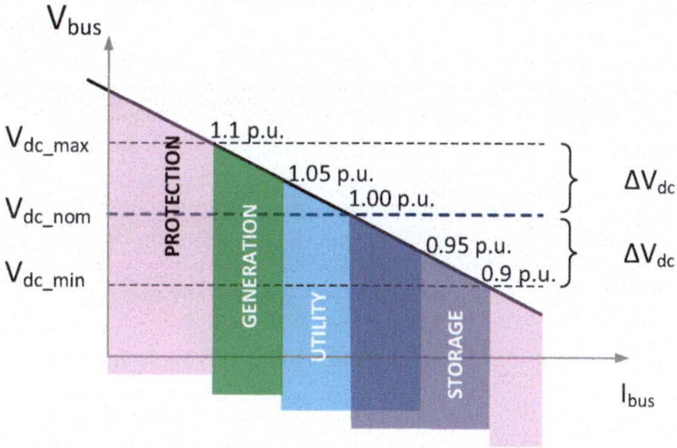

Fig. 7 Typical DC bus voltage in predefined zones

bus voltage that has predefined zones. In each zone, a certain mode will be activated. In DC micro- and nanogrids the following modes are commonly used [23]:

- Generation dominating mode. PV and wind energy converters regulate the DC bus voltage.
- Utility dominating mode. Active front-end converter has the priority and regulates the DC bus voltage. This mode is only available in grid connected operation.
- Storage dominating mode. The storage converter regulates the DC bus voltage.
- Protection mode. The bus voltage is either too high or low. Some protection measures must be applied.

The zones are defined corresponding to the DC bus voltage. The main idea behind DC bus signaling is to prioritize the converter groups connected to the dc nanogrid. If all sources regulate the bus voltage at the same time, the result could be instabilitiy or collapse of the nanogrid. Therefore, in each zone, only the corresponding group of converters can regulate the bus voltage, i.e., work in the droop control mode. All the other converters will work in the current control mode. Depending on the chosen voltage values some zones could also overlap (Fig. 7) e.g., utility and storage mode.

The order of the zones defines the priorities of the energy sources. Depending on the application the order of the zones could be different; typical order for a residential DC nanogrid is presented in [30]. Renewable sources in the generation zone have the highest priority i.e., their output is limited by the droop curve at a very narrow range (Fig. 8) and MPPT operation is implemented most of the time. On the other hand, the storage is shifted into the lowest zone; thus, it will be utilized if all the other sources have been depleted.

The DC bus voltage can fluctuate in certain limits. The typical fluctuation is $\Delta V_{dc} = \pm 10\%$ [30–3233]. If the bus voltage grows over this limit, overvoltage protection will be activated. Similarly, if the bus voltage drops lower than that limit, the undervoltage protection will be activated.

All the converters connected to the the DC nanogrid continuously monitor the DC bus voltage and establish the scorresponding mode. The utility dominating mode $0.95V_{dc} \ldots 1.05V_{dc}$, is managed by the GIC converter, as depicted in Fig. 8. The GIC

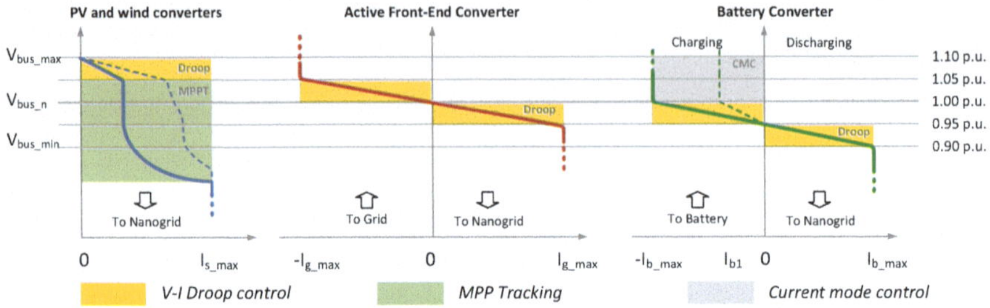

Fig. 8 Principle of droop control with dc bus signaling

is using the V-I droop control to regulate the DC bus voltage. In the first quadrant the DC bus voltage is lower than the nominal value, which refers to the lack of energy. The generation from renewable energy sources (RES) cannot cover consumption. Thus, the GIC draws additional energy from the utility grid and supports the nanogrid. In the second quadrant, the DC bus voltage is higher than the nominal value, which means over generation in the nanogrid. Thus, the GIC injects energy back to the utility grid.

In the generation dominating mode $1.05V_{dc} \ldots 1.1V_{dc}$, the GIC has reached its maximal current limit and goes into a constant current mode while still injecting energy to the grid. But it cannot increase its power level and the DC bus voltage must be limited by other means. It is done by limiting the power generation from RES via the droop control. Starting from the bus voltage value $1.05V_{dc}$ RES will stop MPPT and start the V-I droop control, as shown in Fig. 8. The droop control increases the DC bus voltage linearly while proportionally also reducing the output current of RES. By the time the DC bus voltage reaches its maximum value, the current from RES will be reduced to zero. Thus, there will be no more generation from RES at that point. In the storage and utility mode (DC bus voltage $< 1.05V_{dc}$), the RES will be controlled by the MPPT algorithm maximizing its power output. This operation range is presented by an exponential curve, as shown in Fig. 8. However, due to stochastic nature of wind and solar irradiation the power output of RES is continuously fluctuating, and that curve could have any kind of shape.

Storage control covers the lowest part of the DC bus voltage $0.9V_{dc} \ldots 1.0V_{dc}$.. Note that it partially overlaps with the utility mode. This means that the storage and the GIC are working in parallel. Thus, the dc-bus voltage is defined by both sources. There are different strategies for storage management, but the most common method utilizes storage as a backup power source that supports the nanogrid only when other sources are depleted. In the first quadrant when the current is positive, the storage unit will be discharging. The energy is given into the nanogrid. If the current changes to negative (second quadrant) the storage will be charged. Similar to GIC, also the storage converter has maximum and minimum current limits. Between those limits the *V-I* droop control will be applied. Having reached one of the current limits, the storage converter starts working in a constant current mode. In contrast to the GIC, the current limits of the storage converter are adjustable and can be modified by the battery management system. Depending on the SOC, the current limits will be changed from zero to the rated value, as shown in Fig. 8.

3.2 Communication in Nanogrids

The communication networks in the DC microgrids can be divided into three main categories: home area network (HAN), neighborhood area networks (NAN) and wide area networks (WAN) [33]. In general, those categories are differentiated mainly based on the range. HAN covers distances from 0 to 100 m, NAN from 100 m to 10 km and WAN from 10 to 100 km. The scope of this section covers the HAN network solutions.

The main idea of DC bus signaling (DBS) is to avoid an external communication channel between the loads and sources of the microgrid. Avoidance of the external communication channel will increase the reliability and simplify the infrastructure of the microgrid. However, DBS has its limitations that make it unsuitable in some applications, e.g., variable DC-bus voltage could be a problem for some sensitive loads. Relatively small DC-link voltage range could limit the reliable operation modes, the droop control is susceptible to the line impedance that will cause additional voltage drop and the distorted DC-bus voltage could lead to incorrect operation modes. Another possible scenario could be the case, where the nanogrid includes different types of energy storages that require separate energy management. In such cases, some additional communication method would is needed.

Use of an external low-bandwidth communication link is typical for communication with the distribution system operator (DSO). However, there are other applications as well. In [20], an external communication has been proposed for the DC nanogrid to compensate the DC-bus voltage shift caused by the droop control. The master control unit can order new set points from the slaves and thus, shift the droop curves. Low-bandwidth communication has also been proposed for smart battery management. The GIC could shift droop lines of the storage unit corresponding to renewable energy availability, load profile and grid electricity price [30].

The most common communication technologies used in smart grid and smart buildings are compared in Table 1. In general, there are two communication mediums available: wired and wireless communication. Wired communication is preferred if high reliability is required. Many papers describe wired communication implementation in micro- or nanogrids. The power line communication (PLC) has been widely used by DSO to switch off corrupted parts of the network or read data from smart meters. The main benefit of PLC is that it utilizes existing infrastructure and thus, has lower installation cost [34]. However, the power lines are a rather harsh and noisy environment, which will interfere PLC and reduce the signal quality. Instead of DBS, [34] power line signaling (PLS) is proposed instead of DBS in [35]. It is a distributed control strategy in which the units inject sinusoidal signals of specific frequency into the common DC-bus to communicate. Unlike PLC where digitalized data is sent over the powerline, the PLS is using analog signals to trigger transitions between

Table 1 Communication technologies for DC buildings [42]

Wireless communication	Data rate	Range
PLC (for nanogrid)	<3 Mbps	<3 km
Ethernet	<10.000 Mbps	<100 m
LoRaWan	<50 kbps	<3 km
Bluetooth	721 kbps	<100 m
ZigBee	250 kbps	<100 m
Wifi	54 Mbps	<100 m

different modes. The algorithm can be implemented in the primary control level and no additional communication channel will be needed. However, the drawback is the high frequency noise injected into the DC-bus voltage.

In the large-scale systems, e.g., communication with DSO, a central supervisory controller with low-bandwidth communication will be used. [35] Multidrop serial communication via the RS485 standard is proposed; in [36], Ethernet LAN is used for communication and in the study in [37], demonstrates the communication with the supervisory controller over SPI.

In [38, 39], the inherent delay of the low-bandwidth communication was studied. It was shown that the delay could have critical impact on the control stability, but the situation could be improved with the gain scheduling approach.

Wireless communication methods are more sensitive to interferences; on the other hand they do not need an expensive infrastructure and can be easily used in remote areas or adopted in already existing systems without any reconstruction work. Also, the range is comparable with wired communication mediums (Table 1).

ZigBee is a low bandwidth wireless communication that has low power consumption, low implementation cost and it is simple to implement. ZigBee is used for smart meters, lightning control, energy management, and home automation in general. However, it is based on the 2.4 GHz band (standard IEEE 802.15.4) and could be interfered from other appliances that use the same frequency band e.g., Wifi and Bluetooth. Moreover, the vicinity of ZigBee could corrupt the whole communication channel. Thus, schemes of interference avoidance must be applied [34].

Similar to ZigBee, Wifi is also a cost-effective method for low bandwidth communication in smart grid applications. Wifi uses the 2.4 GHz band (standard IEEE 802.15.4) and typically 54 Mbps data rate (Table 1). In [40], Wifi communication link is compared with ZigBee and Ethernet. It is shown that Ethernet performs better than Wifi or ZigBee when end-to-end delays are considered. However, Wifi has smaller delays compared to ZigBee.

The long-range wide area network (LoRaWAN) has been proposed for vehicle-to-grid systems in smart cities [41]. LoRaWAN uses open-source technology and transmits over unlicensed frequency bands. Due to low power consumption and long range, it is an attractive technology for many smart grid applications.

4 Ancillary Services

In addition to its primary role of managing the residential DC microgrid (e.g., using droop control) and basic management of power flow between the AC and DC ports, the GIC is capable of providing support functions and advanced interactions with the AC grid by delivery of various ancillary services (AS) [43, 44]. The AS are commonly provided to transmission and distribution system operators (TSOs/DSOs) to support the grid and keep its operation within the predefined limits. Typically, the main target of the AS provider is to keep the grid frequency and voltage within the predefined range (Fig. 9a) and restore back to nominal values after any imbalance event [45].

Traditionally, the AS are delivered only at the transmission level and are provided by relatively large entities: either third parties or by TSOs/DSOs themselves. The increase in distributed renewable generation steadily displaces the established centralized arrangement of the grid, creating new challenges for the power system and its stability. The aggregation of a large number of small distributed renewable energy generators for provision of AS is identified as one of the options to overcome this problem [46]. Since the residential DC microgrid with renewable (PV) generation and the GIC is typically considered for near zero or positive energy buildings [7, 47], it can be ultimatelyseen as a prosumer or net producer of energy from the AC grid side. Therefore, a DC building could be treated similar to the converter-interfaced distributed energy source (CIDES). The presence of storage or integration with heating, ventilation and air conditioning of a building allows even flexibility to be obtained in terms of demand-response (DR) and demand-side management. The existing and emerging AS that could be provided by converter-interfaced systems, including DC home with the GIC are presented in Fig. 10 and discussed in brief in the following [4744].

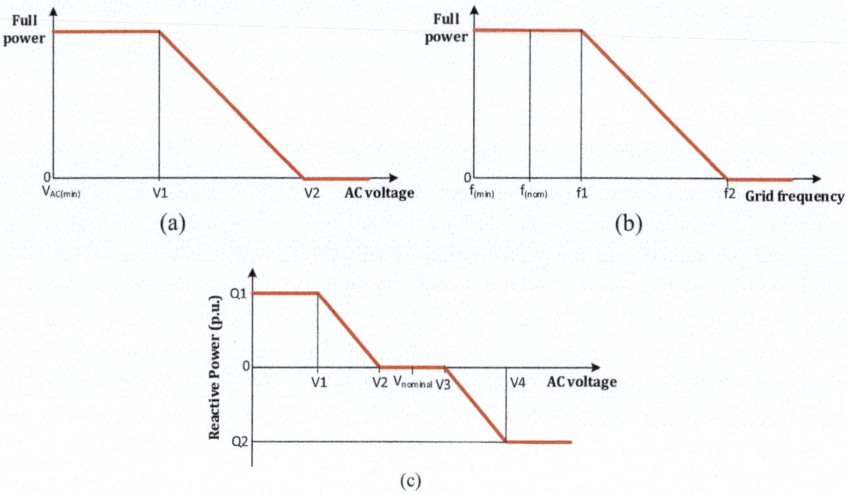

Fig. 9 Generalized representation of output power limitation **a**, frequency response **b** and reactive power regulation **c**

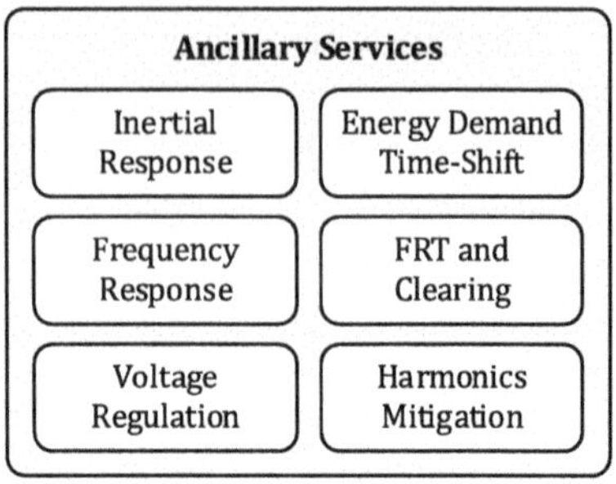

Fig. 10 Potential AS provided by the GIC

Inertial response

The frequency of the power grid is continuously changing in accordance with the balance between the produced and consumed power. A temporary imbalance results in a change in frequency as stored kinetic energy in rotating masses of traditional synchronous generators (SGs) is stored or released. Compared to traditional SGs and some loads (e.g., motors), because of no rotating mass, the converter-interfaced systems inherently provide little or no inertia to the grid. With the increased penetration of such systems and simultaneous withdrawal of traditional fossil fuel generation, the overall inertia of the grid will decrease, potentially leading to issues with frequency deviations and gradients. To address this, significant research efforts are focused on making CIDES behave similar to conventional SGs by means of providing virtual or synthetic inertial response [48]. The coordinated use of fast-acting energy storages (e.g., flywheels, supercapacitors, batteries) enables converter-interfaced systems like DC home GIC to provide inertia to the grid. However, provision of such AS implies additional costs, due to increased stresses on storage. At the same time, as inertial response is currently considered as an inherent support function provided by SGs, and no established methods for measurement and quantification of virtual inertia exist so far. Still, some studies consider inertial response as a commodity to be traded [45].

Frequency response

Unlike virtual inertial response, which mimics the behavior of SGs, the (fast) frequency response provides active power proportional to frequency deviation [49] (9b). Until recently, it was provided by large units automatically or manually, depending on its type (primary, secondary, tertiary). The properly controlled GIC is capable of providing such AS by varying its power flow. For example, the flexible loads can be turned-off and battery storage used in order the house to temporarily inject power to the grid instead of consuming it. Likewise, the consumption can be increased by storing the excess grid energy in the battery or water heater. In some cases, provision of the AS will result in lost opportunity cost (e.g., when the PV generation was clipped) or wear out cost if the storage system exhibited extra charge/discharge cycle.

Newer standards specify the requirement of CIDES assigned as the fast primary frequency control (which is triggered automatically in a decentralized fashion) [5150]. The GIC can potentially follow similar regulations; however, additional research and verification is required since the base power profile of the GIC is different from that of a typical CIDES.

Functioning as a secondary and tertiary reserve with centralized/manual activation is also possible [52]. In this case, the aggregation of GIC is necessary, preferably in conjunction with other grid-connected devices of similar power level.

Voltage regulation

The provision of reactive power for voltage regulation is an established and remunerated AS in transmission networks. At the distribution levels, it is generally under the responsibility of DSO or a customer [47]. Typically, the reactive power is injected locally by automatic voltage regulators or static VAR compensators to achieve the required voltage profile at a specific location (Fig. 9c). As reactive power cannot be transmitted over long distances, there is a big opportunity of small CIDES to provide it as AS. The realization of this function using the GIC, however, would result in additional costs—certain oversizing would be required to allow simultaneous provision of rated active and reactive power. Moreover, the realization of coordinated hierarchical control with good dynamic response still remains an issue that is actively being addressed in various ways [5453]. Other issues are related to proper quantification of the reactive power at the distribution level [55].

Fault ride-through and clearing

Being designed for unidirectional power flow, with increased penetration of CIDES the protection systems for distribution grids are facing new challenges, like the risk of undesirable grid islanding, loss of protection coordination, etc. [56]. To reduce these risks, the CIDES have been forced to disconnect until the fault is cleared. This has led to a situation when simultaneous tripping of CIDES could potentially lead to the undersupply of the European grid that was not possible to compensate using existing reserves [57].

The most recent grid codes now require the fault-ride through (FRT) capability for distribution grids [5051], as shown in Fig. 11. The FRT is defined as the voltage/frequency and time duration profile for the CIDES to follow without disconnection. Potentially, this profile can be adjusted taking into account the various characteristics of the grid, such as short-circuit capacity. Moreover, the CIDES should feature a controlled power feed-in rate—so called "soft start ramp rate", both under nominal conditions and after grid fault (with time delay).

Harmonics mitigation

Various present standards, like IEC 61,000 series, IEEE 519 and IEEE 1547–2018 limit the harmonic pollution of the currents consumed or injected to the distribution grids without considering the voltage distortion at the particular node. With the increased proliferation of non-linear loads and continuous grid expansion, the harmonic content of some nodes may exceed the specified values. The grid-connected

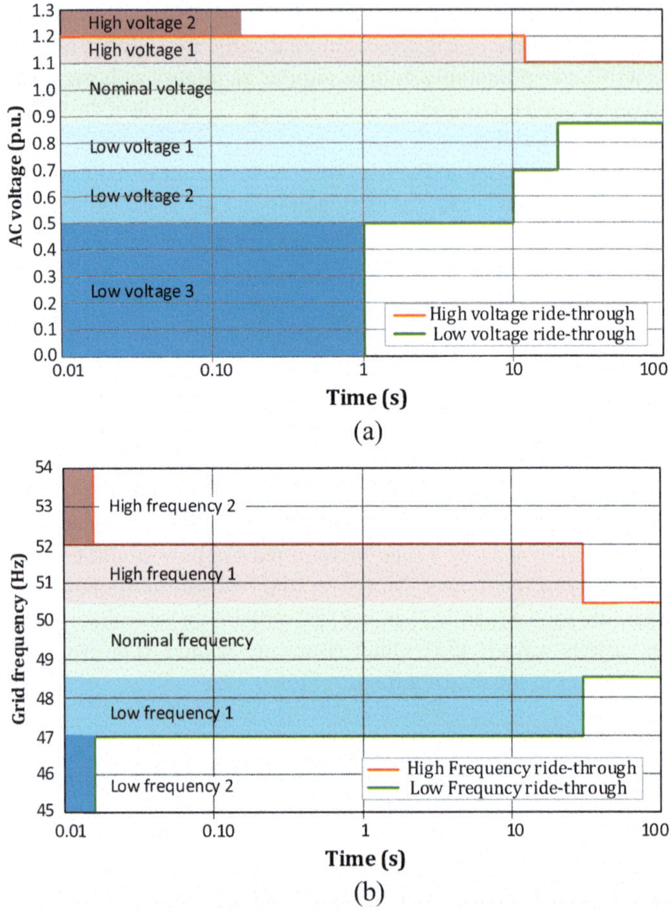

Fig. 11 Generalized low-and high-voltage ride through characteristics **a** and low- and high-frequency ride through characteristics **b**

converter can operate as an active harmonic filter by injecting/sourcing an appropriate harmonic current in addition to the rated sinusoidal one [58]. To enable this as an AS, the existing grid code needs to be modified to define and differentiate these compensating currents. The capability of harmonic order compensation is associated with the switching frequency of the GIC. Similar to voltage regulation, provision of this AS for the GIC may result in certain power dissipation and operational cost increase.

Energy demand time-shift

More complex energy generation patterns that come together with CIDES, create challenges for conventional generators to maintain the grid stability. For a more smooth and predictable generation/load profile, many of the CIDES are equipped with storage systems. Assuming that the home DC grid features energy storage and/or utilizes droop control of flexible loads (e.g., water heater), the GIC can perform adjustments in its operation based on energy arbitrage or other external signals. In order to get the financial benefits, the scheduling should take into account battery degradation and other operational costs [59]. With the installation of utility-scale batteries, some markets will experience large revenue drop for energy arbitrage, thus provision of other AS for small CIDES may end up being more profitable.

Generalizations

Thus, some of the potential AS of small converter-interfaced distributed generators are complementary to those traditionally present on the market. The capability of provision of most of these AS would result in added certain costs for the CIDES [60]. The remuneration models currently lack established measurement and quantification methods and need to solve a range of technical, regulatory and financial barriers. The perspective business models are identified in [44].

Alternatively, it is possible that in the future the provision of certain AS would be required as a condition for interconnection rather than a remunerated service. For example, some of the recent grid codes are already regulating the provision of various services as system support functions. The examples include the updated IEEE Standard 1547–2018 and the EU Standard EN50549-1:2019. These regulations describe such features of grid-connected inverters such as fault ride-through (FRT) capability, follow frequency-droop specifications, ramp rate limitation and are requested to provide the reactive power support.

In general, there are no insurmountable barriers to GICs to provide any of the above-mentioned main ancillary services. Therefore, it should share the functionality and features with other state-of-the-art grid-connected power electronic devices of similar power range: battery storages, PV inverters (including hybrid) and, possibly, electric vehicle chargers. The set of supported AS will depend on the configuration of the particular DC microgrid: presence of local generation and its type, storage and available load flexibility.

To participate in the ancillary service market, such distributed small-scale systems are generally grouped to higher power level clusters and coordinated via an aggregator [616263]. Thus, the decision-making role for provision of certain ancillary services can be external to GIC, which itself has to follow the received operating instructions. Since the GIC incorporates the functions of PV and BESS, it is identified as one of the most perspective aggregated AS providers [47]. Until recently, aggregation of residential devices was mostly limited to DR programs based on air conditioning and water heaters. Recently, several experimental and commercial residential "hybrid" inverters and battery storages started to participate in the frequency response and voltage regulation markets. Sonnen company implements one of the known operational solutions. Together with Next Kraftwerke they are using commercial residential batteries to provide ancillary services primary and secondary frequency response. The virtual power plant is made via aggregation of small (<25 kW) storage units into MW blocks, which are participating in the energy market [64].

5 Power Electronic Solutions for AC Grid Interfacing

The power electronics is a key enabling technology for future electrical systems featuring capability of informed decisions, collective actions and advanced interactions with the grid 103. To satisfy the safety requirements, the GIC should feature galvanic isolation between its AC and DC terminals [104] . This imposes substantial limitations on the configuration of the GIC topology. In fact, one can find some similarities with electric vehicle (bidirectional) chargers, which feature the AC to DC conversion with galvanic isolation as an established industrial practice [65]. Although the isolation requirement can be obtained by using a grid-frequency transformer, this would result in an excessively large and heavy system; therefore, most of the modern isolated converters feature compact high-frequency (HF) transformers.

Typically, the HF-isolated grid-connected converter is composed of two stages—AC-DC and isolated DC-DC, interfaced by an intermediate DC-link capacitor [656667] (Fig. 12). At the same time, some emerging topologies also offer single- and two- stage configurations without (evident) DC-link that can bring benefits in terms of volume and cost [68, 697071].

5.1 Non-isolated AC-DC Stage

Considering the residential application voltage levels, the baseline converter for the first bidirectional rectifier AC-DC stage is a two-level three-phase bridge with a low-pass filter (*L*, *LCL*) per phase at the grid side (Fig. 13a). If connected to a stabilized DC link, the corresponding diagonal devices synthesize the sine wave current using pulse-width modulation (PWM) with 1/6 third harmonic injection or space-vector modulation (SVM).

A range of research works and technology demonstrators has shown that the use of a three-level topology can in fact result in increased efficiency of the stage, particularly at higher switching frequencies [7273]. This is related to reduced commutation voltages and the possibility to use lower voltage devices with smaller on-state voltage drop. Moreover, due to additional voltage levels, dimensions of the low-pass filter can be reduced.

The neutral point clamped (NPC) converter (Fig. 13b) is one of the most widespread three-level topologies, traditionally used at medium voltage levels. The additional diodes clamp the transistor voltages to the capacitor divider mid-point.

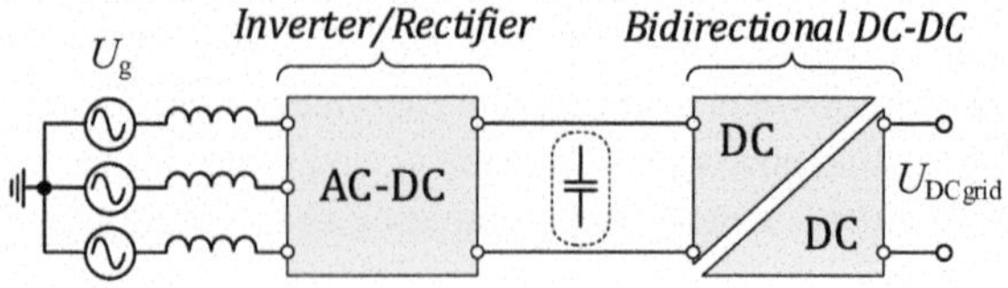

Fig. 12 Two-stage GIC configuration with intermediate DC link

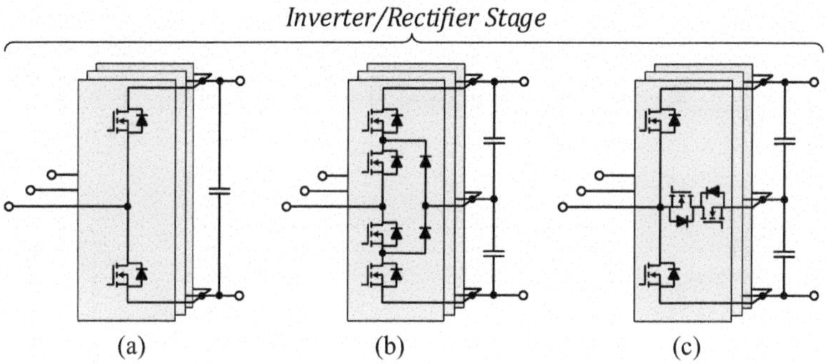

Fig. 13 Bidirectional AC-DC stage configurations: two-level bridge **a**, three-level NPC **b** and three-level T-type **c**

Therefore, all the devices can be rated for half the voltage level when compared to the two-level bridge topology. To achieve better flexibility in the switching states and uniform power losses, the clamping diodes can be substituted with transistors, forming an active neutral point clamped topology (A-NPC) [7475]. Another notable three-level topology is a flying-capacitor converter [76]. It uses a reduced number of semiconductor devices, but needs additional measures to provide voltage balancing.

The Vienna-rectifier (T-type)-based converter is another known three-level topology [77] (Fig. 13c). When compared to the NPC topology, it features a reduced number of semiconductor devices and gate drive supplies. However, half of the devices should be rated for full voltage. Still, the advantages of additional voltage levels can make this topology a potential candidate for residential GIC. In particular, it has been shown that the T-type converter achieves better efficiency at higher switching frequencies than other three-level topologies [7978].

5.2 *Isolated DC-DC Stage*

The DC-DC stage is responsible for provision of galvanic isolation and droop control/regulation of the home DC microgrid voltage. Although there is a range of topologies applicable to this stage, for the target power levels (5…20 kVA?) bridge-type topologies are typically applied.

The dual active bridge (DAB) (Fig. 14a) is an extensively used DC-DC topology that can reach very high maximal efficiency values, thanks to zero voltage switching (ZVS) of the semiconductor devices [80]. It is achieved by introducing a certain phase shift between the commutations of primary and secondary bridges. Due to increased energy circulation, the operation outside the nominal point resulted in significant efficiency reduction, therefore, more comprehensive modulation methods were proposed—dual- and triple phase-shift [81, 82], combined phase shift and frequency [83] as well as asymmetric duty cycle control [84]. In general, these

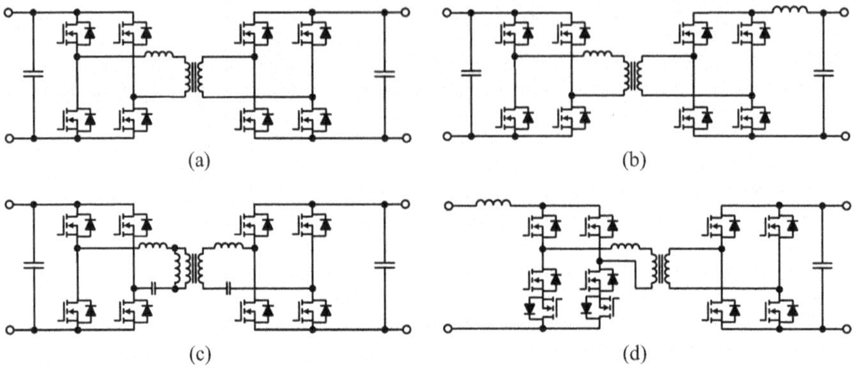

Fig. 14 Various DC-DC stage configurations: DAB **a**, PSFB **b**, CLLC **c** and ASMC **d**

advanced modulations require significant computational efforts, may be sensitive to component parameter variation, and therefore, can be difficult of take full advantage of in some practical applications [85]. Moreover, the DAB inherently imposes a relatively large current ripple on the DC-link capacitor [86].

The CLLC (or CLLLC) resonant converter is a widespread topology used in bidirectional applications, thanks to its symmetrical structure [87] (Fig. 14b). It achieves the highest efficiency when operated near the resonant frequency. Thanks to its sinusoidal current shape, it reduces stress on the HF isolation transformer and requires relatively a low value of series inductance. In contrast to the DAB converter, the regulation is provided by varying the switching frequency and as it moves away from the resonant one, and the efficiency is reduced due to increased energy circulation. There are also certain concerns regarding the controllability of the the converter at light loads [88]. Moreover, some studies show that its peak efficiency is lower than that of a DAB [88, 89].

The phase-shifted full-bridge (PSFB) is another broadly used topology that is capable of bidirectional operation (Fig. 14c). Unlike the previously described solutions, it features an asymmetrical structure: one of its ports is coupled directly to an inductor (of the LC filter). The presence of the inductor significantly reduces the current ripple and RMS currents through the converter components [90]. Other advantages include fixed operating frequency, ZVS/ZCS of the semiconductors as well as the relatively simple design and control. On the other hand, the transistors at the inductor side port are hard-switched and exhibit additional voltage stress due to the current mismatch at the end of the boost (freewheeling) interval. A range of solutions has been proposed to mitigate the problem of overshoots—ranging from passive [91] or regenerative [92, 93] circuits to special modulation methods that eliminate the current mismatch [94, 95]. Another approach lies in certain topology modifications for reducing the current circulation [95] (Fig. 14d). Still, the oscillations between the device output capacitance and transformer leakage inductance remain, which together with the boost-mode operation results in an increased blocking voltage requirement of semiconductors at the inductor side bridge [95].

5.3 Isolated AC-DC Solutions

To achieve further improvements in power density, a range of single- and quasi-stage isolated AC-DC solutions have been proposed. Such converters use a relatively small pulsating DC-link or omit the DC-link completely. This potentially allowsa more compact and possibly less expensive design to be achieved. Although most of such systems are addressing single-phase applications, several concepts have been proposed for the three-phase systems.

One group of converters uses a three-phase unfolder based on a three-level topology (see Fig. 15) to form two variable voltages that constitute parts of the sinusoidal waveforms [71, 96–100]. These variable voltages are then sourced by the two DC-DC converters that are responsible for synthesizing the sinusoidal current waveforms. Such principle was realized by series-resonant version of DAB (Fig. 14a), current-source (Fig. 14d) and three-port converters [99]. A variation of this method based on the "soft" DC-link and synergetic control based on partial PWM operation of the three-phase rectifier was proposed in [100].

The second group is a single-stage solution (Fig. 16a) based on isolated matrix-type converters [101] that incorporate bidirectional switches connected between the grid and the transformer to perform direct low frequency AC to HF AC conversion (Fig. 16b). The second stage then rectifies the HF voltage into the DC one. The matrix stage can be voltage- or current-source converter based. In either case, to achieve a satisfactory performance in terms of stresses and power loss, the modulation strategy of this type of converters should be relatively complex.

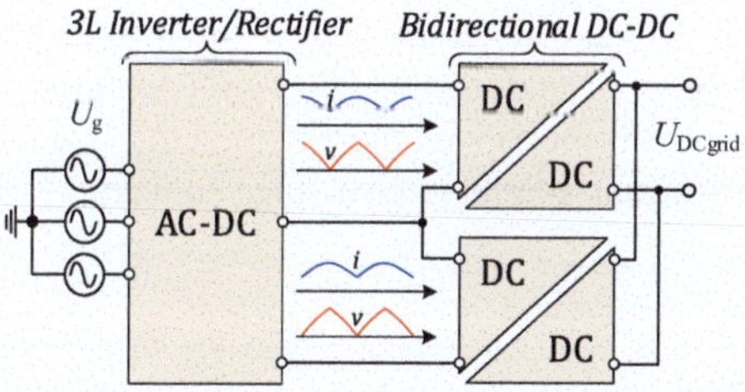

Fig. 15 Quasi single-stage GIC configuration with two pulsating DC links

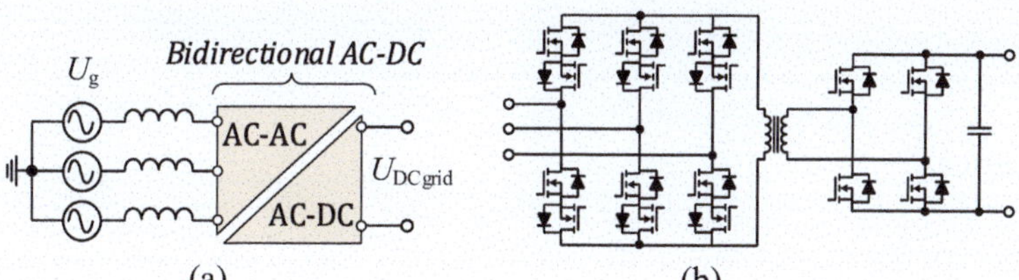

Fig. 16 Single-stage GIC configuration **a**; three-phase isolated matrix converter topology **b**

Despite the potential advantages, the single-stage systems have not yet achieved wide industrial adoption in the kilowatt power range. In addition to complex control, due to the use of bidirectional devices, their overall efficiency does not exceed one of two-stage solutions [102, 103,104].

6 Conclusions

The green transition, backed up by growing penetration of renewable energy sources is continuously shifting the existing centralized power generation to that decentralized, with wide utilization of power electronic converters. he nZEBs, being a part of this process, will become an integral part and an active participant of the future distribution grid. Supplied by PV generation, battery storage and a large number of loads, which are inherently DC, the transition to DC in a building may be one of the remaining sources of increasing the energy savings of nZEBs and one of the key steps towards practical PEBs. However, there is still a lack of knowledge, standards and products that would enable widespread adoption of the residential DC technology.

In this chapter, aspects associated with the transition to DC buildings were analyzed from multiple directions, including DC infrastructure, configurations, safety requirements and existing standards. Dominated by traditional AC technology, in the near term, the DC nZEBs and upcoming PEBs would require grid connection to be realized by means of the GIC. To take full potential of this power electronic interface, such buildings should be treated similar to other CIDES. Therefore, in addition to managing the DC grid with its generation, storage and loads, the GIC should provide a range of advanced iterations with the AC grid. Various aspects of DC grid control in a building were analyzed together with potential services and support functions for the AC grid. Further, given the potential topological realizations of the GIC are provided and various single- or two-stage converter configurations that are addressed.

Acknowledgements This work was supported in part by the Estonian Research Council under Grant PRG1086, in part by the European Union's Horizon 2020 research and innovation programme under the Marie Sklodowska-Curie grant agreement no. 955614 and in part by the Estonian Centre of Excellence in Zero Energy and Resource Efficient Smart Buildings and Districts under Grant 2014-2020.4.01.15-0016, funded by the European Regional Development Fund.

References

1. IEA (2021) Renewable energy market update 2021. IEA, Paris. [Online] https://www.iea.org/reports/renewable-energy-market-update-2021
2. Energy efficiency trends and policies in the household and tertiary sectors—an analysis based on the ODYSSEE and MURE databases. [Online] https://www.odyssee-mure.eu/publications/archives/energy-efficiency-trends-policies-buildings.pdf

3. Directive (EU) 2018/844 of the European Parliament and of the Council of 30 May 2018 amending Directive 2010/31/EU on the energy performance of buildings and Directive 2012/27/EU on energy efficiency, PE/4/2018/REV/1. [Online] https://eur-lex.europa.eu/
4. Ala-Juusela M, ur Rehman H, Hukkalainen M, Reda F (2021) Positive energy building definition with the framework, elements and challenges of the concept. Energies 14(19):6260. https://doi.org/10.3390/en14196260.
5. Magrini A, Lentini G, Cuman S, Bodrato A, Marenco L (2020) From nearly zero energy buildings (NZEB) to positive energy buildings (PEB): the next challenge—the most recent European trends with some notes on the energy analysis of a forerunner PEB example. Develop Built Environ 3: 100019. https://doi.org/10.1016/j.dibe.2020.100019
6. Kumar GMS, Cao S (2021) State-of-the-art review of positive energy building and community systems. Energies 14(16):5046. https://doi.org/10.3390/en14165046
7. Cole RJ, Fedoruk L (2015) Shifting from net-zero to net-positive energy buildings. Build Res Inform 43(1):111–120
8. Santos AFB, Duggan GP, Lute CD, Zimmerle DJ (2018) An efficiency comparison study for small appliances operating in DC and AC in minigrids. IEEE Glob Human Technol Conf (GHTC) 2018:1–2. https://doi.org/10.1109/GHTC.2018.8601861
9. Pantano S, May-Ostendorp P, Dayem K (2016) Demand DC. Accelerating the introduction of DC power in the home. CLASP
10. Strategen Consulting and ARUP Group (2014) Direct-current scoping study: opportunities for direct current power in the built environment. US Department of Energy, Building Technologies Office, Technical Report
11. Patterson BT (2012) DC, come home: DC microgrids and the birth of the 'Enernet.' IEEE Power Energy Mag 10(6):60–69
12. Vossos V, Gerber D, Bennani Y, Brown R, Marnay C (2018) Technoeconomic analysis of DC power distribution in commercial buildings. Appl Energy 230:663–678. [Online]. http://www.sciencedirect.com/science/article/pii/S0306261918312339
13. Directive 2014/35/EU of the European Parliament and of the Council of 26 February 2014 on the harmonisation of the laws of the Member States relating to the making available on the market of electrical equipment designed for use within certain voltage limits. [Online]. http://data.europa.eu/eli/dir/2014/35/oj
14. NL: DC Installations for Low Voltage, Standard NPR 9090:2018 (2018) Royal Dutch Standardization Institute (NEN), pp 1–50
15. Rivera S, Lizana RF, Kouro S, Dragicevic T, Wu B (2021) Bipolar DC power conversion: state-of-the-art and emerging technologies. IEEE J Emerg Select Top Power Electron 9(2):1192–1204
16. Bryan J, Duke R, Round S Decentralized generator scheduling in a nanogrid using DC bus signaling. pp 1–6
17. Hamatwi E, Davidson IE, Agee J, Venayagamoorthy G (2016) Model of a hybrid distributed generation system for a DC nano-grid. In: 2016 Clemson University power systems conference (PSC). pp 1–8
18. Gonzalez-Longatt F, Rajpurohit BS, Singh SN (2015) Smart multi-terminal DC μ-grids for autonomous zero-net energy buildings: implicit concepts. In: 2015 IEEE innovative smart grid technologies–Asia (ISGT ASIA). pp 1–6
19. Schönberger J, Duke R, Round SD (2006) DC-bus signaling: a distributed control strategy for a hybrid renewable nanogrid. IEEE Trans Ind Electron 53(5):1453–1460
20. Wu D, Tang F, Dragicevic T, Guerrero JM, Vasquez JC (2015) Coordinated control based on bus-signaling and virtual inertia for islanded DC microgrids. IEEE Trans Smart Grid 6(6):2627–2638. https://doi.org/10.1109/TSG.2014.2387357
21. Vandoorn TL, Vasquez JC, De Kooning J, Guerrero JM, Vandevelde L (2013) Microgrids: hierarchical control and an overview of the control and reserve management strategies. IEEE Ind Electron Mag 7(4):42–55
22. Guerrero JM, Chandorkar M, Lee T-L, Loh PC (2013) Advanced control architectures for intelligent microgrids-part I: decentralized and hierarchical control. IEEE Trans Ind Electron 60(4):1254–1262

23. Meng L, Shafiee Q, Trecate GF, Karimi H, Fulwani D, Lu X, Guerrero JM (2017) Review on control of DC microgrids. IEEE J Emerg Select Top Power Electron 1–1
24. Bidram A, Davoudi A (2012) Hierarchical structure of microgrids control system. IEEE Trans Smart Grid 3(4):1963–1976
25. Shahid MU, Khan MM, Hashmi K, Boudina R, Khan A, Yuning J, Tang H (2019) Renewable Energy Source (RES) Based Islanded DC Microgrid with Enhanced Resilient Control. Int J Electr Power Energy Syst 113:461–71
26. Boroyevich D, Cvetkovic I, Dong D, Burgos R, Wang F, Lee F (2010) Future electronic power distribution systems a contemplative view. In: 2010 12th International conference on optimization of electrical and electronic equipment. pp 1369–1380
27. Nguyen TL, Guerrero JM, Griepentrog G (2020) A self-sustained and flexible control strategy for Islanded DC nanogrids without communication links. IEEE J Emerg Select Top Power Electron 8(1): 877–892. https://doi.org/10.1109/JESTPE.2019.2894564
28. Liu G, Khodamoradi A, Mattavelli P, Caldognetto T, Magnone P (2018) Plug and play DC-DC converters for smart DC nanogrids with advanced control ancillary services. In: 2018 IEEE 23rd international workshop on computer aided modeling and design of communication links and networks (CAMAD). pp 1–6
29. Ito Y, Zhongqing Y, Akagi H (2004) DC micro-grid based distribution power generation system. In: Power electronics and motion control conference. pp 2352–2360
30. Zhang W, Lee FC, Huang P-Y (2014) Energy management system control and experiment for future home. In: 2014 IEEE energy conversion congress and exposition (ECCE). IEEE, Pittsburgh, PA, USA, pp 3317–3324
31. Wunder B, Ott L, Kaiser J, Gosses K, Schulz M, Fersterra F, Marz M, Lavery M, Han Y (2017) Droop controlled cognitive power electronics for DC microgrids. In: 2017 IEEE international telecommunications energy conference (INTELEC). IEEE, Broadbeach, QLD, Australia, pp 335–342. https://doi.org/10.1109/INTLEC.2017.8214158
32. Katiraei F, Iravani MRR, Lehn PWW (2005) Micro-grid autonomous operation during and subsequent to islanding process. IEEE Trans Power Deliv 20(1):248–257
33. Kumar D, Zare F, Ghosh A (2017) DC microgrid technology: system architectures, ac grid interfaces, grounding schemes, power quality, communication networks, applications, and standardizations aspects. IEEE Access 5:12230–12256. https://doi.org/10.1109/ACCESS.2017.2705914
34. Gungor VC, Sahin D, Kocak T, Ergut S, Buccella C, Cecati C, Hancke GP (2011) Smart grid technologies: communication technologies and standards. IEEE Trans Industr Inf 7(4):529–539. https://doi.org/10.1109/TII.2011.2166794
35. Dragičević T, Guerrero JM, Vasquez JC (2014) A distributed control strategy for coordination of an autonomous LVDC microgrid based on power-line signaling. In: IEEE Trans Industr Electron (61)7:3313–3326. https://doi.org/10.1109/TIE.2013.2282597
36. Kim S, Jeon J, Cho C, Ahn J, Kwon S (2008) Dynamic modeling and control of a grid-connected hybrid generation system with versatile power transfer. IEEE Trans Industr Electron 55(4):1677–1688. https://doi.org/10.1109/TIE.2007.907662
37. Werth A, Kitamura N, Tanaka K (2015) Conceptual study for open energy systems: distributed energy network using interconnected DC nanogrids. IEEE Trans Smart Grid 6(4):1621–1630. https://doi.org/10.1109/TSG.2015.2408603
38. Pradhan R, Chirayath M, Thale S (2016) Coordinated control strategy for a DC microgrid with low bandwidth communication. In: 2016 IEEE international conference on power electronics, drives and energy systems (PEDES). pp 1–6. https://doi.org/10.1109/PEDES.2016.7914276
39. Liu S, Wang X, Liu PX (2015) Impact of communication delays on secondary frequency control in an islanded microgrid. IEEE Trans Industr Electron 62(4):2021–2031. https://doi.org/10.1109/TIE.2014.2367456
40. Ahmed MA, Kim Y (2013) Communication Networks of Domestic Small-Scale Renewable Energy Systems. In: 2013 4th International conference on intelligent systems, modelling and simulation. pp 513–518. https://doi.org/10.1109/ISMS.2013.21

41. Klaina H et al (2020) Aggregator to electric vehicle LoRaWAN based communication analysis in vehicle-to-grid systems in smart cities. IEEE Access 8:124688–124701. https://doi.org/10.1109/ACCESS.2020.3007597
42. Mandic-Lukic J, Milinkovic B, Simic N (2016) Communication solutions for smart grids, smart cities and smart buildings. In: Mediterranean conference on power generation, transmission, distribution and energy conversion (MedPower 2016). pp 1–7. https://doi.org/10.1049/cp.2016.1109
43. Kaushal A, Hertem DV (2019) An overview of ancillary services and HVDC systems in European context. Energies 12(18):3481
44. Demoulias CS, Malamaki K-ND, Gkavanoudis S, Mauricio JM, Kryonidis GC, Oureilidis KO, Kontis EO, Martinez Ramos JL (2020) Ancillary services offered by distributed renewable energy sources at the distribution grid level: an attempt at proper definition and quantification. Appl Sci 10(20):7106
45. Thiesen H, Jauch C, Gloe A (2016) Design of a system substituting today's inherent inertia in the European continental synchronous area. Energies 9(8):582
46. ENTSO-E; Wind Europe; Solar Power Europe and T&D Europe (2020) High penetration of power electronic interfaced power sources and the potential contribution of grid forming converters. Technical Report. 2020. https://eepublicdownloads.blob.core.windows.net/public-cdn-container/cleandocuments/Publications/SOC/High_Penetration_of_Power_Electronic_Interfaced_Power_Sources_and_the_Potential_Contribution_of_Grid_Forming_Converters.pdf. Accessed 22 July 2021
47. Holmberg D, Omar F (2021) "Characterization of residential distributed energy resource potential to provide ancillary services", NIST SP 1900–601, 2018. . https://doi.org/10.6028/NIST.SP.1900-601
48. Zhong Q (2016) Virtual synchronous machines: a unified interface for grid integration. IEEE Power Electr Magaz 3(4):18–27
49. Eriksson R, Modig N, Elkington K (2018) Synthetic inertia versus fast frequency response: a definition. IET Renew Power Gener 12:507–514
50. Commission Regulation (EU) 2016/631 of 14 April 2016, establishing a network code on requirements for grid connection of generators. 2016. https://eur-lex.europa.eu/legal-content/EN/TXT/PDF/?uri=CELEX:32016R0631&from=EN. Accessed 16 June 2021
51. EN50549–2 (2019) Requirements for generating plants to be connected in parallel with distribution networks—Part 2: connection to a MV distribution network—generating plants up to and including Type B. European Committee for Electrotechnical Standardization (CENELEC), Brussels, Belgium
52. Goebel C, Hesse H, Schimpe M, Jossen A, Jacobsen H (2017) Model-based dispatch strategies for lithium-ion battery energy storage applied to pay-as-bid markets for secondary reserve. IEEE Trans Power Syst 32(4):2724–2734
53. Calderaro V, Galdi V, Lamberti F, Piccolo A (2015) A smart strategy for voltage control ancillary service in distribution networks. IEEE Trans Power Syst 30(1):494–502
54. Kryonidis GC, Demoulias CS, Papagiannis GK (2020) A two-stage solution to the bi-objective optimal voltage regulation problem. IEEE Trans Sustain Energy 11(2):928–937
55. De Oro Arenas L, e Melo GA, Canesin CA (2017) FPGA-based power meter implementation for three-phase three-wire and four-wire power systems, according to IEEE 1459–2010 standard. In: 2017 Brazilian Power Electronics Conference (COBEP). pp 1–6
56. Telukunta V, Pradhan J, Agrawal A, Singh M, Srivani SG (2017) Protection challenges under bulk penetration of renewable energy resources in power systems: a review. CSEE J Power Energy Syst 3(4):365–379
57. Brundlinger R, Strasser T, Lauss G et al (2015) Lab tests: verifying that smart grid power converters are truly smart. IEEE Power Energy Mag 13(2):30–42
58. Liang X, Andalib-Bin-Karim C (2018) harmonics and mitigation techniques through advanced control in grid-connected renewable energy sources: a review. IEEE Trans Indus Appl 54(4):3100–3111

59. Kumar GVB, Palanisamy K (2020) A review of energy storage participation for ancillary services in a microgrid environment. Inventions 5(4):63
60. Liu Y et al (2020) Aging effect analysis of PV inverter semiconductors for ancillary services support. IEEE Open J Indus Appl 1:157–170
61. Oureilidis K, Malamaki K-N, Gallos K, Tsitsimelis A, Dikaiakos C, Gkavanoudis S, Cvetkovic M, Mauricio JM, Maza Ortega JM, Ramos JLM, Papaioannou G, Demoulias C (2020) Ancillary services market design in distribution networks: review and identification of barriers. Energies 13(4):917
62. Liu K, Chen Q, Kang C, Su W, Zhong G (2018) Optimal operation strategy for distributed battery aggregator providing energy and ancillary services. J Modern Power Syst Clean Energy 6(4):722–732
63. Dolatabadi M, Siano P (2020) A scalable privacy preserving distributed parallel optimization for a large-scale aggregation of prosumers with residential PV-battery systems. IEEE Access 8:210950–210960
64. Stiefenhofer T (2019) Remote battery management system, management device, and remote battery management method. US Patent US 10,516,547 B2, Sonnen GmbH, 2019
65. Rivera S, Kouro S, Vazquez S, Goetz SM, Lizana R, Romero-Cadaval E (2021) Electric vehicle charging infrastructure: from grid to battery. IEEE Ind Electron Mag 15(2):37–51
66. Wei C, Shao J, Agrawal B, Zhu D, Xie H (2019) New surface mount SiC MOSFETs enable high efficiency high power density bi-directional on-board charger with flexible DC-link voltage. IEEE Appl Power Electron Conf Exposit (APEC) 2019:1904–1909
67. Wei C, Zhu D, Xie H, Shao J (2019) A 6.6kW high power density bi-directional EV on-board charger based on SiC MOSFETs. In: PCIM Europe 2019; International Exhibition and Conference for Power Electronics, Intelligent Motion, Renewable Energy and Energy Management. pp. 246–252
68. Baranwal R, Iyer KV, Basu K, Castelino GF, Mohan N (2018) A reduced switch count single-stage three-phase bidirectional rectifier with high-frequency isolation. IEEE Trans Power Electron 33(11):9520–9541
69. Wu F, Li X, Yang G, Liu H, Meng T (2020) Variable switching periods based space vector phase-shifted modulation for dab based three-phase single-stage isolated AC–DC converter. IEEE Trans Power Electron 35(12):13725–13734
70. Das D, Weise N, Basu K, Baranwal R, Mohan N (2019) A bidirectional soft-switched DAB-based single-stage three-phase AC–DC converter for V2G application. IEEE Trans Transp Electrificat 5(1):186–199
71. Rabkowski J, Blinov A, Zinchenko D, Wrona G, Zdanowski M (2020) Grid-frequency Vienna rectifier and isolated current-source DC-DC converters for efficient off-board charging of electric vehicles. In: 2020 22nd European conference on power electronics and applications (EPE'20 ECCE Europe). pp 1–10
72. Teichmann R, Bernet S (2005) A comparison of three-level converters versus two-level converters for low-voltage drives, traction, and utility applications. IEEE Trans Indus Appl 41(3):855–865
73. Schweizer M, Friedli T, Kolar JW (2013) Comparative evaluation of advanced three-phase three-level inverter/converter topologies against two-level systems. IEEE Trans Industr Electron 60(12):5515–5527
74. Bruckner T, Bemet S (2001) Loss balancing in three-level voltage source inverters applying active NPC switches. In: 2001 IEEE 32nd annual power electronics specialists conference (IEEE Cat. No.01CH37230). vol. 2, pp. 1135–1140
75. Kim Y-J, Kim S-M, Lee K-B (2021) Improving DC-link capacitor lifetime for three-level photovoltaic hybrid active NPC inverters in full modulation index range. IEEE Trans Power Electron 36(5):5250–5261
76. Davoodnezhad R, Holmes DG, McGrath BP (2014) A novel three-level hysteresis current regulation strategy for three-phase three-level inverters. IEEE Trans Power Electron 29(11):6100–6109

77. Wang Q, Zhang X, Burgos R, Boroyevich D, White AM, Kheraluwala M (2018) Design and implementation of a two-channel interleaved vienna-type rectifier with >99% efficiency. IEEE Trans Power Electron 33(1):226–239
78. Stempfle M, Fischer M, Nitzsche M, Wölfle J, Roth-Stielow J (2016) Efficiency analysis of three-level NPC and T-Type voltage source inverter for various operation modes optimizing the overall drive train efficiency by an operating mode selection. In: 2016 18th European conference on power electronics and applications (EPE'16 ECCE Europe). pp 1–10
79. Anthon A, Zhang Z, Andersen MAE, Holmes DG, McGrath B, Teixeira CA (2017) Comparative Evaluation of the Loss and Thermal Performance of Advanced Three-Level Inverter Topologies. IEEE Trans Indus Appl 53(2):1381–1389
80. Akagi H, Kinouchi S, Miyazaki Y (2016) Bidirectional isolated dual-active-bridge (DAB) DC-DC converters using 1.2-kV 400-A SiC-MOSFET dual modules. CPSS Trans Power Electron Appl 1(1):33–40
81. Bai H, Mi C (2008) Eliminate reactive power and increase system efficiency of isolated bidirectional dual-active-bridge DC–DC converters using novel dual-phase-shift control. IEEE Trans Power Electron 23(6):2905–2914
82. Huang J, Wang Y, Li Z, Lei W (2016) Unified triple-phase-shift control to minimize current stress and achieve full soft-switching of isolated bidirectional DC–DC converter. IEEE Trans Industr Electron 63(7):4169–4179
83. Hiltunen J, Väisänen V, Juntunen R, Silventoinen P (2015) Variable-frequency phase shift modulation of a dual active bridge converter. IEEE Trans Power Electron 30(12):7138–7148
84. Mou D et al (2021) Optimal asymmetric duty modulation to minimize inductor peak-to-peak current for dual active bridge DC–DC converter. IEEE Trans Power Electron 36(4):4572–4584
85. Kasper M, Bortis D, Deboy G, Kolar JW (2017) Design of a highly efficient (97.7%) and very compact (2.2 kW/dm ^{3}) isolated AC–DC telecom power supply module based on the multicell ISOP converter approach. IEEE Trans Power Electron 32(10):7750–7769
86. Huang P-Y, Ohta T, Fujii M, Kado Y (2019). Bidirectional isolated ripple cancel dual active bridge DC-DC converter. In: 2019 IEEE Third International Conference on DC Microgrids (ICDCM). pp 1–6
87. Chen W, Rong P, Lu Z (2010) Snubberless bidirectional DC–DC converter with new CLLC resonant tank featuring minimized switching loss. IEEE Trans Industr Electron 57(9):3075–3086
88. Siebke K, Mallwitz R (2020) Comparison of a dual active bridge and CLLC converter for on-board vehicle chargers using GaN and time domain modeling method. In: 2020 IEEE Energy Conversion Congress and Exposition (ECCE) 2020. pp 1210–1216
89. Mukherjee S, Kumar A, Chakraborty S (2021) Comparison of DAB and LLC DC–DC converters in high-step-down fixed-conversion-ratio (DCX) applications. IEEE Trans Power Electron 36(4):4383–4398
90. Escudero M, Meneses D, Rodriguez N, Morales DP (2020) Modulation scheme for the bidirectional operation of the phase-shift full-bridge power converter. IEEE Trans Power Electron 35(2):1377–1391
91. Zhu L (2006) A novel soft-commutating isolated boost full-bridge ZVS-PWM DC–DC converter for bidirectional high power applications. IEEE Trans Power Electron 21(2):422–429
92. Wu T, Yang J, Kuo C, Wu Y (2014) Soft-switching bidirectional isolated full-bridge converter with active and passive snubbers. IEEE Trans Industr Electron 61(3):1368–1376
93. Miura Y, Kaga M, Horita Y, Ise T (2010) Bidirectional isolated dual full-bridge dc-dc converter with active clamp for EDLC. In: 2010 IEEE energy conversion congress and exposition. pp 1136–1143
94. Xuewei P, Rathore AK (2013) Novel interleaved bidirectional snubberless soft-switching current-fed full-bridge voltage doubler for fuel-cell vehicles. IEEE Trans Power Electron 28(12):5535–5546
95. Blinov A, Kosenko R, Vinnikov D, Parsa L (2020) Bidirectional isolated current-source DAB converter with extended ZVS/ZCS range and reduced energy circulation for storage applications. IEEE Trans Industr Electron 67(12):10552–10563

96. Chen W, Zane R, Seltzer D, Corradini L (2014) Isolated bidirectional DC/AC and AC/DC three-phase power conversion using series resonant converter modules and a three-phase unfolder. In: 2014 IEEE 15th Workshop on Control and Modeling for Power Electronics (COMPEL)
97. Chen WW, Riar B, Zane R (2017) A three-port series resonant converter for three-phase unfolding inverters. In: 2017 IEEE 18th workshop on control and modeling for power electronics (COMPEL). pp 1–7
98. Chen W, Riar B, Zane R (2018) Battery integrated modular multifunction converter for grid energy storage. In: 2018 IEEE Energy Conversion Congress and Exposition (ECCE). 2157–2163
99. Yelaverthi DB, Hatch R, Mansour M, Wang H, Zane R (2019) 3-Level asymmetric full-bridge soft-switched PWM converter for 3-phase unfolding based battery charger topology. In: 2019 IEEE energy conversion congress and exposition (ECCE). 2737–2743
100. Antivachis M, Anderson JA, Bortis D, Kolar JW (2020) Analysis of a synergetically controlled two-stage three-phase DC/AC buck-boost converter. CPSS Trans Power Electron Appl 5(1):34–53
101. Korkh O, Blinov A, Vinnikov D, Chub A (2020) Review of isolated matrix inverters: topologies, modulation methods and applications. Energies 13(9):2394
102. Blinov A, Verbytskyi I, Peftitsis D, Vinnikov D (2021) Regenerative passive snubber circuit for high-frequency link converters. IEEE J Emerg Select Top Indus Electron. https://doi.org/10.1109/JESTIE.2021.3066897.
103. Martins JF, Romero-Cadaval E, Vinnikov D, Malinowski M (2022) Transactive Energy: Power Electronics Challenges. IEEE Power Electronics Magazine 9(1):20–32. https://doi.org/10.1109/MPEL.2022.3140981
104. Carvalho EL, Blinov A, Chub A, Emiliani P, de Carne G, Vinnikov D (2022) Grid Integration of DC Buildings: Standards Requirements and Power Converter Topologies. IEEE Open J Power Electron 3:798–823. https://doi.org/10.1109/OJPEL.2022.3217741

Chapter 4
Dynamic Voltage Restorer–A Custom Power Device for Power Quality Improvement in Electrical Distribution Systems

Vinay Kumar Awaar, Praveen Jugge, S. Tara Kalyani, and Mohsen Eskandari

Abstract Power quality (PQ) is an important issue in power systems that reflects a measure of quality of energy supply to the consumers. Power electronics-based equipment is increasing in power systems that cause voltage distortions which may damage a sensitive appliance in the distribution grid. Besides, many electrical and electronic equipment is susceptible to power supply disturbances in modern industrial establishments. The most vital power quality issues affecting large commercial and industrial customers are momentary power interruptions and voltage sags. Usually, these PQ issues are associated with the faults that occur in the supplying power systems. Voltage sags are very common since they can be associated with the faults that happen remotely to the customer. Further, these PQ problems with about 4–5 cycles may cause an extensive malfunction to the sensitive equipment. Therefore, an appropriate solution is necessary to handle this problem. A Dynamic voltage restorer (DVR) is a custom power device (CPD) that can protect sensitive equipment against voltage disturbances and improve power quality in electrical distribution systems. In recent years, many CPDs have been developed to meet the need of the industries and their critical load establishment. The DVR is one of the most promising CPDs used to improve the power quality before feeding it to the sensitive load to protect electric appliances from damage. DVR is the combination of a voltage source inverter (VSI), which can impose the voltage in series through a series injection transformer (SIT)

V. K. Awaar (✉) · P. Jugge
Gokaraju Rangaraju Institute of Engineering and Technology, Hyderabad, India
e-mail: vinaykumaar.a@gmail.com

P. Jugge
e-mail: drpraveen@griet.ac.in

S. T. Kalyani
Jawaharlal Nehru Technological University Hyderabad, Hyderabad, India

M. Eskandari
University of New South Wales, Sydney, Australia
e-mail: m.eskandari@unsw.edu.au

A. K. Giri et al. (eds.), *Power Quality: Infrastructures and Control*,
Studies in Infrastructure and Control,
https://doi.org/10.1007/978-981-19-7956-9_4

with the power distribution system during voltage sag. The required voltage compensation can be achieved by adopting the pulse width modulation (PWM) technique for VSI. The DVR's performance mainly depends upon the control system that includes voltage disturbance detection, reference generation, voltage & current control and PWM switching strategy. The main objective of the control system is to compensate for the disturbance at the earliest using a fast detection technique, optimum control variables for the control circuit and a suitable PWM switching methodology. Many such compensation methods and control techniques are proposed in the literature along with various DVR topologies which are analysed and studied in this chapter.

Keywords Power quality · Dynamic voltage restorer · Voltage source inverter · PI controller · Harris hawk's optimization (HHO) algorithm

1 Introduction

The power system should ascertain the good quality of the electric power supply, which ensures the waveforms of the voltage and current are sinusoidal. Also, voltage and current should be balanced in 3-phase grids. Further, it must maintain the levels of voltage within the permissible band (about $100 \pm 5\%$ of its rated value). Otherwise, the system performance may be degraded and hence there is a need for voltage compensation. To regulate the voltage at the load bus many conventional compensating methods are used such as shunt/series capacitors, static synchronous capacitor, tap changing/booster transformer, and dynamic voltage restorer (DVR). DVR is used at the distribution level whereas other devices are also used at the transmission networks [1].

A DVR is a series-connected power device that can be used to compensate for the voltage disturbances at the load bus when it is connected to a critical or sensitive load. As described in Fig. 1, DVR can control the load voltage by compensating the voltage excursion in case of a voltage sag/dip. A DVR can also be described as a solid-state inverter that is based on a series injection voltage with a power distribution system. Hence, by injecting the voltage in series, it can restore the load voltage for a sensitive load by compensating the voltage sag at the supply side [2].

2 DVR Components

The basic components of a DVR are described in Fig. 1. The main parts include a voltage source inverter (VSI), filter, series injection transformer (SIT), and an energy source with a DC-link capacitor. Since this DC-link capacitor gets discharged during voltage sag compensation and it would require increasing the size of the DC-link capacitor for longer voltage sags. But this can be avoided if the energy source is replaced with a grid-connected rectifier.

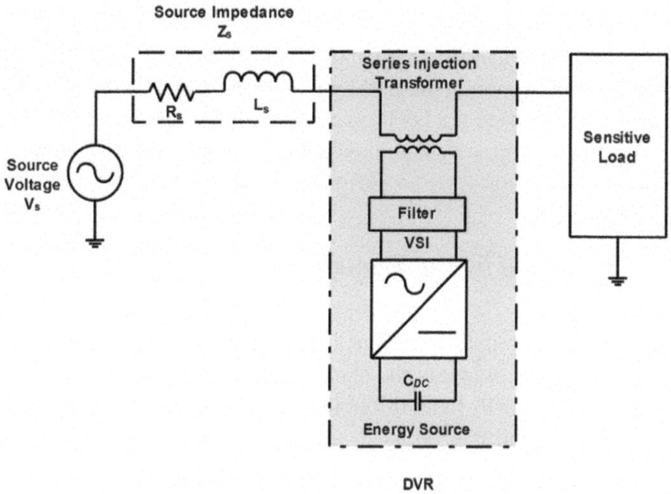

Fig. 1 Schematic diagram of the DVR

Voltage source inverter: It is a power electronics-based device where the semiconductor switches are controlled via the PWM method to modulate DC voltage (from DC-link) to AC voltages injected into the system. Therefore, generally, the inverter performs as a voltage source.

Filter: It is placed to reduce the harmonic content produced by the PWM-based VSI.

Series injection transformer (SIT): is inserted mainly to allow electric isolation for the converter configuration and thereby provide protection to the equipment.

DC-link/Energy source: A DC-link voltage is provided to the VSI to synthesize the AC voltage into the grid and injects the necessary power into the supply during most voltage sags.

3 DVR Topologies

The location of the DVR is intended either at the medium voltage (MV) level or the low-voltage (LV) level. The main advantage of a large centralized MV level connected DVR is its lower cost per MVA compared with the capital and maintenance costs associated with decentralized low-voltage units. However, it requires a high isolation level and a high short-circuit level. Whereas this short-circuit level can be reduced by the distribution transformer in case of LV level connected DVR and it is possible to target specifically more at load buses with sensitive loads to voltage sag. But, if there is a voltage sag with a zero-sequence component, then the DVR controller should generate positive, negative and zero-sequence voltages to compensate for the load voltages connected between phase and neutral. Further, it increases the complexity of the DVR controller. Hence, the location of the DVR is selected as per the requirement, which further depends upon the method by which DVR has connected to the grid.

Many topologies are used for DVR connection in the distribution system, which is mainly categorized in the way that, how a converter is connected to the DVR.

3.1 DVR Connected by the Transformer

In this connection, the transformer ratio is arbitrarily chosen, which provides the scaling of the converter voltage to the standard industrial voltages, thus increasing the performance of the DVR. In addition, the SIT is used to ensure the DVR's basic insulation level (BIL) and can be used as a part of the filter that can reduce the harmonic content produced by the VSI. However, with the insertion of SIT, the losses increase, and the size, volume and cost also increase with the low-frequency injection transformers.

3.2 Transformerless or Directly Connected DVR

By removing SIT, the size of the DVR can be reduced and a compact solution with less weight and low volume can be developed. Since the voltage drops and non-linearity effects, (caused by the transformer) are avoided, the performance of DVR is expected to improve. However, the removal of SIT may lead to the requirement for a complex converter topology, which further increases the total number of components. Also, high isolation to the ground must be ensured; otherwise, the protection of power electronic switches would be complicated.

DVRs are also categorized in the perspective of power exchange between the grid and DVR, i.e., when a DVR injects the voltage during voltage sag and restores the supply voltage, then the DVR exchanges active and reactive power with the system. So, when DVR supplies active power to the load, it needs a source for the energy. Therefore, this energy source can be provided by energy storage elements like batteries, supercapacitors or double-layer capacitors, etc. In the case of no-storage DVR topology, the energy is directly taken from the supply during voltage sag. Various topologies for DVR are described in the following section (only pictorial representation is presented). Interested readers can refer [3] for more details (Figs. 2 and 3).

3.3 DVRs Based on the Type of the Power Converter

1.1 Current Source Converter (CSC) based DVR
1.2 Voltage Source Converter (VSC) based DVR

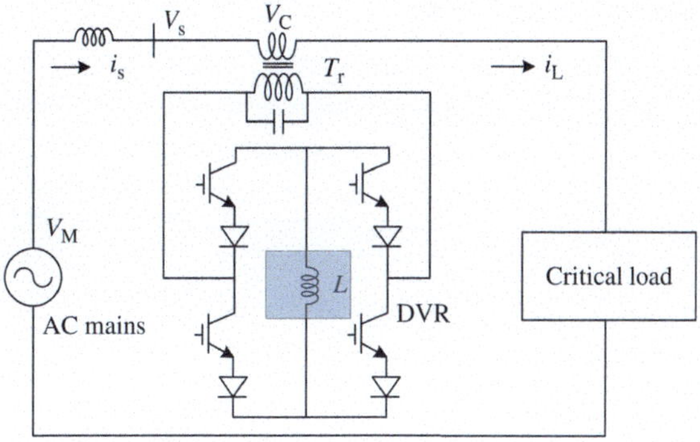

Fig. 2 Single-phase DVR (CSC-based)

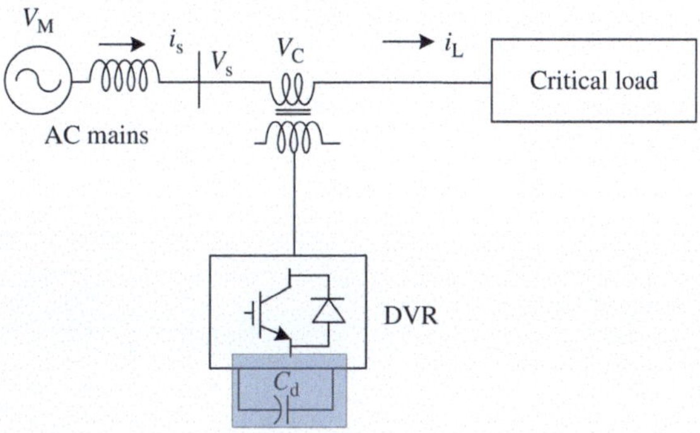

Fig. 3 Single-phase DVR (VSC based)

3.4 DVRs Based on the Converter Topology

1.3 Single-phase Half-Bridge VSC based DVR
1.4 Single-phase Full-Bridge VSC based DVR

See (Figs. 4, 5 and 6).

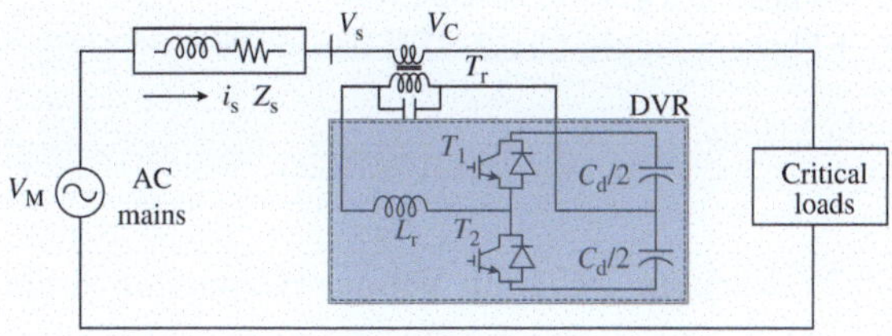

Fig. 4 Single-phase DVR (Half-bridge VSC based)

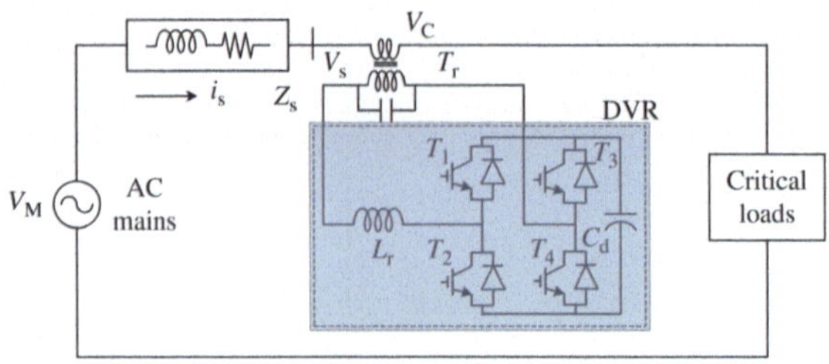

Fig. 5 Single-phase DVR (Full-bridge VSC based)

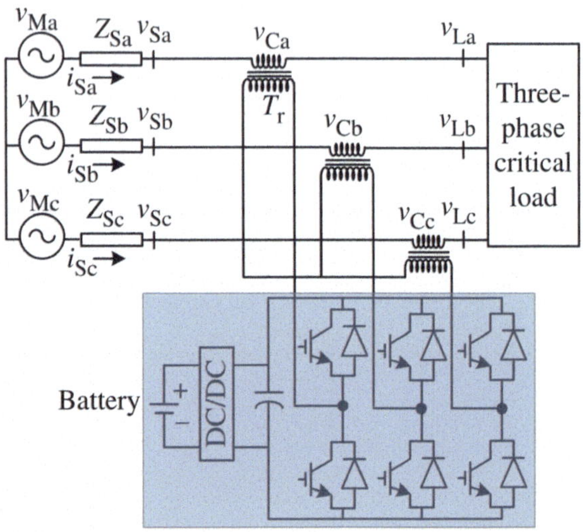

Fig. 6 Battery supported based DVR (3-phase 3-leg VSC)

3.5 *Energy Storage-Based DVR*

3.5.1 Constant DC-Link Voltage (Battery Supported)

- 3-Phase 3-leg VSC based DVR
- 3-Phase 2-leg VSC based DVR
- 3-Phase 3-single-phase VSC based DVR

See (Figs. 7, 8 and 9).

3.5.2 Variable DC-Link Voltage (Capacitor Supported)

- 3-Phase 3-leg VSC based DVR
- 3-Phase 2-leg VSC based DVR
- 3-Phase 3-single-phase VSC based DVR

See (Figs. 10 and 11).

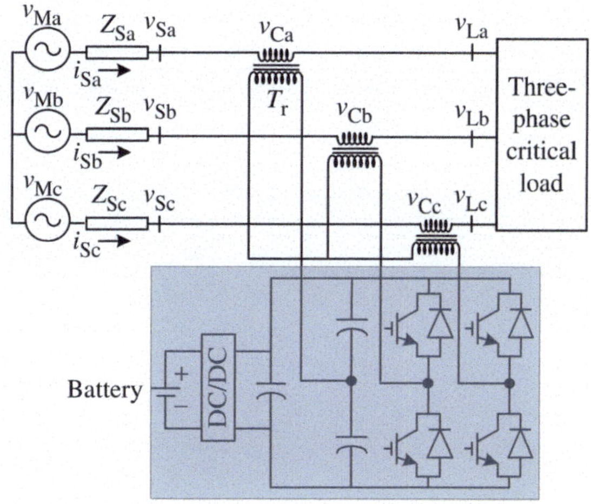

Fig. 7 Battery supported based DVR (3-phase 2-leg VSC)

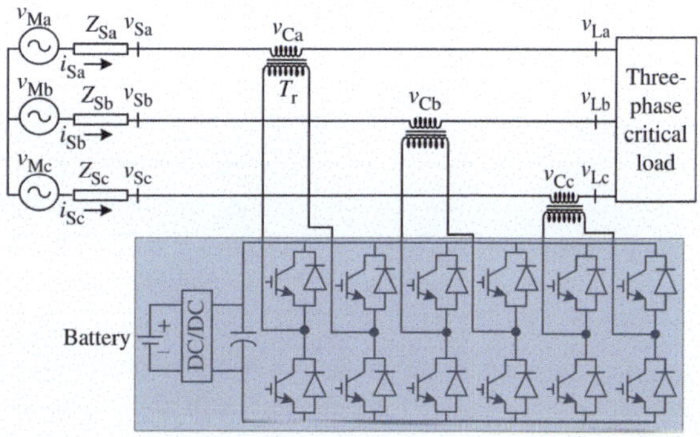

Fig. 8 Battery supported based DVR (3-phase 3-single phase VSC)

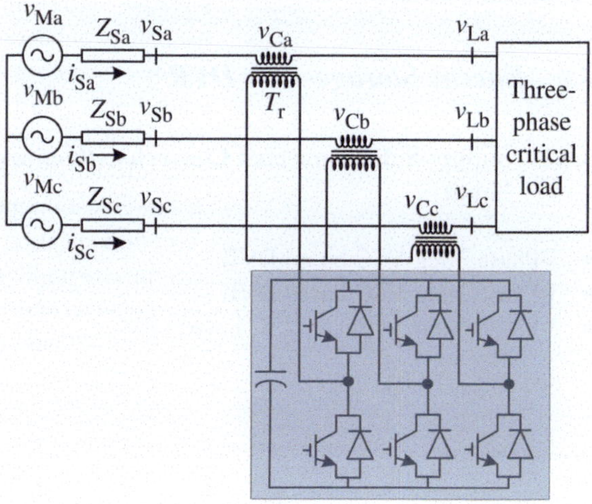

Fig. 9 Capacitor supported based DVR (3-phase 3-leg VSC)

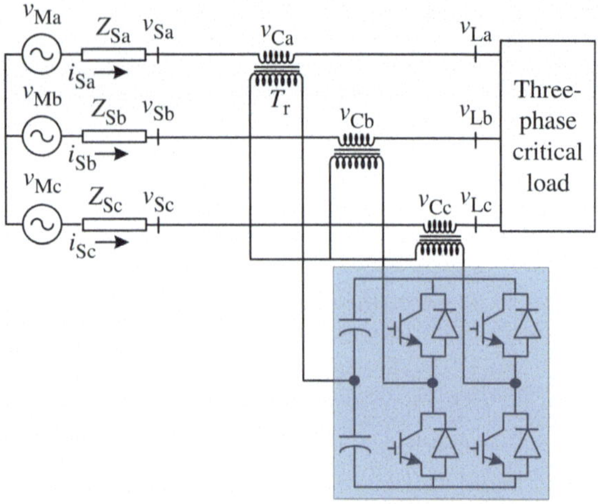

Fig. 10 Capacitor supported based DVR (3-phase 2-leg VSC)

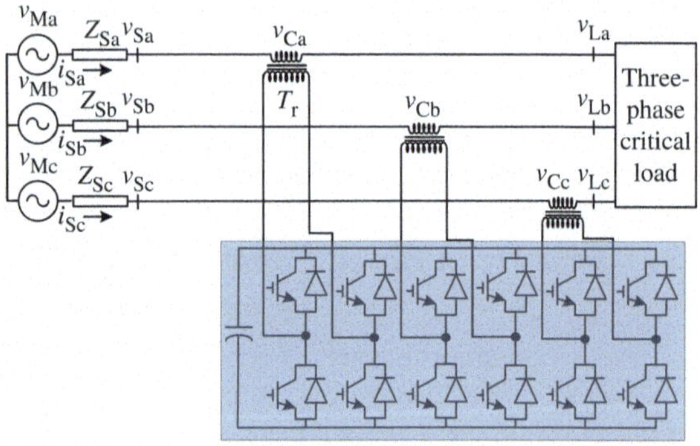

Fig. 11 Capacitor supported based DVR (3-phase 3-single phase VSC)

3.6 Energy Storage-Less DVR

3.6.1 Supply-Side Connected Converter (Rectifier Supported-Left Shunt)

- 3-Phase 3-leg VSC based DVR
- 3-Phase 2-leg VSC based DVR
- 3-Phase 3-single-phase VSC based DVR

See (Figs. 12 and 13).

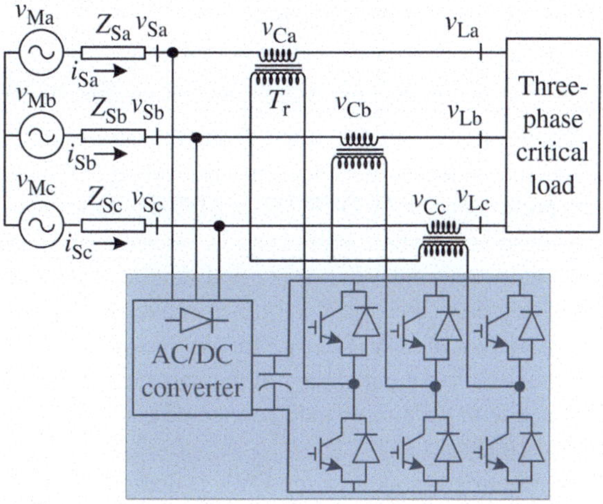

Fig. 12 Supply-side connected converter DVR (3-phase 3-leg VSC)

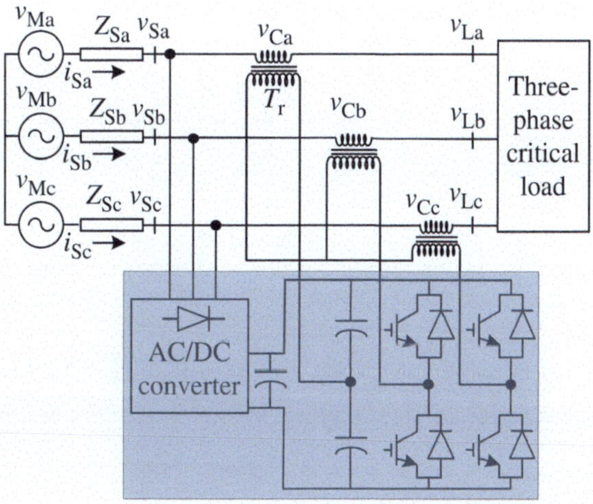

Fig. 13 Supply-side connected converter DVR (3-phase 2-leg VSC)

3.6.2 Load Side Connected Converter (Rectifier Supported-Right Shunt)

- 3-Phase 3-leg VSC based DVR
- 3-Phase 2-leg VSC based DVR
- 3-Phase 3-single-phase VSC based DVR

The selection from the above topologies can be made by focusing on the following parameters.

- DC-link voltage level and its controllability during voltage sag;
- Its ability to control the voltage sags of various characters;
- The rating of the voltage source converter;

- Its effect on the source and their neighboring loads;
- The complete system and its complexity;
- Cost estimations related with respect to other topologies.

Therefore, it can be noted from the above discussion that a DVR topology connecting with a series injection transformer (SIT) provides a charging circuit along with protection for the DVR. Also, a DVR without any specific energy storage is suitable for the grid where there is a low intensity of severe voltage sags. To get a better performance by the VSC, it is good practice to connect it at the load side (but this would increase the cost for higher converter capacity). Further, it is also important to protect the DVR during the downstream short-circuiting and to maintain the continuous current path to avoid uncontrolled charging of the DC-link, which can be obtained with the supply side connected converter-based DVR as shown in Figs. 14, 15 and 16 (Fig. 17).

A simplified model for the DVR is depicted in Fig. 18.

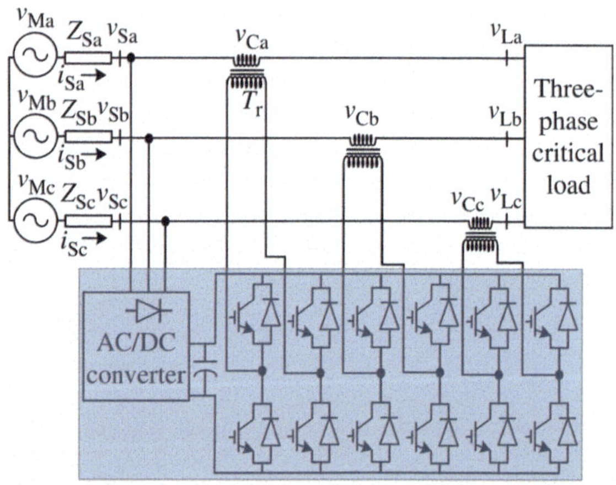

Fig. 14 Supply-side connected converter-based DVR (3-phase 3-single phase VSC)

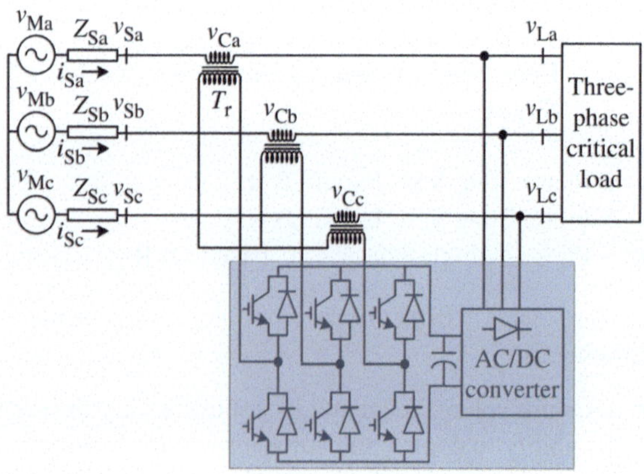

Fig. 15 Load-side connected converter-based DVR (3-phase 3-leg VSC)

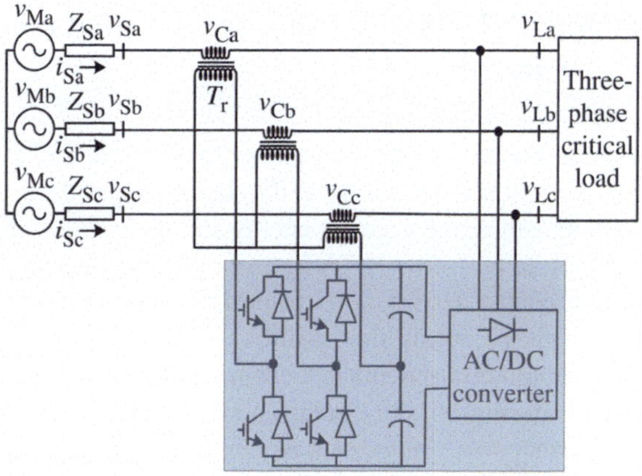

Fig. 16 Load-side connected converter-based DVR (3-phase 2-leg VSC)

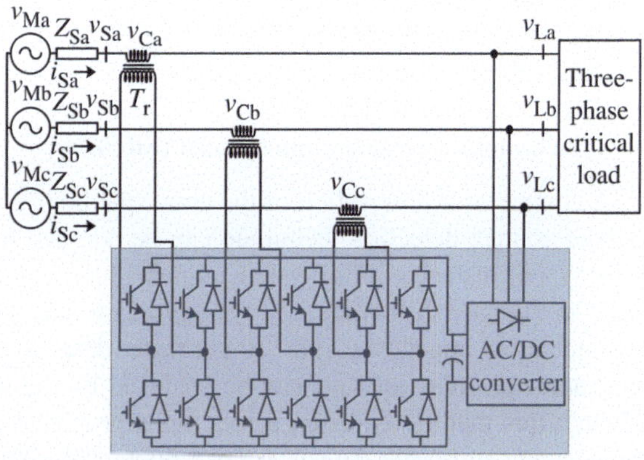

Fig. 17 Load-side connected converter-based DVR (3-phase 3-single phase VSC)

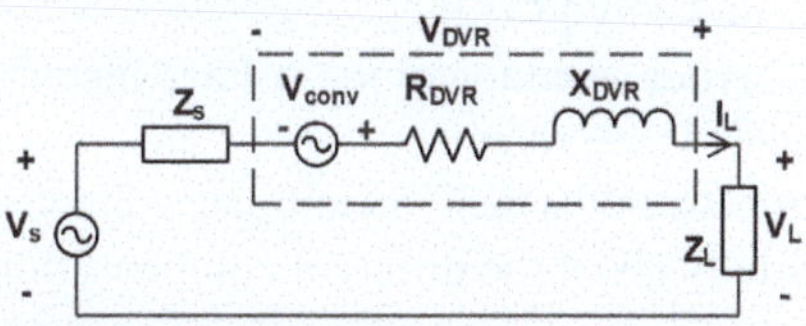

Fig. 18 1-Φ Simplified model of DVR

A simplified model of DVR is supplied with a controllable voltage source. The voltage source V_s represents a single-phase supply system in series with the source impedance Z_s with a fixed resistance R_{DVR}, as a lumped model of power losses in DVR along with the reactive elements in the DVR, i.e., fixed reactance X_{DVR}.

4 Sag Detection Techniques Adopted for the Commercial DVRs

The requirement of the real power injection is minimal in the case of the in phase compensation technique as well as in the case of the energy-optimized compensation technique thus, the need for energy storage is reduced. Owing to this advantage, the commercially available DVRs use either of these techniques. The main objective of any control technique is to describe the method to quantify the compensating DVR voltage during the disturbance. In addition to the compensation technique, there should be a scheme that can track the magnitude and phase angle of the supply voltage during normal operation and should detect the occurrence of the voltage disturbance like voltage sag, its duration, and phase angle [4–7].

The following are the commonly used voltage sag detection techniques.

- Peak value detection (PVD);
- Fourier transform (FT);
- Phase-locked loop (PLL);
- Wavelet transform method (WT);
- Synchronous reference frame model (SRF).

A comparative description of the above methods is given in Table 1.

The voltage detection technique can be adopted based on the requirement and priority. For instance, if the system contains harmonics and data needed associating with depth/phase shift of sags, and phase/magnitude of supply; then the FT technique is suitable but at a slower rate. To get faster detection, WT is preferable. Since the start and end points of sags can be analysed by the wavelet transform coefficient. Though this method is faster, it requires a proper selection of wavelet prototypes. An accurate and faster detection can be obtained with the SRF technique, which is more suitable for three-phase systems. The only constraint with this method is its complexity [8–12].

After detection of the voltage disturbance, the reference voltage should be generated. These reference voltage data may vary depending upon the type of compensation method which can mitigate voltage disturbances. The following section describes briefly about the different compensation methods.

5 Methods of Compensation

The main objective of the compensating device is that the amplitude of the load voltage must be compensated without tripping off the load. There are different strategies studied by various researchers and a few standards are presented here.

Table 1 Voltage disturbance detection methods and their comparison

Criterion/Detection method	*Type of Approach*	*Parameters can be obtained*	*Advantages*	*Dis-advantages*
PVD	Calculates the instant at which the voltage gradient of the input signal becomes zero and thereby obtains the peak value	The only peak value	Simple to implement	Takes more time in detection about 500 ms
FT	The input signal can be represented in the Frequency domain and integral transform can be done in both directions	Magnitude and frequency of each phase	Other than fundamental component magnitude, possible to get the magnitudes of each frequency component like 5th and 7th Easy to implement in the real-time control system	Since the calculation is the averaging method, the sag detection is slower, about 20 ms
PLL	Closed-loop feedback control system	Locks the magnitude and phase of the input signal	The controlled voltage output of PLL is a function of frequency	The noise from the feedback signal, the tracking of the phase of the signal may not be accurate and difficult to implement in real-time
WT	Signal can be represented both in the Frequency and Time domain	Magnitude and phase of the input signal	Faster compared to Fourier Transform	Integral transform of the signal can be done in only one direction and difficult to interpret the results in real-time operations

(continued)

5.1 Pre-sag Compensation

This method is one of the standard solutions for compensating the voltage sag. In this method, both the amplitude and phase of the voltage will be restored before the sag occurs. This method of compensation leads to the lowest distortions at the

Table 1 (continued)

Criterion/Detection method	*Type of Approach*	*Parameters can be obtained*	*Advantages*	*Dis-advantages*
SRF	An improved method of PLL principle, combining the technique of sag detection using the principal synchronous reference frame voltage quantities	Positive sequence components	Fast detection and easy to implement with the software tools like LabVIEW and CCS in real-time operations	This control technique is complex

load since the phase of the voltage at the load will not be affected during the sag. Thus, this method needs to synchronize the load voltage with the phase-locked loop (PLL) to restore the phase angle. And the DVR can restore the exact load voltage only when it delivers a higher voltage amplitude depending on the phase angle of the grid voltage during the voltage sag. So, with this method, the DVR must be designed for maximum voltage with an addition of low energy extraction from the DC-link [13, 14].

5.2 In-Phase Compensation

This strategy can be realized for the minimized voltage amplitude since the voltage will be compensated in phase to the grid voltage after the sag occurs. But this method cannot restore the phase angle, which leads to phase jumps and further may cause the tripping of the loads [15]. Thus, this strategy also PLL must be synchronized with the grid voltage.

5.3 Energy-Optimized Compensation

The other existing control strategy is based on drawing more active power from the grid so that the amount of active power needed from the DC-link can be reduced. But this method of compensation with pure reactive power is suitable only for shallow sags. And also, this method causes phase jumps and necessitates high voltage amplitude DVR [16, 17].

Once the compensation method is adopted, this is required to develop the control strategy for the compensation. There are many control strategies that have been developed by the researchers with the use of the conventional methods like predictive, repetitive, feed-forward and feedback, feedback linearization and metaheuristic algorithms etc. [18]. One of the metaheuristic algorithms is described in the following section.

6 Control Strategy for DVR

6.1 The Harris Hawks Optimization (HHO)

Harris Hawks Optimization (HHO) algorithm is a metaheuristic up-gradation practice that resembles the supportive action of a good rushing form of Harris Hawks referred to as "surprise pounce" [19]. Same as the different metaheuristic algorithms, the HHO rule conjointly consists of exploration and exploitation stages. Moreover, HHO is a population-based, gradient-free improvement practice. So, it is often made applicable with a correct design aimed at any improvement problem. The HHO comprises two exploration phases and four exploitative steps. Besides, the scientific illustration of this supportive movement recommends a brand-new random method for numerous optimization problems [20]. The following section explains the connection of Harris Hawks Optimization to the proposed control system.

The selection of parameters of PI controller is made using the metaheuristic optimization algorithm called Harris Hawks Optimization (HHO) Algorithm that offers the most optimum PI parameters. The readers are suggested to refer [21–23] for more explanation on HHO.

7 Simulation Results

The three-phase dynamic voltage restorer's closed-loop control arrangement is simulated in MATLAB/Simulink with HHO algorithm for PI parameters estimation. Figure 19 shows the system performance when tested against 10% voltage sag. Figure 19 shows the voltage of the point of common coupling (PCC) in an ideal situation from t = 0 s to 0.05 s. At t = 0.05 s to 0.10 s, VPCC experiences a 10% voltage sag of its rated value to reach a maximum value of 294 V. For the equivalent duration, DVR generates a counter voltage of 33 V that nullifies voltage sag.

In Fig. 20 the PCC voltage experiences a voltage sag of 20% from 0.05 to 0.10 duration, which produces a counter voltage of reduced voltage 65.5 V in demand to return the load voltage to its rated peak of 327.

In Fig. 21, the PCC voltage suffers a voltage sag of 30% from 0.05 to 0.10 duration, which produces a counter voltage of reduced voltage 98 V to restore the load voltage to its rated peak of 327 V.

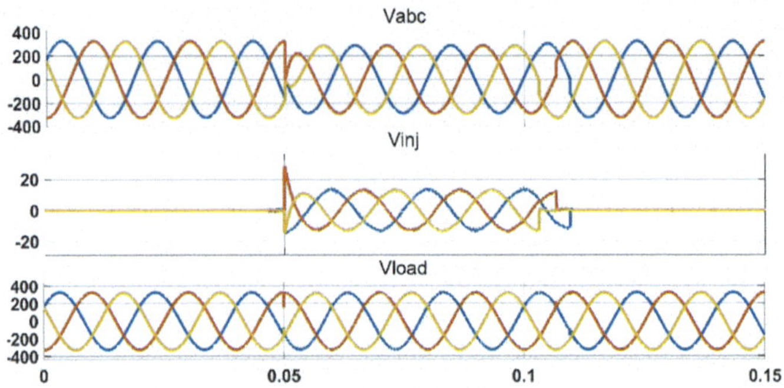

Fig. 19 Voltage compensation for 10% sag

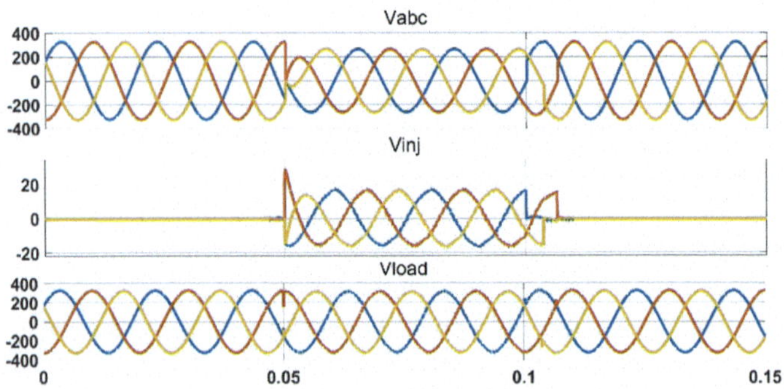

Fig. 20 Voltage Compensation for 20% sag

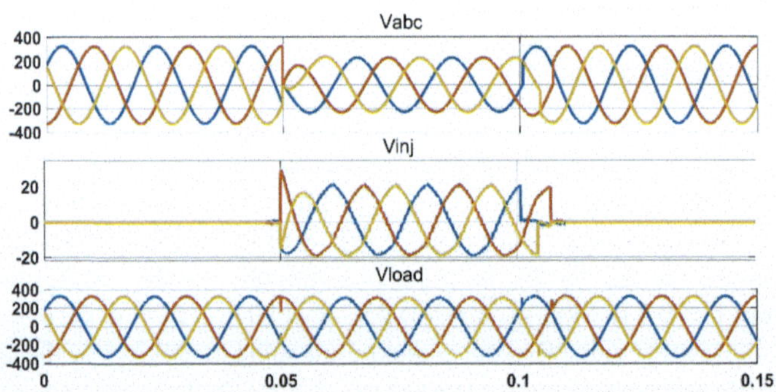

Fig. 21 Voltage Compensation for 30% sag

In Fig. 22, the PCC voltage suffers a voltage sag of 40% from 0.05 to 0.10 duration, which produces a counter voltage of reduced voltage 131 V in demand to re-establish the load voltage to its rated peak of 327 V.

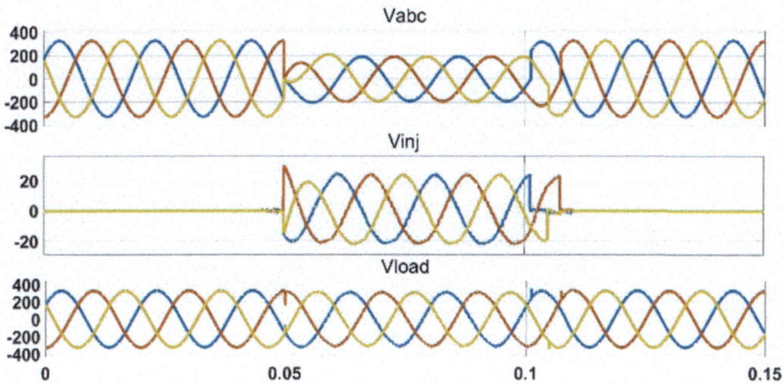

Fig. 22 Voltage compensation for 40% sag

Figure 23, the PCC voltage suffers a voltage sag of 50% from 0.05 to 0.10 duration, which produces a counter voltage of reduced voltage 163.5 V in demand to bring back the load voltage to its rated peak of 327 V (Table 2).

In Fig. 24, voltage sag happens in phases *a* and *b* but not in phase *c*. In that case, DVR can also counter the sag and produces the required voltage to nullify the sag. In Fig. 25, voltage sag is noticed in only one phase (i.e., phase *a*). DVR produces the required voltage in phase with the phase *a* to compensate for the voltage loss. Results added showcase that the control system could address any disturbance at PCC and the outstanding dynamic response.

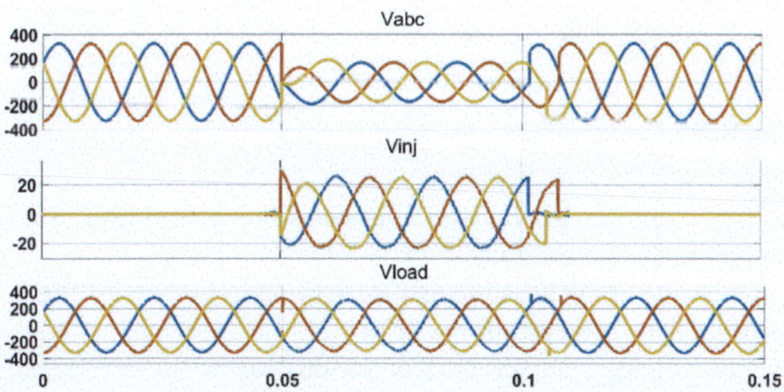

Fig. 23 Voltage compensation for 50% sag

Table 2 Voltage compensation for different sag levels

% of Sag	V_{abc} (V)	V_{DVR} (V)	V_{Load} (V)
10	294	33	327
20	261.5	65.5	327
30	229	98	327
40	196	131	327
50	163.5	163.5	327

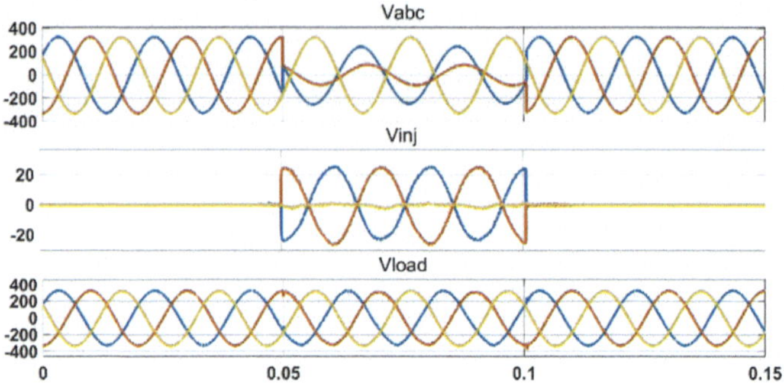

Fig. 24 Unbalanced Voltage compensation

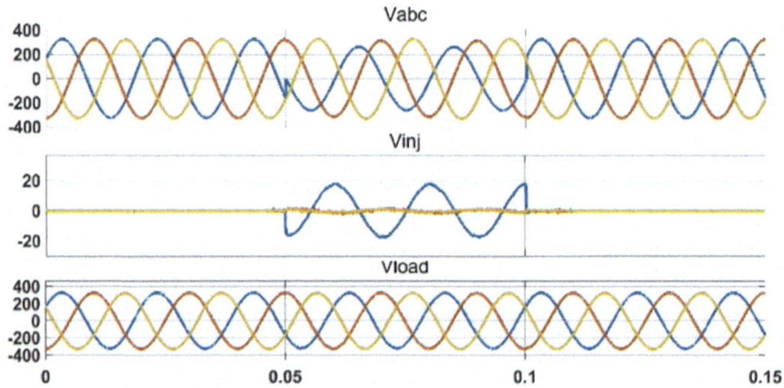

Fig. 25 Voltage Compensation in one phase

8 Conclusions

In this chapter, a comprehensive review of various topologies of DVR was presented and different compensation methods and possible control strategies were discussed. The main components of DVR and their functions were explained. The compensation process of a DVR as a custom power device was presented with an example. In particular, the voltage compensation at various sag levels was explained with the simulation results. The performance of DVR using a metaheuristic algorithm such as Harris Hawks Optimization (HHO) was evaluated with the help of simulation results. The combination of feed-forward upstream detection and closed-loop feedback management signal greatly helps in removing the momentary response and temporary oscillations. Since these oscillations are undesirable during voltage compensation, calibration of PI parameters is supported by the HHO algorithm to obtain the best values. Thus, this control strategy improves the performance of the DVR and hence the power quality of the electrical distribution system.

References

1. Woodley NH (2000) Field experience with Dynamic Voltage Restorer (DVRTMMV) systems. Power Engineering Society Winter Meeting, 2000
2. Kim H, Lee S-J, Sul S-K (2004) A calculation for the compensation voltages in dynamic voltage restorers by use of PQR power theory. In: Applied power electronics conference and exposition, vol 1, pp 573–579
3. Singh B, Chandra A, Al-Haddad K (2014) Power quality: problems and mitigation techniques. Wiley
4. Choi SS, Wang TX, Vilathgamuwa DM (2005) A series compensator with fault current limiting function. IEEE Trans Power Del 20(3):2248–2256
5. Moghassemi A, Padmanaban S (2020) Dynamic voltage restorer (DVR): a comprehensive review of topologies, power converters, control methods, and modified configurations. Energies 13(16):4152
6. Awaar VK, Jugge P, Tara Kalyani S (2019) Validation of control platform using TMS320F28027F for dynamic voltage restorer to improve power quality. J Control Autom Electr Syst 30:601
7. Ryan M1, Lorem RD (1995) A high perfomance Sine waye inverter controller with capacitor cumnt feedback and BackEMF decoupling. In: Weedings of IEEE power electronics specialists conference, pp 507–513
8. Awaar VK, Jugge P, Tara Kalyani S (2016) Optimal design and testing of a dynamic voltage restorer for voltage sag compensation and to improve power quality. In: IECON 2016-42nd annual conference of the IEEE industrial electronics society, Florence, pp 3745–3750
9. Kanjiya P, Singh B, Chandra A, Al-Haddad K (2013) SRF theory revisited: to control self-supported dynamic voltage restorer (DVR) for unbalanced and nonlinear loads. IEEE Trans Ind Appl 49(5):2330–2340
10. Awaar VK, Jugge, P, Tara Kalyani S, Thota S (2021) Validation of SRFPI control methodology for voltage sag compensation with a dynamic voltage restorer to improve power quality. In: 2021 international conference on sustainable energy and future electric transportation (SEFET). IEEE, pp 1–6
11. Sánchez PR, Acha E, Calderon JEO, Feliu V, Cerrada AG (2009) A versatile control scheme for a dynamic voltage restorer for power-quality improvement. IEEE Trans Power Del 24(1):277–284
12. Katole DN, Daigavane MB, Gawande SP, Daigavane P (2017) Modified single phase SRF d-q theory based controller for DVR mitigating voltage sag in case of nonlinear load. J Electromag Anal Appl
13. Nielsen JG, Newman M, Nielsen H, Blaabjerg F (2004) Control and testing of a dynamic voltage restorer (DVR) at medium voltage level. IEEE Trans Power Electron 19(3):806–813
14. Awaar VK, Jugge P, Tarakalyani S (2015) PQ Improvement by moderation of multi-level inverter controlling techniques and intensifying the performance of DVR. In: Advances in electrical and electronic engineering, vol 13, no 2
15. Ramachandaramurthy K, Fitzer C, Arulampalam A, Zhan C, Barnes M, Jankins N (2002) Control of a battery supported dynamic voltage restorer. Proc Inst Elect Eng Gen Transm Distrib 149(5):533–542
16. Zmwd DN, Holmes DG (1999) Stationary frame current regulation of PWM inverters with zero steady-state. In: Proceedings of IEEE power electronics specialists conference, pp 1185–1190
17. Vilathgamuwa DM, Perera AADR, Choi SS (2003) Voltage sag compensation with energy optimized dynamic voltage restorer. IEEE Trans Power Deliv 18(3):928–936
18. Coban R, Ercin O (2012) Multi-objective bees algorithm to optimal tuning of PID controller. Cukurova Univ J Fac Eng Arcchit 27(2):13–26
19. EKinci S, Hekimoglu B (2019) Improved kidney-inspired algorithm approach for tuning of PID controller in AVR systems. IEEE Access 7:39935–39947
20. Bao X, Jia H, Lang C (2019) 'A novel hybrid harris hawks optimization for color image multilevel thresholding segmentation.' IEEE Access 7:76529–76546

21. Heidari AA, Mirjalili S, Faris H, Aljarah I, Mafarja M, Chen H (2019) Harris hawks optimization: algorithm and applications. Future generation computer systems, vol 97, pp 849–872
22. Abbasi A, Firouzi B, Sendur P (2021) On the application of Harris hawks optimization (HHO) algorithm to the design of microchannel heat sinks. Eng Comput 37(2):1409–1428
23. Sobhy MA, Ezzat M, Hasanien HM, Abdelaziz AY (2019) Harris hawks algorithm for automatic generation control of interconnected power systems. In: Proceedings of 21st International Middle East Power System Conference (MEPCON), Dec 2019, pp 575–582

Chapter 5
Unified Power Quality Conditioner

Priyank Shah and Xiaowei Zhao

1 Introduction

In the distribution network, the power quality of an electrical supply plays a vital role for the satisfactory operation of home appliances. The electrical power network faces manifold challenges in the system as various kinds of local loads (e.g., constant impedance load, constant power load, constant current load, etc.) are tied at the common coupling point. In addition, the nonlinear loads pollute the supply currents by drawing the harmonic components from the source [1, 2]. These harmonics hamper the voltage quality at the coupling point. Likewise, the typical value of coupling point voltage is also increased and decreased as a huge reactive load is disconnected or connected from the system, respectively. Therefore, the series compensated device should be coupled in the system in order to improve the voltage profile at the load side network. The shunt compensated device should be connected in the system in order to improve the current quality of the grid side network. The series compensated device [3, 4] improves the voltage quality under manifold circumstances such as voltage sag, harmonically polluted grid voltages, voltage swell, and unbalanced grid voltages. Henceforth, several researchers have investigated the distribution static compensator (DSTATCOM) and unified power quality conditioner (UPQC) configurations to improve the power quality of the distribution network. This chapter deals with the classification, basic working principle, detailed discussion on various control strategies, and simulation results and analysis. The basic MATLAB model of the UPQC system is also provided in this chapter.

P. Shah · X. Zhao (✉)
Intelligent Control and Smart Energy (ICSE) Research Group, School of Engineering, The University of Warwick, Coventry CV4 7AL, UK
e-mail: Xiaowei.Zhao@warwick.ac.uk

A. K. Giri et al. (eds.), *Power Quality: Infrastructures and Control*,
Studies in Infrastructure and Control,
https://doi.org/10.1007/978-981-19-7956-9_5

2 Classification of UPQC System

The UPQC system is classified into four distinct categories, namely converter-based classification, topology-based classification, supply system-based classification, and rating-based classification. The detailed chart is depicted in Table 1 [1, 2].

3 Working Principle of UPQC System

In this section, the working principle of the UPQC system is discussed as follows. Figure 1 shows the schematic of the solar photovoltaic (PV) coupled unified power quality conditioner. Figure 2 shows the phasor diagram of the right shunt compensating device-based UPQC system [1]. In a nominal operating scenario, the load voltage (V_{L1}) and coupling voltage (V_{s1}) is the same. The load current is typical lag the load voltage by load angle (ϕ) and solar current (I_{pvg1}) is injected into the grid. As voltage is reduced from the nominal value, series VSC injects the compensating voltage (V_{inj}) to sustain the load voltage (V_{L2}) as per the reference nominal value (V_{S1}). The shunt VSC injects the compensating currents (I_{sh1} and I_{sh2}). These compensating currents contain load reactive power and active solar power current (I_{pvg1} and I_{pvg2}). It is worth noticing that the solar PV system is considered to grasp the whole understanding of the integration process of renewable energy sources with the UPQC system and works satisfactory even in absence of solar insolation. Hence, the overall discussion will be focused on the UPQC system.

Series Voltage Source Converter: The series voltage source converter is often known as a dynamic voltage restorer. The series compensating device compensates the voltage profile related problems such as voltage sag-swell, harmonically polluted voltages, unbalanced voltage, flicker, notches, etc. The pole of the series converter is connected with the transformer, which is a series connection between the grid and local load. The ripple filter is connected across the terminal of the transformer to suppress the switching harmonics [1, 2]. The pulse width modulation technique is used herein to perform an optimal operation of the series converter.

Shunt Voltage Source Converter: The shunt voltage source converter compensates for the current related issues such as local reactive power requirement, nonlinear current, unbalanced load currents, etc. It provides unity power factor operation to the grid and compels the balanced grid currents. The ripple filter is connected after the interfacing inductor to suppress the switching harmonics, generated from the shunt voltage source converter. The pulse width modulation technique is used herein to perform an optimal operation of the shunt converter.

Table 1 Detailed classification of UPQC systems

Converter based classification	The voltage source converter (VSC) and current source converter (CSC) based configurations are available for the UPQC system, however, the VSC based UPQC system has distinct and solid advantages over CSC based UPQC system, therefore, it is more popular and adopted in the real-time application
Topology based classification	Based on the location of the shunt compensation device (e.g., connected at the grid side network or load side network), it is classified into the left shunt and right shunt configurations. The right shunt compensation is widely adopted in the UPQC system due to having certain advantages in contrast with the left shunt compensation device
Supply system based classification	Two-wire UPQC: The single-phase configuration is known as a two-wire UPQC system. This configuration has two-wire feeds for the single-phase loads (e.g., one as supply wire and the second as return wire/neutral wire)
	Three-wire UPQC: In this configuration, a total of two three-leg voltage source converters are connected as a series compensating device and shunt compensating device, respectively. This configuration has one limitation such as it can only feed the three-phase loads. To overcome this issue, the four-wire UPQC system is designed for the reliable operation of single-phase load
	Four-wire UPQC: The drawback of the three-wire UPQC is alleviated using a four-wire UPQC system. In this configuration, the additional fourth wire is connected to the three-phase circuit in order to supply a three-phase load as well as a single-phase load. The three-leg and four leg voltage source converters are used as series compensating devices and shunt compensating devices to design a four-wire UPQC system, respectively. Here, it is worth noticing that the fourth leg of the shunt voltage source converter helps to alleviate the neutral current
Rating based classification	UPQC-P: In this configuration, the shunt voltage source converter ameliorates the current-profile related problems in order to oblige the unity power factor (UPF) operation of the grid. The series voltage source converter injects the compensating voltages in in-phase with the grid voltages for both voltage sag and voltage swell scenarios. As in-phase compensating voltages are provided, the rating of the series converter is quite low

(continued)

Table 1 (continued)

	UPQC-Q: In this configuration, the shunt voltage source converter improves the current-profile related issues in order to oblige the UPF operation of the grid. The series voltage source converter injects the compensating voltages in quadrature with the grid voltages for the voltage sag scenario. As quadrature compensating voltages are provided (e.g., either 90º lag or 90º lead), it cannot be able to work satisfactorily under voltage swell scenario (i.e., it cannot reduce the voltage level)
	UPQC-S: In this configuration, the series and shunt device ratings are very optimum, however, the control strategy is quite complex, unlike UPQC-P and UPQC-Q systems. The shunt converter provides partially reactive power compensation and the rest of the reactive power is provided by the series converter. Therefore, the optimum rating can be archived by sharing the reactive power. It is capable to work for both voltage sag and voltage swell phenomena

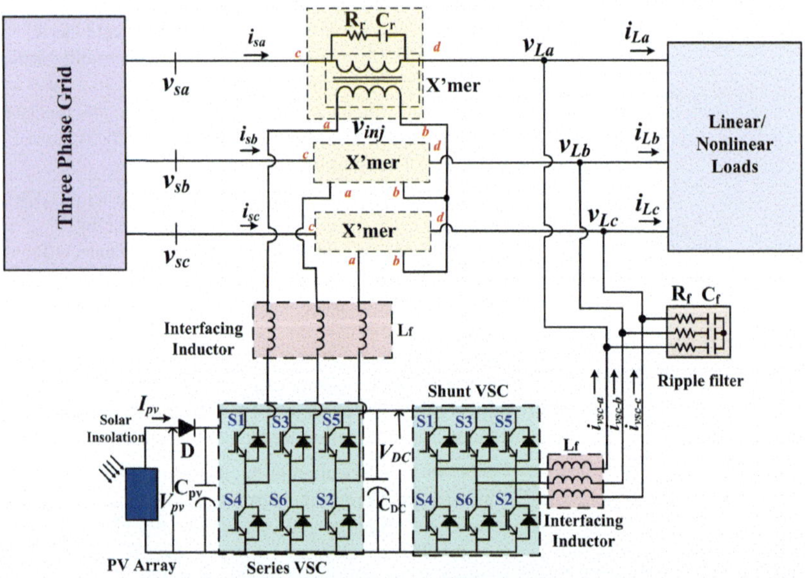

Fig. 1 Schematic of solar PV coupled unified power quality conditioner

Fig. 2 Graphical phasor representation of solar PV fed unified power quality conditioner

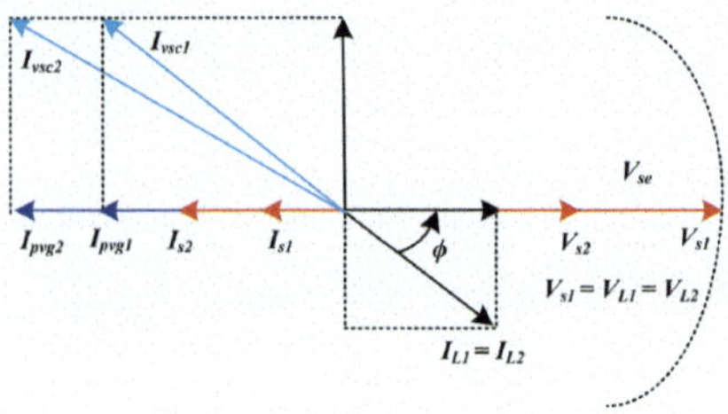

4 Control Strategy

The manifold control strategies are available in the state-of-art to achieve optimal operation of the UPQC system. The list of several control strategies [5–15] is mentioned as follows.

- Synchronous reference frame theory (e.g., phase-locked loop (PLL) based algorithm)
- Least mean algorithm (e.g., least mean square, least mean fourth, leaky least mean square, etc.)
- Generalized integrator-based algorithm (e.g., second-order, third-order, multiple second-order, etc.)
- Lyapunov controller
- Repetitive controller
- Neural network-based algorithm
- Fuzzy logic controller
- Kalman filter-based algorithm (e.g., Unscented Kalman filter, Cubature Kalman filter, etc.)

The precise implementation of selected control strategies is described as follows.

4.1 Synchronous Reference Frame Theory-Based Control Strategy (PLL Based Algorithm)

The distinct advantages of synchronous reference frame theory are explained as follows.

(1) The measurements are transformed into the *dq*-domain, which is the DC domain, unlike the *abc*-domain of AC quantities. Here, it is worth noticing that two proportional-integral (PI) controllers are required to control the system. However, a total of three proportional-integral (PR) controllers are required to control the *abc*-quantities of the system.

(2) The frequency is reflected as the system gain in the *dq*-domain, therefore, it is easy to achieve the zero steady-state error (i.e., even in case of fluctuations in the frequency). In the case of the PR controller, it is difficult to achieve zero steady-state error as it is explicitly adjusted to extract the specific frequency component. Hence, most industry application has adopted the *dq*-domain for their commercial applications.

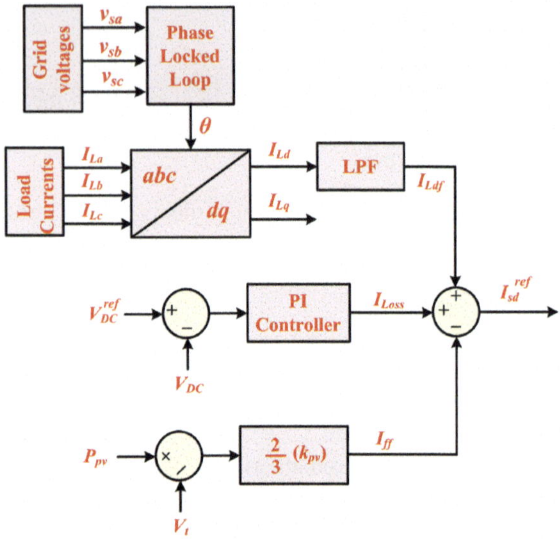

Fig. 3 Detailed implementation of synchronous reference frame-based control strategy to obtain reference grid current

The detailed explanation of the shunt and series VSC control configuration is described as follows.

A. *Shunt VSC Control Strategy*

Figure 3 depicts the overall structure of the control strategy of the shunt voltage source converter. The overall controller is divided into five sections namely estimation of the fundamental active weight of load current, evaluation of loss component, evaluation of feed-forward term, generation of reference currents, and generation of gating signals of shunt converter.

(1) *Estimation of Fundamental Active Weight of Load Current*

To estimate the active weight of the fundamental load component, the phase-locked-loop is required to transform the *abc*-domain quantities to *dq*-domain using Park transformation [9]. The detailed schematic of an implementation of this structure [9] is shown in Fig. 3. The grid voltages are fed as input of the phase-locked loop, and frequency and phase are the output of it. Using this phase angle, in-phase and quadrature vectors are generated and help to transform DC quantities of the load currents. Furthermore, these components are used later to generate reference grid currents. The d-axis and q-axis component is employed further to compute the net amplitude of the reference grid signals. The d-axis and q-axis components (I_{Ld} and I_{Lq}) are assigned as an active and reactive weight of the fundamental load current [9].

(2) *Evaluation of Loss Term*

To accommodate the losses in converter and filter, the instantaneous difference amongst reference command (V_{DC}^{ref}) and sensed voltage (V_{DC}) is fed to the controller and the corresponding output is assigned as a loss component (I_{Loss}). In case of availability of solar PV array, the DC bus reference is chosen based on maximum power point tracking operation. In contrast, it is chosen based on the amplitude of line voltages. It plays an important role to regulate the DC bus voltage. It is expressed as [6, 7],

$$I_{\mathrm{Loss}} = (K_p + \frac{K_i}{s})(V_{DC}^{\mathrm{ref}} - V_{DC}) \quad (1)$$

where, K_p and K_i are proportional and integral gains of the PI controller. The loss component is processed further to estimate the amplitude of the reference grid currents.

(3) *Evaluation of Feed-Forward Term*

In the case of sustainable energy sources, solar power (P_{pv}) and amplitude of terminal voltage (V_t) are accommodated as a feed-forward component included in the controller. The feed-forward term (I_{ff}) is computed as [7, 8],

$$I_{ff} = \frac{2k_{pv}P_{pv}}{3V_t} \quad (2)$$

where, k_{pv} is accommodated herein to consider the losses in the converter and filter. The typical value of this intermediate constant lies in the range of 0.95–0.97. The feed-forward component is processed further to estimate the amplitude of the reference currents.

(4) *Generation of Reference Grid Currents*

The estimated loss component (I_{Loss}), feed-forward component (I_{ff}), and active weight of fundamental load component (I_{Load}) are used herein to acquire the amplitude of the reference component (I_{sd}^{ref}). It is expressed as [8, 9],

$$I_{sd}^{\mathrm{ref}} = I_{\mathrm{Loss}} - I_{ff} + I_{\mathrm{Load}} \quad (3)$$

The three-phase reference grid currents (i_{sa}^{ref}, i_{sb}^{ref}, and i_{sc}^{ref}) are obtained through this direct and quadrature axis current components. The instantaneous error is obtained as,

$$e_{sa} = i_{sa}^{\mathrm{ref}} - i_{sa};\ e_{sb} = i_{sb}^{\mathrm{ref}} - i_{sb};\ e_{sc} = i_{sc}^{\mathrm{ref}} - i_{sc} \quad (4)$$

This error is fed to the hysteresis regulator to obtain the gating pulses.

(5) *Generation of Switching Pulses of Shunt Converter*

Figure 4 exhibits the detailed schematics of the obtained gating pulses for the shunt VSC. The error amongst the reference and sensed grid currents are fed to the hysteresis regulator. As sensed grid currents cross the upper band of the hysteresis width, the lower switch of the respective leg of the converter is turned on. In contrast,

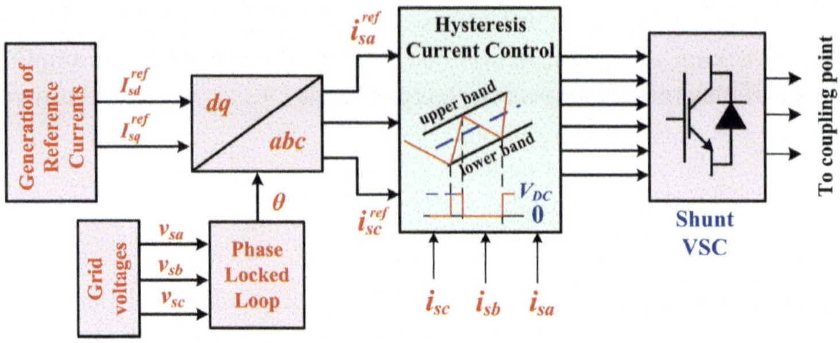

Fig. 4 Generation of switching pulses for shunt VSC

if sensed grid currents cross the lower hysteresis width, then the upper switch of the respective leg of the converter is turned on to regulate the current dynamics within the band. The detailed design and operation are explained in the literature [10].

B. *Series VSC Control Strategy*

The detailed schematic of the series VSC controller [9] is depicted in Fig. 5. The load voltages and grid voltages are transformed into the *dq*-domain from the *abc*-domain. The error between active and reactive components of the load and grid voltages are processed with the reference voltages. Thereafter, it is fed to the PI regulator. The outcome of the PI controller is transformed into *abc*-domain (v_{inj}^{ref}) using the inverse Park transformation. Thereafter, the pulse width modulated pulses are generated for the series voltage source converter. The detailed implementation to generate switching pulses for the series converter is explicitly illustrated in Fig. 6.

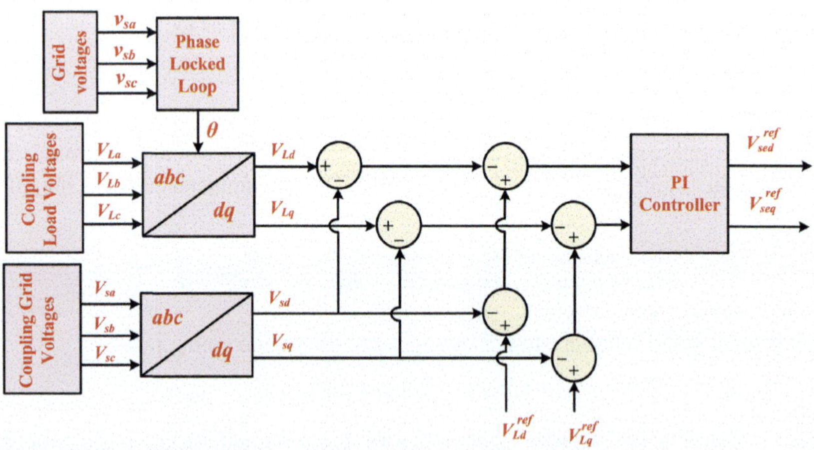

Fig. 5 Detailed implementation of synchronous reference frame-based control strategy to obtain reference compensation voltages

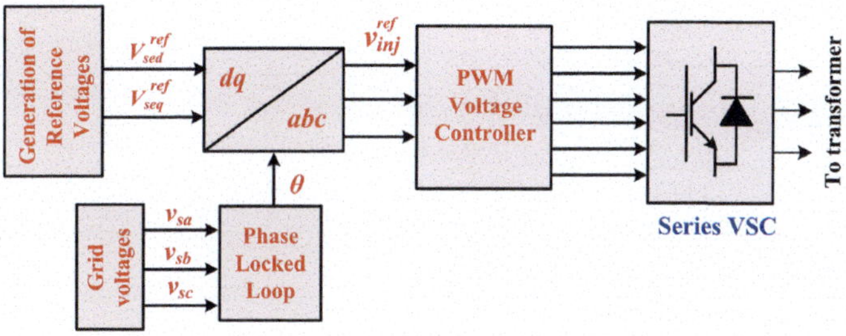

Fig. 6 Generation of switching pulses for series voltage source converter

Table 2 Design of system parameters

Parameter	Expression
DC link voltage	$V_{DC} = \frac{2\sqrt{2}}{\sqrt{3}} \times v_{sab}$
DC link capacitor	$C_{DC} = \frac{2\times k\times a\times P_{pv}\times t}{(V_{DC}^2 - V_{DC\,\min}^2)}$
Interfacing inductor	$L_{vsc} = \frac{\sqrt{3}\times m\times V_{DC}}{12\times a\times f_{sw}\times I_{vsc}^{\text{ripple}}}$
Voltage rating of VSC switch	$V_{sw} = sf \times V_{DC}^{\max}$
Current rating of VSC switch	$I_{sw} = sf \times I_{vsc}$

(where, $v_{sab}, k, a, t, m, sf, i_{vsc}$ represents line voltage, variation of energy during dynamics in percentage, overloading factor, time-span to recover the DC bus voltage, modulation index, safety factor, and rated converter current, respectively)

The design of the system parameters is given in Table 2. The selection of the system components is given in Appendix. The performance of the system is explained as follows.

Case-I Response of UPQC with Load Current Unbalancing Scenario

Figure 7a depicts the behaviour of the system at an event of unbalancing in the load. Figure 7a depicts the behaviour of v_{sabc}, i_{sabc}, i_{Labc}, i_{vsc}, and v_{sa}-i_{sa}. The grid voltages are symmetrical and sinusoidal as shown in Fig. 7a. The grid currents are also symmetrical and sinusoidal even with the presence of nonlinear load. The dynamics of the local load are exhibited in Fig. 7a. The current dynamics of the load network are not hampered as the constant admittance load is tied at the coupling point. The compensating currents are provided by the voltage source converter to ensure the unity power factor. The unity power factor operation is demonstrated using phase voltage and current in Fig. 7a. Henceforth, it is proved that it complies to the IEEE std. 519 and IEC std. 61,727. Figure 7b presents the dynamics of the V_{DC}, I_{pv}, P_{pv}, P_{grid}, and Q_{grid}. The DC bus voltage is not affected even under unbalancing phenomena in the load side network, which means the loss component is not much varied from its nominal value. The dynamics of solar photovoltaic are not affected as maximum power point tracking is maintained for the solar power generation. The behaviour of the solar PV array power is illustrated in Fig. 7b. The variation of active power in

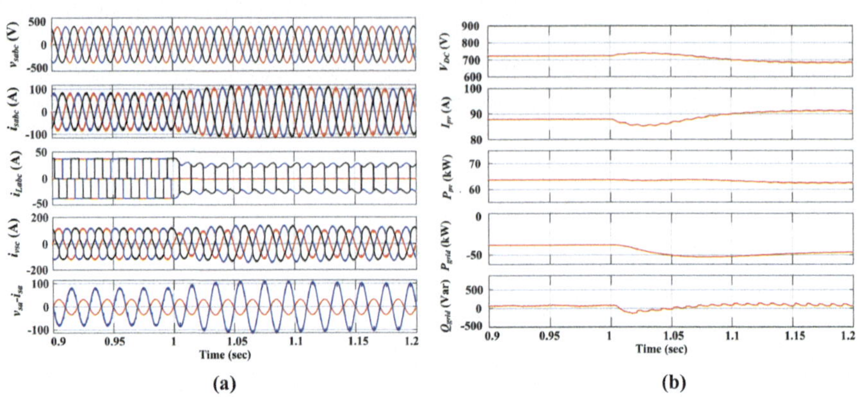

Fig. 7 System dynamics in case of load current unbalancing scenario

the grid side network is depicted in Fig. 7b. The reactive power flow in the grid side network is null, which also indicates the unity power factor operation. Thereby, the presented system also follows the IEEE std. 1159.

Case-II Performance of System at Voltage Sag

Figure 8a outlines the system dynamics in an event of voltage sag phenomena. Figure 8a shows the waveform of v_{sabc}, v_{inj}, v_{Labc}, i_{Labc}, i_{vsc} and i_{sabc}. As voltage sag occurs at 1.60 s, the performance of the grid side network is depicted in Fig. 8a. The series converter injects the compensating voltage to regulate the voltage at the load side network. The typical behaviour of the load side network is depicted in Fig. 8a. The load voltage is regulated as per the reference command. The current dynamics of the grid, load, and shunt VSC network are shown in Fig. 8a. The grid currents are balanced and symmetrical as exhibited in Fig. 8a. The nominal value of grid currents is increased (i.e., the injecting power is almost constant) as the typical value of grid voltage is decreased as depicted in Fig. 8b. One can notice that the performance of the load side network is not compromised even with the presence of voltage sag phenomena. It is worth noticing that the presented system complies to the IEEE std. 1159. Figure 8b shows the dynamics of the V_{DC}, I_{pv}, P_{pv}, P_{grid}, and Q_{grid}. The dynamics of the loss component are shown in Fig. 8b. The DC bus voltage is regulated and maintained as per the reference voltage. Solar energy is not affected, which can be illustrated using significant signals such as solar PV array current and power. The active and reactive power dynamics of the grid side network are illustrated in Fig. 8b. The null reactive power flow in the grid side network represents that the unity power factor operation is successfully achieved even with the presence of voltage sag phenomena. The presented system simultaneously complies to the IEEE stds. 519 and 1159, and IEC std. 61,727.

The distinct disadvantages of SRFT or PLL-Based Algorithm are explained as follows.

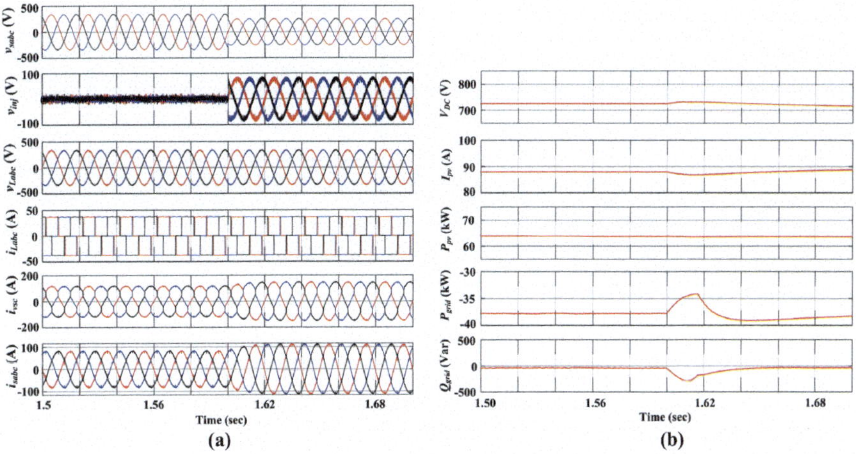

Fig. 8 Performance of the system with voltage sag phenomena

- The response of the UPQC is hampered by the participation of abnormal grid scenarios as the state-of-art PLL fails to extract phase angle under abnormalities in the grid side network.
- The implementation of PLL based algorithm is quite complicated as several transformations and trigonometric functions are required.
- The dynamic performance can be hampered as the low-pass filter is connected in the path of tracing the *dq*-component.
- The performance of the system is poor under abnormal grid scenarios (e.g., distorted grid voltages, unbalanced grid voltages, etc.).

To overcome these issues, the detailed implementation of a least mean algorithm is explained as follows.

4.2 Least Mean Based Control Strategy

The least mean algorithm works based on the typical value of instantaneous error between the sensed load current and estimated load current. The distinct advantages of this algorithm are mentioned as follows.

- The realization of least mean algorithm is quite easy and very computation efficient as compared to the phase-locked loop-based methodology.
- The steady-state performance of the UPQC system is satisfactory.

The basic MATLAB/Simulink implementation with detailed step by step procedure is explicitly given in [16, 17]. The detailed implementation of shunt and series converter controls are explained as follows.

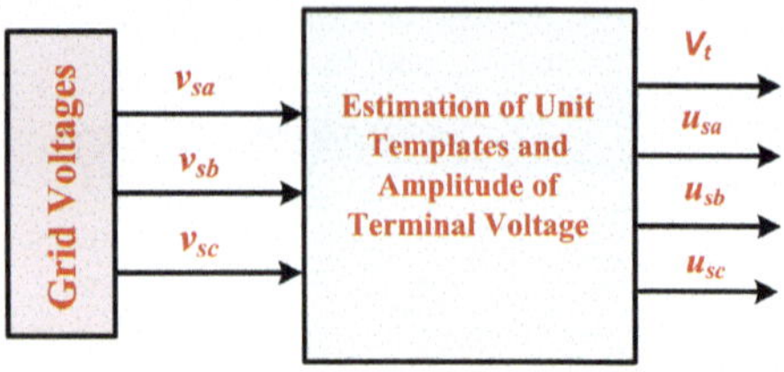

Fig. 9 Schematic to extract unit templates

A. *Shunt VSC Controller*

Figures 9, 10, 11 and 12 demonstrate the overall structure of the presented least mean square algorithm for the UPQC system. It consists of the calculation of loss and feed-forward terms, computation of fundamental active load component, generation of reference grid currents, and switching pulses for shunt VSC. The detailed computation of each term is described as follows.

(1) *Computation of Unit Templates*

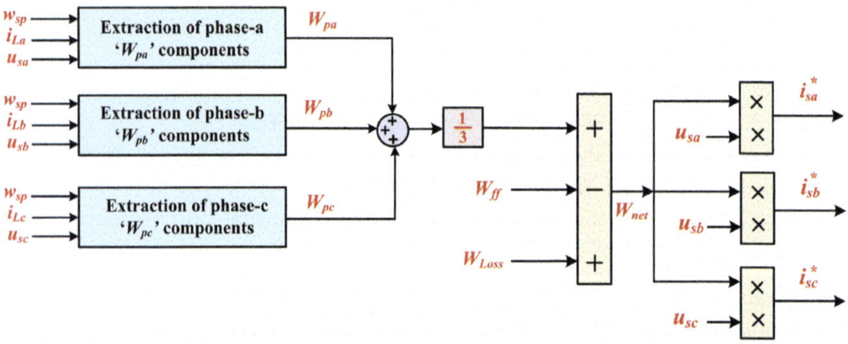

Fig. 10 Detailed implementation of least mean square-based control strategy to acquire reference grid currents

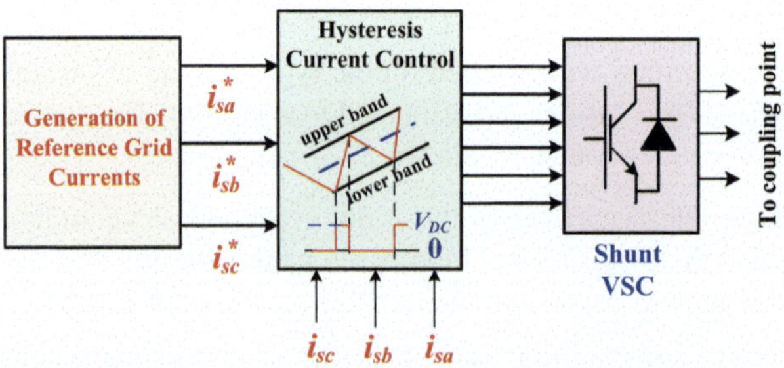

Fig. 11 Generation of switching signals for shunt voltage source converter

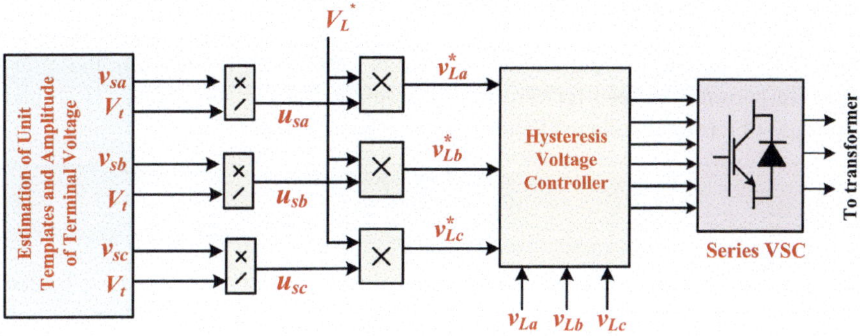

Fig. 12 Detailed implementation of control strategy for series voltage source converter

The phase voltage of the grid side network is sensed and processed further to compute the terminal voltage of the coupling point. The in-phase unit templates are computed to obtain the unit vectors of the phase voltages. The amplitude of the terminal voltage is (V_t) computed as [6, 7],

$$V_t = \sqrt{(2/3)\left(v_{sa}^2 + v_{sb}^2 + v_{sc}^2\right)} \tag{5}$$

where, v_{sa}, v_{sb}, and v_{sc} are three-phase voltages of the grid voltages. Figure 9 shows the realization to extract the unit vectors. The unit templates (u_{pa}, u_{pb}, and u_{pc}) are computed as [6, 7],

$$u_{sa} = v_{sa}/V_t;\ \ u_{sb} = v_{sb}/V_t;\ \ u_{sc} = v_{sc}/V_t \tag{6}$$

These unit vectors are processed further to calculate the reference currents.

(2) *Estimation of Loss Component*

The active power component is required to regulate the DC bus voltage as per the reference command and the corresponding active component is known as the loss component. In the presence of the solar power generation, the active component is provided by the solar power itself, and the rest of the power is fed to the VSC. In contrast, the active component is supplied from the grid in absence of the solar power generation. The reference DC bus voltage is chosen as explicitly mentioned in subsection A-2 of Sect. 4.1. The loss term is computed using instantaneous error is passed through the PI controller and it is expressed as [6, 7],

$$W_{\mathrm{Loss}} = (K_p + \frac{K_i}{s})(V_{DC}^{\mathrm{ref}} - V_{DC}) \tag{7}$$

where, K_p and K_i are proportional and integral gains of the PI controller. The loss term is proceeded further to compute the reference amplitude of the grid currents.

(3) *Estimation of PV Feedforward Component*

In order to compute the contribution of the renewable energy sources in the reference grid current, the feed-forward term is computed using solar power (P_{pv}) and amplitude of terminal voltage (V_t). It is expressed as [7, 8],

$$W_{ff} = \frac{2k_{pv}P_{pv}}{3V_t} \tag{8}$$

where, k_{pv} is constant and its typical value falls between 0.95 and 0.97. It is adjusted in the feed-forward term to adapt the losses in the semiconductor devices and filters. This feed-forward term is employed further to acquire the reference amplitude of the grid currents.

(4) *Extraction of Fundamental Weight of Load Current*

The least mean square-based algorithm is facilitated herein to compute the amplitude of the fundamental component of the load current. The least mean square algorithm is realized as follows. The fundamental active weight is expressed as [11, 12],

$$W_{pa}(n+1) = W_{pa}(n) + \mu e_{pa}(n) \tag{9}$$

where, μ represents intermediate tuning gain and $e_{pa}(n)$ is an instantaneous error between sensed and estimated load current. The estimation error for phase 'a' is obtained as [11, 12],

$$e_{pa}(n) = i_{La}(n) - W_{pa}(n)u_{pa}(n) \tag{10}$$

Likewise, the active weight of load current for phases 'b' and 'c' are expressed as [11, 12],

$$W_{pb}(n+1) = W_{pb}(n) + \mu e_{pb}(n) \tag{11}$$

$$W_{pc}(n+1) = W_{pc}(n) + \mu e_{pc}(n) \tag{12}$$

These magnitudes of fundamental load current are processed further to compute the net magnitude of load current (W_{Load}).

(5) *Generation of Reference Grid Currents*

Figure 10 shows the detailed block diagram to estimate the amplitude of reference grid currents [12]. The total active component of the reference current is calculated using a fundamental magnitude of the load component, feed-forward component, and loss component. It is computed as follows.

$$W_{\text{net}} = W_{\text{Loss}} - W_{ff} + W_{\text{Load}} \tag{13}$$

The reference grid currents are obtained using in-phase unit templates and the net amplitude of the reference grid current. These reference grid currents are processed further to acquire gating pulses for the voltage source converter as follows.

$$i_{sa}^{*} = u_{sa} W_{\text{net}}; \; i_{sb}^{*} = u_{sb} W_{\text{net}}; \; i_{sc}^{*} = u_{sc} W_{\text{net}} \tag{14}$$

Here, it is worth noticing that the in-phase unit vectors are used as the unity power factor operation is a key objective of this work.

(6) *Generation of Gating Pulses*

Figure 11 depicts the realization to generate the optimum switching pulses for the shunt voltage source converter. For UPF operation, the reference currents are calculated by taking product of an in-phase unit vector with the calculated net active grid current term. The instantaneous current error is fed to the current controller. The typical band of the hysteresis bandwidth is chosen in the range of 0.2. The detailed design steps of the hysteresis regulator is given in the literature [10].

B. *Series VSC Controller*

Figure 12 shows the control structure of the series converter to regulate the load voltages. The reference load voltages are acquired in order to provide gate pulses for the series VSC [8, 9]. The in-phase unit templates are multiplied by the amplitude of the load side network (V_L^*) to obtain the reference load voltages (v_{La}^*, v_{Lb}^*, and v_{Lc}^*). It is expressed as,

$$v_{La}^{*} = u_{sa} V_L^{*}; \qquad v_{Lb}^{*} = u_{sb} v_{La}^{*}; \qquad v_{Lc}^{*} = u_{sc} V_L^{*} \tag{15}$$

where, V_L^* is the amplitude of reference load voltages. The instantaneous error between reference load voltages and sensed load voltages are fed to the hysteresis current controller. The hysteresis bandwidth is decided based on the literature [10]. The performance of the system is explained as follows.

Case-I Behaviour of UPQC System with Absence of Sustainable Energy Sources

Figure 13a shows the response of the system in the absence of solar irradiation (e.g., night-time case). It demonstrates the waveforms of v_{sabc}, i_{sabc}, i_{Labc}, i_{vsc}, and v_{sa}-i_{sa}. As solar energy is lessened to zero, the phase reversal is detected in the grid currents. The dynamics of load currents are illustrated in Fig. 13a. The VSC coupled system only provides compensating currents instead of active components to the grid side network. The UPF operation of the grid is presented using an in-phase operation between phase voltage and grid current. Figure 13a shows the dynamics of the V_{DC}, I_{pv}, P_{pv}, P_{grid}, and Q_{grid}. The dynamics of DC bus voltage are depicted in Fig. 13b. It is stable and regulated as per reference command. The dynamics of solar power current and power are depicted in Fig. 13a. The steady-state value of an active power in the grid side network is varied and it can be observed in Fig. 13b. There is not any change in the reactive power flow in the grid, (i.e., as solar power generation is not available) which exhibits the VSC compensates for the reactive power requirement

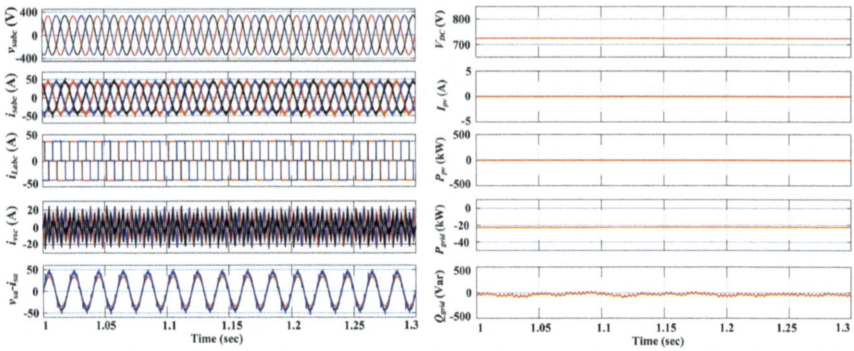

(a) (b)

Fig. 13 Performance of the presented system with absence of solar power generation

of the local loads. Therefore, the presented UPQC system complies the revised IEEE std. 519 and IEC std. 61,727.

Case-II Behaviour of UPQC System Under Voltage Swell

Figure 14a shows the response of the system in an event of voltage swell phenomena. It demonstrates the waveforms of v_{sabc}, v_{inj}, v_{Labc}, i_{Labc}, i_{vsc} and i_{sabc}. As voltage swell occurs at second, the dynamics of grid voltages is presented in Fig. 14a. The series VSC injects the counter compensating voltages in order to regulate the voltage at the load side network. The dynamics of the load side network are depicted in Fig. 14a. The load current dynamics are smooth even with the presence of voltage swell phenomena. One can notice that the presented system adheres to the IEEE std. 1159. The voltage source converter provides compensation to generate balanced currents as demonstrated in Fig. 14a. As the typical value of grid voltage is increased, the typical value of grid currents is decreased (i.e., the injecting power is almost the same) as depicted in Fig. 14a. The presented system follows the IEEE std. 519 and IEC std. 61,727. Figure 14b shows the waveform of V_{DC}, I_{pv}, P_{pv}, P_{grid}, and Q_{grid}. The loss component is not much varied from its nominal value as exhibited in Fig. 14b. The DC bus voltage is controlled based on its nominal value even though the typical value of grid voltage is higher than the nominal value. The solar current and power are extracted based on maximum power point operation. Hence, the surplus power is fed to the grid and its dynamics of active and reactive powers are shown in Fig. 14b.

The several disadvantages of the least mean methodology are expressed as follows.

- The estimated amplitude of load current has an oscillatory response, which generates oscillations while generating the reference grid currents, thereby, the THDs of grid current are relatively high.
- It is not capable to estimate the amplitude under biases in the load current (e.g., soldering machine, old television, etc.).
- The convergence rate of the least mean methods is quite poor, which affects the system performance under load unbalancing scenarios.

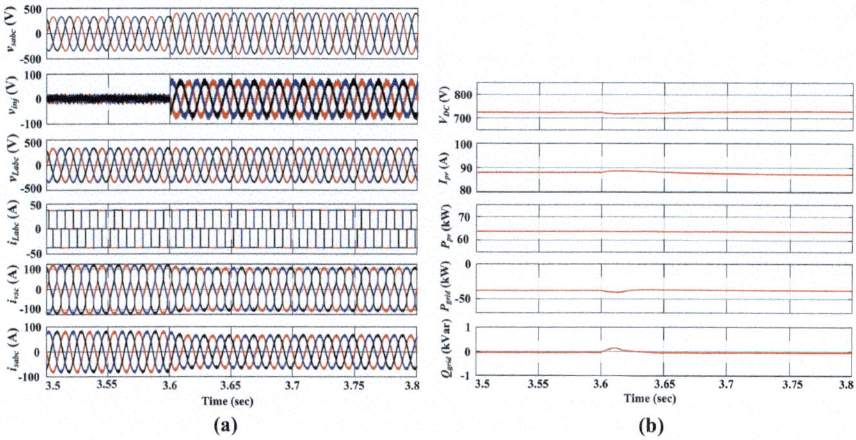

Fig. 14 Performance of the presented system with voltage swell phenomena

- The performance of the system is poor under abnormal grid scenarios (e.g., distorted grid voltages, unbalanced grid voltages, etc.).

To overcome these issues, the generalized integrator-based algorithm is analyzed in the following section.

4.3 Generalized Integrator Based Methodology

The generalized integrator has certain advantages such as good accuracy, ease of implementation, good convergence rate, behaving as a low-pass filter or band-pass filter, etc. The distinct advantage of this work is explained as follows.

- It can work as low-pass filter and band-pass filter for the extracted quadrature and in-phase components of an input signal, respectively.
- The realization of this method is quite simple, and the computational burden is also low as compared with the phase-locked loop-based methodology.
- The steady-state and dynamic performance is noticeably improved as compared with the state-of-art least mean and phase-locked loop-based methodologies.

The detailed implementation of a generalized integrator-based strategy is explained as follows.

A. *Shunt VSC Controller*

Figures 15, 16 and 17 demonstrate the overall structure of the presented least mean square algorithm for the UPQC system. It consists of the calculation of loss and feed-forward terms, computation of fundamental active load term, generation of reference grid currents, and gating pulses for shunt converter. The comprehensive details for computation of each term are described as follows.

(1) *Estimation of Unit-Templates*

The amplitude of the terminal voltage (V_t) is expressed as [6, 7],

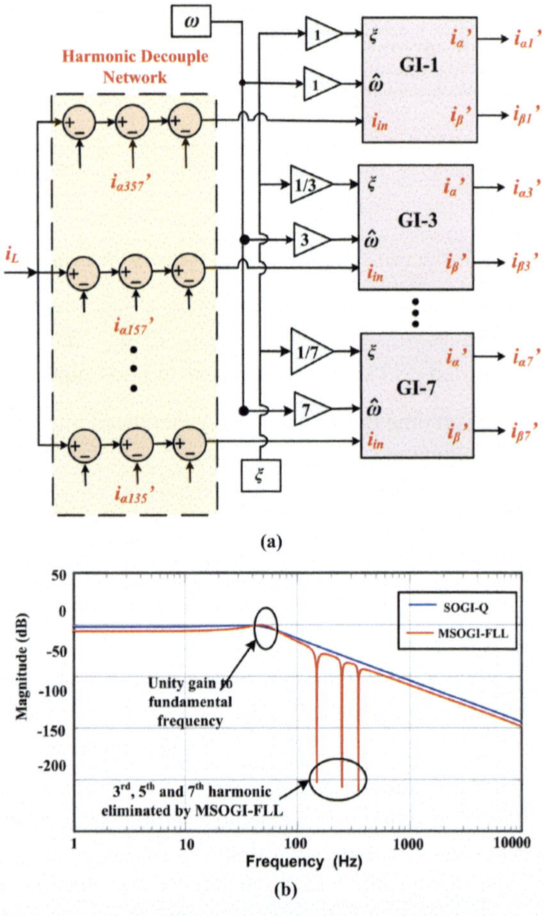

Fig. 15 Detailed implementation of multiple second-order generalized integrator

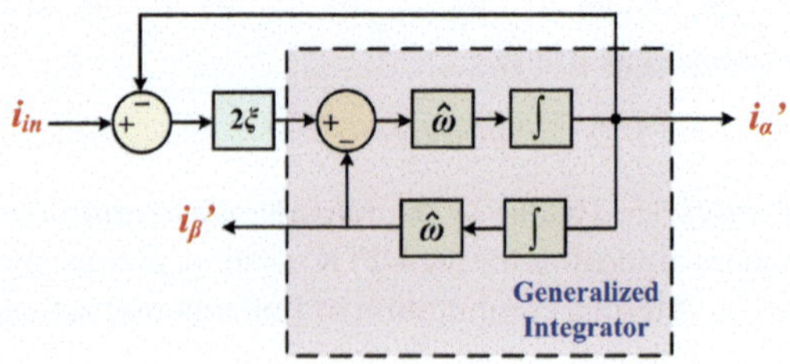

Fig. 16 Detailed implementation of generalized integrator

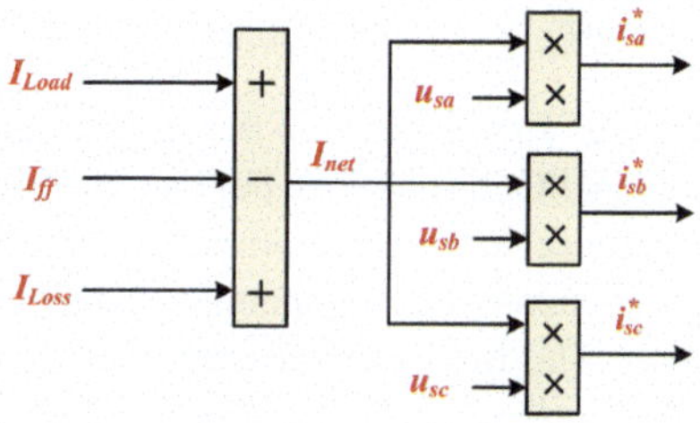

Fig. 17 Schematic of control strategy to obtain the reference currents

$$V_t = \sqrt{\frac{2}{3}\left(v_{sa}^2 + v_{sb}^2 + v_{sc}^2\right)} \tag{16}$$

where, v_{sa}, v_{sb}, and v_{sc} are phase voltages of the grid side network. The in-phase unit-templates (u_{sa}, u_{sb}, and u_{sc}) are computed using phase voltages and amplitude of terminal voltages. It is computed as [6, 7],

$$u_{sa} = v_{sa}/V_t; \qquad u_{sb} = v_{sb}/V_t; \qquad u_{sc} = v_{sc}/V_t \tag{17}$$

These templates play a vital role to generate in-phase reference grid currents in order to ensure the unity power factor operation.

(2) *Estimation of Loss Term*

In order to regulate the DC bus voltage, the PI controller is used herein to adjust the losses in the voltage source converter. The in-between instantaneous error is fed to the PI controller and the corresponding outcome is processed further to acquire the amplitude of reference currents. The reference DC bus voltage is chosen as explicitly mentioned in subsection A-2 of Sect. 4.1. The mathematical formulation is expressed as [6, 7],

$$I_{\text{Loss}} = (K_p + \frac{K_i}{s})(V_{DC}^{\text{ref}} - V_{DC}) \tag{18}$$

where, K_p and K_i are proportional and integral gains of the controller, V_{DC}^{ref} and V_{DC} are a reference and sensed DC bus voltages.

(3) *Estimation of Feed-Forward Term*

The deviation in solar energy and amplitude of terminal voltage are adjected in the control strategy whole evaluating the amplitude of the reference grid currents. It is evaluated as [7, 8],

$$I_{ff} = \frac{2k_{pv}P_{pv}}{3V_t} \tag{19}$$

where, P_{pv}, V_t, and k_{pv} are solar power, the amplitude of terminal voltage, and intermediate constant, respectively. The typical value of constant falls between 0.95 and 0.97.

(4) *Estimation of Amplitude of Fundamental Load Component*

Figures 15 and 16 show the realization of multiple second-order generalized integrators (MSOGI) [13, 14]. For ease of implementation and understanding. The three sets of generalized integrators are connected in parallel to estimate the selective harmonic components. Each generalized integrator is tuned with help of a frequency locked loop. The fundamental frequency component is estimated with help of estimation error in load component and quadrature fundamental component. The generalized integrator behaves as low- pass filter and band-pass filter for quadrature fundamental component and in-phase fundamental component, respectively. The transfer function for in-phase and quadrature components ($i_d^{'}(s)$ and $i_q^{'}(s)$) is expressed as [13, 14],

$$D(s) = \frac{i_d^{'}(s)}{i_{in}^{'}(s)} = \frac{k\widehat{\omega}s}{s^2 + k\widehat{\omega}s + \widehat{\omega}^2} \tag{20}$$

$$Q(s) = \frac{i_q^{'}(s)}{i_{in}^{'}(s)} = \frac{k\widehat{\omega}^2}{s^2 + k\widehat{\omega}s + \widehat{\omega}^2} \tag{21}$$

where, k is a tuning parameter for the designed generalized integrator. To estimate the selective harmonics, damping coefficient (ξ), intermediate gain block of harmonic frequency (K_f) and gain of damping factor are mandatory to be designed. Figure 15b depicts the frequency domain analysis of MSOGI method [14].

Design Procedure of Damping Factor and Its Gain: The damping factor plays an important role to extract selective harmonics. The typical value is designed based on settling time [13, 14],

$$t_s = \frac{4}{\xi\omega_n} \tag{22}$$

The transfer function for the in-phase component for hth harmonic is expressed as,

$$\left|D(jh\widehat{\omega})\right| = \frac{2\xi h}{\sqrt{\left(1 - h^2\right)^2 + (2\xi h)^2}} \tag{23}$$

where, h is an odd integer and the typical value of damping factor is in the range of zero to one. The typical value of h can be three, five and seven and it can be expressed as,

$$\left|D(jh\widehat{\omega})\right| = \frac{2\xi}{h} \tag{24}$$

Therefore, the choice of intermediate gain is selected as 0.3333, 0.2, and 0.1428, respectively for 3rd, 5th and 7th harmonic extraction of a generalized integrator.

Design Procedure of Gain (K_f): The gain K_f is tuned based on the requirement to extract the specific harmonic components from the given input signal. In this work, third, fifth and seventh harmonics are extracted. The frequency of the third, fifth and seventh harmonic components are 150 Hz (50 × 3 = 150 Hz), 250 Hz (50 × 5 = 250 Hz), and 350 Hz (50 × 7 = 350 Hz). Therefore, the values of K_f for GI-2, GI-3, and GI-4 are 3, 5, and 7, respectively.

The in-phase and quadrature component of fundamental load current is expressed in terms of third, fifth and seventh harmonic as [13, 14],

$$i'_{d1}(s) = \left[D_1(s)\left(\frac{1 - D_5(s)}{1 - D_5(s)D_1(s)}\right)\left(\frac{1 - D_7(s)}{1 - D_7(s)D_1(s)}\right)\right]i_L \tag{25}$$

Likewise, the transfer function for quadrature fundamental component is expressed as,

$$i'_{q1}(s) = \left[Q_1(s)\left(\frac{1 - Q_5(s)}{1 - Q_5(s)Q_1(s)}\right)\left(\frac{1 - Q_7(s)}{1 - Q_7(s)Q_1(s)}\right)\right]i_L \tag{26}$$

After obtaining quadrature components of load currents (i'_{aq1}, i'_{bq1} and i'_{cq1}), zero-crossing indicators and sample and hold circuits are employed herein to acquire the magnitude of fundamental load components (I_{pa}, I_{pb}, and I_{pc}) by using in-phase vectors (u_{sa}, u_{sb}, and u_{sc}), respectively.

(5) *Computation of Reference Currents*

The estimated loss component, feed-forward term, and amplitude of fundamental load current are used herein to evaluate the amplitude of reference grid currents as depicted in Fig. 17. The reference grid currents (i^*_{sa}, i^*_{sb}, and i^*_{sc}) are evaluated using a magnitude of reference signals (I_{net}) and in-phase templates (u_{sa}, u_{sb}, and u_{sc}) as follows [10].

$$I_{\text{net}} = I_{\text{Loss}} - I_{ff} + I_{\text{Load}} \tag{27}$$

$$i^*_{sa} = u_{sa}I_{\text{net}};\;\; i^*_{sb} = u_{sb}I_{\text{net}};\;\; i^*_{sc} = u_{sc}I_{\text{net}} \tag{28}$$

These reference grid currents are processed further to generate switching pulses for the shunt voltage source converter.

(6) *Generation of Gate Pulses*

Figure 11 exhibits the block diagram to acquire the gating signals for the shunt converter using the hysteresis current controller. The instantaneous difference between the reference and sensed source currents is fed to the hysteresis current regulator to ensure the unity power factor operation of the system. The mathematical formulation of these reference currents is computed as [13, 14],

$$e_{sa} = i^*_{sa} - i_{sa};\;\; e_{sb} = i^*_{sb} - i_{sb};\;\; e_{sc} = i^*_{sc} - i_{sc} \tag{29}$$

where, e_{sa}, e_{sb}, and e_{sc} are instantaneous errors in grid currents. The shunt VSC has variable switching frequency as e_{sabc} changes from time to time. The detailed procedure to acquire the hysteresis band is explained in the literature [10].

B. *Series VSC Controller*

Figure 12 demonstrates the detailed implementation of series VSC control of the UPQC system. To ensure the voltage at the load side network, the error amongst reference load voltage and sensed load voltage is fed with the hysteresis regulator. The switching frequency of the hysteresis current controller is variable as it depends upon the hysteresis width and DC bus voltage. The detailed design procedure is explicitly discussed in the literature [10]. The performance of the overall system is explained as follows.

Case-I Performance of System with DSTACOM to PV Mode

Figure 18 presents the response of the UPQC at the transition from DSTATCOM to PV mode. Figure 18a demonstrates the waveforms of v_{sabc}, i_{sabc}, i_{Labc}, i_{vsc}, and v_{sa}-i_{sa}. The variation in grid side voltage does not occur for nominal operating scenario as depicted in Fig. 18a. The grid currents are symmetrical and sinusoidal as depicted in Fig. 18a. The load currents are nonlinear as constant admittance load is coupled at the coupling point. Initially, the VSC provides compensating current to compel the balanced grid currents. As solar power generation regained at second, the typical value of VSC currents is increased as depicted in Fig. 18a. The unity power factor operation is depicted using significant signals such as phase voltage and grid current. Figure 18 shows the waveforms of V_{DC}, I_{pv}, P_{pv}, P_{grid}, and Q_{grid}. The DC bus voltage is regulated using PI controller. The detailed tuning procedure is explicitly explained in the literature [6–8]. The dynamics in solar PV array current and power are illustrated in Fig. 18b. The variation in active and reactive powers are illustrated in Fig. 18b. The more negative value of active power indicates that the solar energy is fed to the grid as demonstrated in Fig. 18b.

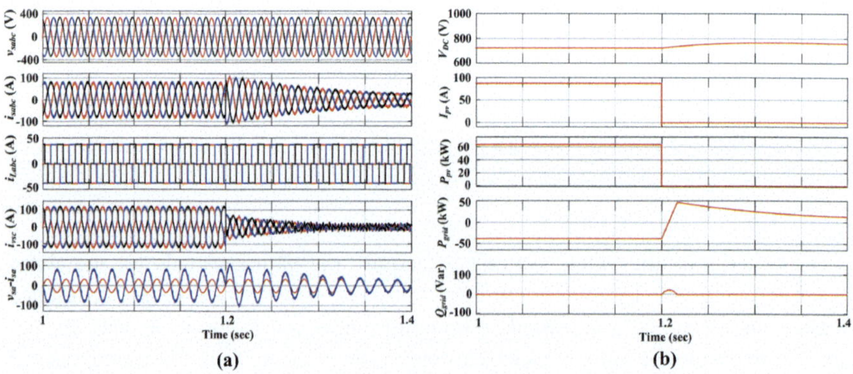

Fig. 18 Performance of the presented system with DSTATCOM to PV Mode

Case-II Performance of System Under Voltage Sag Phenomena

Figure 19 exhibits the behaviour of the UPQC under voltage sag phenomena. Figure 19a demonstrates the waveforms of v_{sabc}, v_{inj}, v_{Labc}, i_{Labc}, i_{vsc} and i_{sabc}. As the voltage of the grid side network is reduced from its nominal rating, the series VSC provides compensating voltages to make balanced and sinusoidal voltages. Hence, the performance of the load side network is not compromised by this sag phenomenon. The typical voltage of the load side network is regulated as per the reference voltage command as depicted in Fig. 19a. The shunt VSC improves the current quality by providing compensating currents at the coupling point. As the typical value of grid voltages is decreased, the amplitude of grid currents is increased due to having almost the same power is injected from the solar PV array side network. The grid currents are balanced and sinusoidal as illustrated in Fig. 19a. Figure 19b shows the waveforms of V_{DC}, I_{pv}, P_{pv}, P_{grid}, and Q_{grid}. The dynamics in DC bus voltage are analyzed by examining the loss component. The DC bus voltage is regulated and sustained as per reference voltage. The solar PV array power is not hampered even under voltage sag phenomena. Likewise, the active-reactive power flow in the grid side network is depicted in Fig. 19b.

The distinct disadvantage of a generalized integrator-based algorithm is explained as follows.

- The realization of this algorithm is quite cumbersome as several generalized integrator blocks are required.
- The performance of the system is poor under abnormal grid scenarios (e.g., distorted grid voltages, unbalanced grid voltages, etc.).

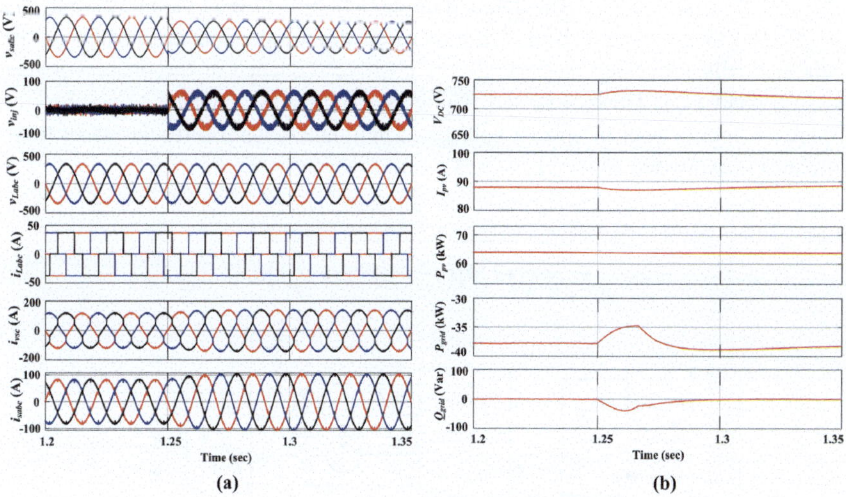

Fig. 19 Performance of the presented system with voltage sag phenomena

To overcome these issues, the Lyapunov controller is analyzed in the following section.

4.4 Lyapunov Technique Based Control Strategy

The Lyapunov controller has certain advantages such as exponential convergence rate and ease of implementation, and it has the capability to work satisfactorily under abnormal grid scenarios. The detailed implementation of the Lyapunov controller is explained as follows.

A. *Shunt VSC Controller*

The Lyapunov controller is presented herein to identify the fundamental component of the load current. The overall control structure is divided into five sections: (1) Estimation of a +ve seq. of phase voltages and templates, (2) Estimation of loss term, (3) Estimation of feed-forward term, (4) Extraction of fundamental weight of load current, (5) Generation of reference currents, and (6) Generation of gate pulses. The detailed explanation of each section is described as follows.

(1) *Estimation of a Positive Sequence of Phase Voltages and Templates*

Figure 20 demonstrates the detailed implementation +ve sequence of unit templates. The three-phase grid voltages are processed in order to transform the *abc*-quantity into the $\alpha\beta$-domain [6, 7]. These quantities are passed through the Lyapunov controller to suppress the harmonics and unbalance from the phase voltages to generate balanced voltages. The detailed implementation of the Lyapunov controller is explained in the next section. The processed αβ-components ($v_{p\alpha}$ and $v_{p\beta}$) are mathematically processed to generate the positive sequence of the phase voltages (v_{pa}, v_{pb}, and v_{pc}). These +ve seq. voltages are processed further to acquire the in-phase component of the phase voltages. The mathematical formulation of these templates is expressed as [6, 7],

$$V_t = \sqrt{(2/3)\left(v_{pa}^2 + v_{pb}^2 + v_{pc}^2\right)} \quad (30)$$

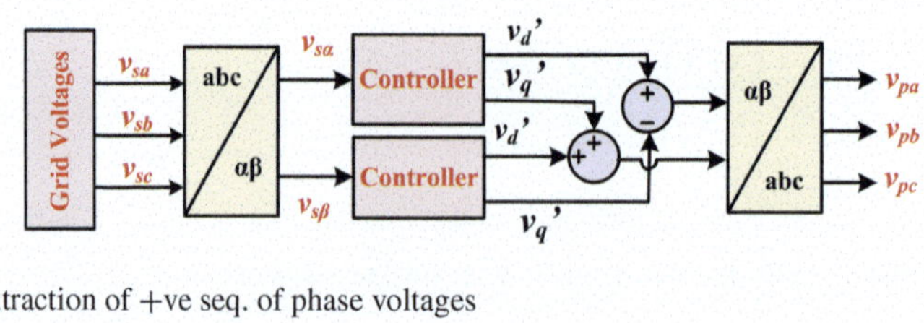

Fig. 20 Extraction of +ve seq. of phase voltages

where, v_{pa}, v_{pb}, and v_{pc} are positive sequences of phase voltages of the grid side network. The in-phase unit templates (u_{pa}, u_{pb}, and u_{pc}) are computed using phase voltages and amplitude of terminal voltages. It is computed as [6, 7],

$$u_{pa} = v_{pa}/V_t; \quad u_{pb} = v_{pb}/V_t; \quad u_{pc} = v_{pc}/V_t \tag{31}$$

Here, it is worth noticing that the smooth and balanced unit templates are obtained even with the presence of distorted and unbalanced grid voltages. These templates play a vital role to generate in-phase reference grid currents for the purpose of UPF correction.

(2) *Estimation of Loss Component*

The DC bus voltage is regulated with help of a PI controller. The error between reference DC bus voltage and sensed DC bus voltage is fed to the controller to acquire the loss component. The reference DC bus voltage is chosen as explicitly mentioned in subsection A-2 of Sect. 4.1. The typical value of the loss component is estimated as [6, 7],

$$I_{\text{Loss}} = (K_p + \frac{K_i}{s})(V_{DC}^{\text{ref}} - V_{DC}) \tag{32}$$

where, K_p and K_i are proportional and integral gains of the controller, V_{DC}^{ref} and V_{DC} are a reference and sensed DC bus voltages. This loss term is fed further to acquire the amplitude of reference grid currents.

(3) *Estimation of PV Feedforward Component*

The stochastic nature of the solar power generation is accommodated herein as a feed-forward component in the control strategy. The typical value of a feed-forward component is estimated as [7, 8],

$$I_{ff} = \frac{2k_{pv}P_{pv}}{3V_t} \tag{33}$$

where, P_{pv}, V_t, and k_{pv} are solar power generation, the amplitude of terminal voltage, and intermediate constant, respectively. The typical value of constant falls between 0.95 and 0.97. This value is used further to compute the amplitude of the reference grid current.

(4) *Extraction of Fundamental Weight of Load Current*

The Lyapunov controller is used herein to estimate the amplitude of the fundamental load current. The load current is expressed with sum of fundamental and different harmonic terms. The adaptive controller law is given as [15],

$$\hat{\dot{\theta}} = \Gamma \tilde{J}_{La} \Upsilon \tag{34}$$

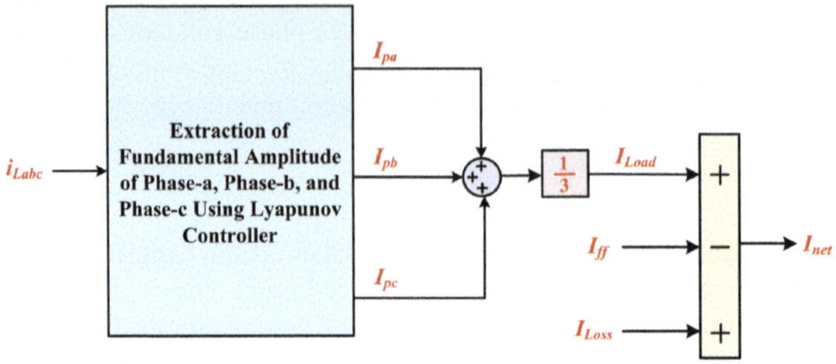

Fig. 21 Detailed schematic of shunt VSC control strategy

where, θ represents the distinct harmonic signal from the given input current (i_{La}), Γ is adaptive gain, $\Upsilon > 0$, $\tilde{J}_{La}$ represents the estimation error of the cost function. The sample and hold circuit is used to extract the amplitude of load current (I_{pa}). Likewise, the amplitude for phase 'b' and 'c' (I_{pb} and I_{pc}) is obtained as depicted in Fig. 21. The detailed implementation of the Lyapunov controller is explained in the literature [15].

(5) *Generation of Reference Currents*

Figure 21 demonstrate the detailed control schematic for the development of reference grid currents. The net amplitude of fundamental load current, feed-forward component, and loss component are used herein to estimate the amplitude of reference grid currents. It is computed by multiplying the in-phase template with the amplitude of the reference grid current. These are expressed as [8, 9],

$$I_{\text{net}} = I_{\text{Loss}} - I_{ff} + I_{\text{Load}} \tag{35}$$

$$i_{sa}^* = u_{sa} I_{\text{net}};\ i_{sb}^* = u_{sb} I_{\text{net}};\ i_{sc}^* = u_{sc} I_{\text{net}} \tag{36}$$

These reference grid currents are processed further to generate switching pulses for the shunt voltage source converter.

(6) *Generation of Gate Pulses*

Figure 11 demonstrates the detailed schematic for the generation switching pulses of the shunt VSC. The instantaneous error between the reference and sensed grid currents is processed to the hysteresis regulator to acquire gating signals. The design of the hysteresis bandwidth and working principle is explicitly explained in the literature [8, 9].

$$e_{sa} = i_{sa}^* - i_{sa};\ e_{sb} = i_{sb}^* - i_{sb};\ e_{sc} = i_{sc}^* - i_{sc} \tag{37}$$

where, e_{sa}, e_{sb}, and e_{sc} are instantaneous errors in grid currents. The shunt VSC has variable switching frequency as e_{sabc} changes from time to time. The detailed procedure to acquire the hysteresis band is explained in the literature [10].

B. *Series VSC Controller*

The reference load voltages are obtained with help of the terminal voltage and in-phase unit templates. The error between the reference and sensed load voltages are fed to the hysteresis current controller to provide gating pulses of series VSC as depicted in Fig. 12. The instantaneous error between reference load voltages and sensed load voltages are fed to the hysteresis current controller. The hysteresis bandwidth is decided based on the literature [10]. The behaviour of the system is explained as follows.

Case-I Response of System Under Unbalanced Source Voltages

Figure 22a, b present the behaviour of the presented system in event of unbalanced grid voltages. Figure 22a shows the waveforms of v_{sabc}, v_{inj}, v_{Labc}, i_{Labc}, and i_{sabc}. The grid voltages are highly unbalanced as depicted in Fig. 22a. However, the load voltages are maintained balanced and sinusoidal based on the reference voltage command. The presented UPQC system complies with the IEEE std. 1159. The harmonic contents in the grid voltages are also restricted based on the IEEE-519 standard. The behaviour of the load currents are also depicted in Fig. 22a. The shunt VSC provides compensating currents to make balanced grid currents. Therefore, the presented UPQC system complies with the IEEE std. 519 and IEEE std. 61,727. Figure 22b shows the waveforms of V_{DC}, I_{pv}, P_{pv}, P_{grid}, and Q_{grid}. The DC bus voltage is also regulated within the permissible range as exhibited in Fig. 22b. The solar power generation is not affected, which is demonstrated using solar PV array current and power. The active and reactive powers are also illustrated in Fig. 22b.

5 Conclusion

In this chapter, the detailed operation of a unified power quality conditioner is discussed. The working principle of the UPQC system is explained with a phasor diagram plot with the integration of solar power generation. Manifold control strategies such as phase-locked loop-based synchronous reference frame theory, least mean algorithm, generalized integrator, and Lyapunov controller, are designed and analyzed for unified power quality conditioner. The distinct operating scenarios are tested, and it shows the satisfactory operation of the presented control strategies. The advantages and disadvantages of each control strategy are also described herein in this chapter.

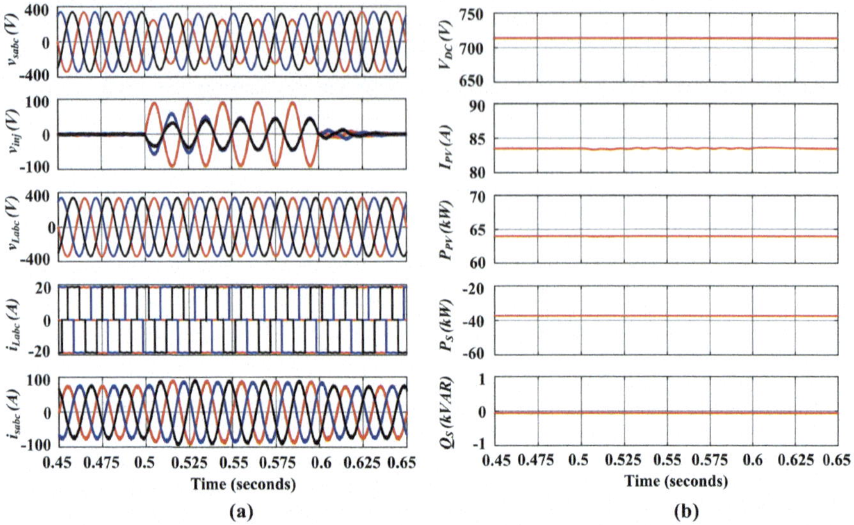

Fig. 22 Performance of the presented system with unbalanced voltages

Acknowledgements P. Shah and X. Zhao are with the Intelligent Control and Smart Energy (ICSE) research group, School of Engineering, The University of Warwick, Coventry, CV4 7AL, United Kingdom. (Corresponding author: X. Zhao). This work was funded by the UK Engineering and Physical Sciences Research Council (EPSRC) under grant EP/S001905/1.

Credit Author Statement **Priyank Shah**: Conceptualization; Methodology; Software; Validation; Formal Analysis; Investigation; Data Curation; Writing—Original Draft; Writing—Review & Editing; Visualization; Project Administration. **Xiaowei Zhao**: Conceptualization; Methodology; Formal Analysis; Investigation; Resources; Writing—Review & Editing; Visualization; Supervision; Project Administration; Funding acquisition.

Appendix

Solar Photovoltaic Array Parameters: $P_{pv} = 64$ kW, $V_{oc} = 36.6$ V, $I_{sc} = 7.97$ A, $V_{DC} = 710$ V; $v_{sab} = 415$ V; $K_p = 0.01$, $K_i = 0.001$; $C_{DC} = 6450$ μF; Shunt Converter Parameters: $L_f = 3$ mH; Shunt Converter Ripple filter: $R_f = 10\ \Omega$, $C_f = 10$ μF; Nonlinear Load: $R_L = 30\ \Omega$, $L_L = 80$ mH; Series Converter Parameters: $L_f = 3.5$ mH; Series Converter Ripple filter: $R_f = 5\ \Omega$, $C_f = 25$ μF; Sampling Time = 10 μs.

References

1. Singh B, Chandra A, Al-Haddad K (2015) Power quality problems and mitigation techniques, 1st edn. Wiley, Chichester, UK
2. Khadkikar V (2012) Enhancing electric power quality using UPQC: a comprehensive overview. IEEE Trans Power Electron 27(5):2284–2297
3. Masoum M, Fuchs EF (2015) Power quality in power systems and electrical machines, 2nd edn. Academic Press, United States
4. Sanjeevikumar P, Sharmeela C, Holm-Nielsen JB, Sivaraman P (Nov 2020) Power quality in modern power systems, 1st edn. Academic Press, United States
5. Weedy B, Cory B, Jenkins N, Ekanayake J, Strbac G (Aug 2012) Electric power systems, 5th edn. Wiley, United States
6. Shah P, Singh B (2019) Robust EnKF with improved RCGA-based control for solar energy conversion systems. IEEE Trans Ind Electron 66(10):7728–7740
7. Shah P, Hussain I, Singh B, Chandra A, Al-Haddad K (2019) GI-based control scheme for single-stage grid interfaced SECS for power quality improvement. IEEE Trans Ind Appl 55(1):869–881
8. Shah P, Hussain I, Singh B (2016) FOGI-QSG based control of multifunctional grid tied SECS. In: Proceeding of the IEEE 7th power India international conference (PIICON), Bikaner, India, pp 1–6
9. Devassy S, Singh B (2018) Design and performance analysis of three-phase solar PV integrated UPQC. IEEE Trans Ind Appl 54(1):73–81
10. Shah P, Singh B (2021) Power quality improvement using three-phase transformer less grid interfaced SECS enabling leakage current suppression. IET Power Electron 14(2):2067–2083
11. Hosseini K, Montazeri A, Alikhanian H, Kahaei MH (2008) New classes of LMS and LMF adaptive algorithms. In: Proceedings of the IEEE 3rd international conference on information and communication technologies: from theory to applications, Damascus, Syria, pp 1–5
12. Shah P, Hussain I, Singh B (2018) Real-time implementation of optimal operation of single-stage grid interfaced PV system under weak grid conditions. IET Gener Transm Distrib 12(7):1631–1643
13. Golestan S, Guerrero JM, Abusorrah AM, Vasquez JC, Al-Turki Y (2022) LTP modeling and stability assessment of multiple second-order generalized integrator-based signal processing/synchronization algorithms and their close variants. IEEE Trans Power Electron 37(5):5062–5077
14. Rodríguez P, Luna A, Candela I, Mujal R, Teodorescu R, Blaabjerg F (2011) Multiresonant frequency-locked loop for grid synchronization of power converters under distorted grid conditions. IEEE Trans Ind Electron 58(1):127–138
15. Srinivas VL, Kumar S, Singh B, Mishra S (2018) A multifunctional GPV system using adaptive observer based harmonic cancellation technique. IEEE Trans Ind Electron 65(2):1347–1357
16. Learn MATLAB Simulink, MATLAB implementation of unified power quality conditioner (UPQC) for power quality improvement, Jan 2021. https://www.youtube.com/watch?v=Zmp35AYKCjI&ab_channel=LearnMATLABSimulink
17. Devi S, Singh B, Devassy S (2020) Modified generalised integrator based control strategy for solar PV fed UPQC enabling power quality improvement. IET Gener Transm Distrib 14(16):3127–3138

Chapter 6
Analysis of Behavior of Variable-Speed Pumped Storage Hydropower Plants—Designing a Vector Controller

Mohsen Alizadeh Bidgoli

Abstract Variable-speed pumped storage power plants (VSPSPP), as opposed to fixed speed pumped storage power plants, use a DFIM in conjunction with a back-to-back converter. Electrical components' nominal equipment parameters are designed for a VSPSPP. Additionally, a traditional vector controller comprises three components: a machine-side controller, a grid-side controller, and a pump-turbine controller. Each path contains an external and an internal loop; therefore, the internal and outer loops are extracted, and the transfer function is extracted to tune the control coefficients; the open- and closed-loop transfer functions of various control sections are extracted from the machine, and grid-side converters, as well as the pump-turbine, and their stability is guaranteed by drawing the root locus diagram.

Keywords Pumped storage power plant · Doubly fed induction machine · Vector control

Nomenclature

Abbreviation

DFIM	Doubly fed induction machine
FRT	Fault ride-through
VS	Variable-speed
FS	Fixed-speed
HPT	Hydro pump-turbine
RPT	Reversible pump turbines

M. Alizadeh Bidgoli (✉)
Department of Electrical Engineering, Yadegar-E-Imam Khomeini (RAH) Shahre-Rey Branch, Islamic Azad University, Tehran, Iran
e-mail: m.alizadeh.b@gmail.com

A. K. Giri et al. (eds.), *Power Quality: Infrastructures and Control*,
Studies in Infrastructure and Control,
https://doi.org/10.1007/978-981-19-7956-9_6

PSPP	Pumped storage power plant
WT	Wind turbines
MSC	Machine side converter
GSC	Grid-side converter
PSS	Power system stabilizer
SMIB	Single machine infinite bus

Parameters

R	Resistance
L	Inductance
i	Current
V	Voltage
s	Slip
S	Laplase transform
ω	Angular speed
δ	Machine power angle
D	Rotor damping constant
f, m	Constant coefficients of state and the input matrix
P	Active power
Q	Reactive power
η	Efficiency
q	Water flow rate of turbine
T	Electric torque
k	Constant control coefficient
e	Error signal
α	Phase of Thevnan voltage
H	The Inertia constant
β	Bandwidth
g	Percentage of opening the gate
F, M	State and input matrix
x, u	State variable and input signal
h	Head
S	Apparent power

Indices

r	Rotor
s	Stator
b	Base
ds, qs	Stator winding on dq axes

dr, qr	Rotor winding on *dq* axes
m	Magnetizing
G	Generator
M	Motor
e	Electric
g	Grid
p	Active power portion
q	Reactive power portion
0	Equilibrium point
tr	Exciting transformer
N	Number of turns of the coil
T, P	Turbine, Pump mode
GSC	Grid-side converter
MSC	Machine side converter
n	Nominal
pp	Proportional part of active power loop
ip	Integral part of active power loop
pv	Proportional part of voltage loop
iv	Integral part of voltage loop
pq	Proportional part of reactive power loop
iq	Integral part of reactive power loop
pg	Proportional part of turbine gate loop
ig	integral part of turbine gate loop
nl, fl	No load, full load
cl	Closed-loop
ol	Open-loop
cr	Current loop
pcr	Proportional part of current loop
icr	Integral part of current loop
pvdc	Proportional part of dc-link voltage loop
ivdc	Integral part of dc-link voltage loop
$p\omega$	Proportional part of speed loop
$i\omega$	Integral part of speed loop

1 Introduction

Traditional pumped storage power plants that utilize synchronous motor-generators consume additional electrical energy as a large load in motor mode to pump water from the lower reservoir to the upper reservoir. Given that power plants are always capable of producing energy at full load, a PSPP can contribute to load curve shaving, i.e., reducing peak load when acting as a generator and shifting it to times of lower load when acting as a motor. Additionally, by incorporating a PSPP into the power grid, significant capacity is added to the required capacity for grid frequency regulation. In comparison, a thermal power plant is limited to contributing up to 5% of its rated power to frequency control. Additionally, a PSPP operating in condenser mode can use its full capacity to improve the power system's dynamic and continuous stability by injecting reactive power or absorption required by the power grid. The capacity required for frequency regulation in the power grid is estimated using the following relationship [1], based on total installed capacity and the accepted frequency adjustment error (±0.1 Hz) reported by Kansai Electric in Japan:

$$S = \sqrt{1.44P + (0.01P)^2} \tag{1}$$

where S is the required capacity for frequency regulation in megawatts and P is the system capacity in megawatts. For example, using the above formula, the required capacity for frequency adjustment (given the tolerance specified) is 597 MW for a system with an installed capacity of approximately 53 GW. By adding a variable-speed pumped storage power plant to the network, the required capacity for frequency adjustment is increased significantly. Meanwhile, a thermal power plant can contribute up to 5% of its nominal power to frequency control.

A DFIM is a critical machine in today's industry, particularly in WT with capacities less than 10 MW and PSPP with capacities greater than 100 MW. Numerous studies and articles have been published on wind turbines; however, less attention has been paid to PSPP used in motor and condenser modes in addition to generator modes. In recent years, and for a variety of reasons discussed later, synchronous machines operating in the FS mode have been capable of being replaced with DFIMs in new projects, thereby improving the performance of power plants. VS-PSPP's electrical components include a wound rotor induction motor and a back-to-back converter. Due to the fact that these machines can also be used in variable-speed wind farms, the only distinction between WTs and PSPPs is the power level of the machine and the voltage level of the converter. Table 1 shows the recently published work based on this chapter.

1.1 Various Types of PSPPs in the World

Due to the development of nuclear power plants between the 1960s and 1990s, PSPPs were the primary solution for converting off-peak energy generated by thermal or nuclear power plants into high-value, on-peak energy. However, by limiting the growth of atomic power plants due to safety concerns, public attention in advanced countries was diverted away from the construction of new cases. Following that, in recent years, the expansion of renewable energy resources has justified the construction of new PSPPs (e.g., wind turbine and solar energy). Without energy storage, dispatching the bulk power system with high WTs and PV penetration is implausible. In other words, from the grid's perspective, utilizing the flexibility of a PSPP

Table 1 Comparing the contribution of recently published papers to that of this chapter

References	Contribute	Contribute not	Year
[2]	To compare VS and FS-PSPP from a technical point of view	To consider the pump mode and provide details of the control system	2010
[3]	To provide a vector controller of VS-PSPP for generator and pump modes	To consider the details of the controller design of a VS-PSPP	2010
[4]	To propose a variable-speed operation for small size Kaplan turbine with fixed wicket gates	To consider for large-size Francis turbine with the application of PSPPs	2012
[5]	To compare VS and FS-PSPPs	To have a technical view and consider details of the controller	2016
[6]	To design and analyze the crowbar system in VS-PSPP to increase transient stability in a weak network connection	To provide controller details in permanent condition, failure to check pump mode	2017
[7]	To investigate the dynamic performance of VS-PSPP in pumping mode	To design controller in generator mode	2017
[8]	To analyze the start-up and braking mode of VS-PSPPs	To provide controller design in the steady-state condition	2018
[9]	To review the use of power electronics in VS-PSPPs	To have a technical view and detail of the control system	2018
[10]	To investigate the benefits of VS-PSPP for reducing wind turbine fluctuations	To explain details of vector controller design for VS-PSPP and consider the pumping mode	2019
[11]	To compare VS and F PSPP with conventional vector control in both generator and motor modes	To consider the details of the controller design of VS-PSPP	2020
[12]	To improve reliability by using the hybrid power system of PSPP and wind turbine simultaneously (as HVDC system)	To design controller and consider pumping mode	2020
[13]	To review published articles of the VS-PSPP	To provide technical specifications and details of the control system	2020
[14]	To design a wide area signal for VS-PSPPs to improve small-signal stability	To explain the vector controller design for PSPP and consider pumping mode	2021
[15]	To compare transient stability for VS and FS-PSPPs	To explain points of vector controller design for PSPP and consider pumping mode	2021
This chapter	To design and analyze vector controller for a variable-speed PSPP for generator and pumping modes and simulation of PSPP	To develop an advanced controller	2022

is necessary to deal with an unexpected increase or decrease in the production of sustainable energies. For instance, the large wind energy ramps down from 1500 to 300 MW in West Texas is one of these unpredictable events that occurred in 2008 in the United States of America [16].

The earliest PSPPs lacked reversible pump turbines (RPTs), meaning they used separate pumps and turbines, though RPTs are now common in new PSPP projects. While some plants utilized physically distinct motor-driven pumps and turbine-driven generators, others utilized a tandem unit consisting of a generator-motor and a separate pump and turbine coupled on a standard shaft. Additionally, a large proportion of the early PSPPs incorporated horizontal shafted motor pumps and turbine generators. However, large reversible vertical shaft machines are preferred in today's PSPPs. It should be noted that these early plants achieved high overall cycle efficiencies because individual pumps and turbines could be optimized for maximum efficiency at the generator/synchronous motor's speed. Despite this, when considering the total cost of the plant, the electrical–mechanical equipment costs for these plants are high due to the separate motor–pump and generator–turbine installations. At the time, it was not straightforward to design a single pump-turbine impeller capable of operating efficiently at a single expected synchronous speed. By contrast, an SM can be rotated in any direction, i.e., motor pump or generator-turbine, by simply reversing two of the three electrical phases. Some statistics are provided in Table 2.

Finally, the modern PSPP epoch in the United States began in the mid-1960s when PSPPs were operated to the benefit of their capability so as to provide frequency control, load following, and spinning reserve. Advanced PSPP technology includes reversible pump-turbine, high technology power electronic converters, improved electric insulation, and electric drive controllers. A PSPP project might utilize a variety of configurations to accomplish the aim of electric energy storage. However, new PSPP projects usually, but not always, are being constructed using three strategies; 1—fixed speed, 2—adjustable speed, and 3—ternary technology, which are described in the following.

1.1.1 Fixed-Speed PSPP

In conventional PSPPs using SM, the rotational speed of the generator and turbine must be constant to produce the grid frequency. However, operation at a constant speed has a flaw in which the pumping power cannot be controlled. The controller of FS-PSPP consists of two parts, as shown in Fig. 1. The former includes the static excitation system regulating the stator winding voltage. Also, an auxiliary input known as a power system stabilizer (PSS) is added to the excitation input. The latter, controller of HPT, in motor and generator operating modes are different. Although the turbine governor can control the active power of the unit in generating operating mode, the active power of SM in the motor operating mode cannot be adjusted; that is, the guide vane opening is optimized to increase the pump's efficiency. Further details can be found in [17].

Table 2 Ten largest PSPPs in the world [1]

Plant/Dam	Country	Rated power (GW)	Turbine type	Max annual generation (TWh)
Three Gorges	China	22.5	32 × 700 MW Francis 2 × 50 MW Francis	84.4
Itaipu	Brazil/Paraguay	14	20 × 700 MW Francis	94.7
Xiluodu	China	13.9		64
Baihetan	China	13.1		
Belo Monte	Brazil	11	20 × 550 + 611 MW Francis 7 × 25.9 MW Kaplan	38.2
Guri	Venezula	10.2	10 × 730 MW Francis + 4 × 180 MW Francis 3 × 400 MW Francis + 3 × 225 MW Francis 1 × 340 MW Francis	53.4
Wudongde	China	8.7		
Tucurui	Brazil	8.4	12 × 350 MW Francis 11 × 375 MW Francis	41.4
Grand Coulee	USA	6.8	27 Francis 26 Pump-turbine	20
Langtan	China	6.4	9 × 714 MW Francis	18.7

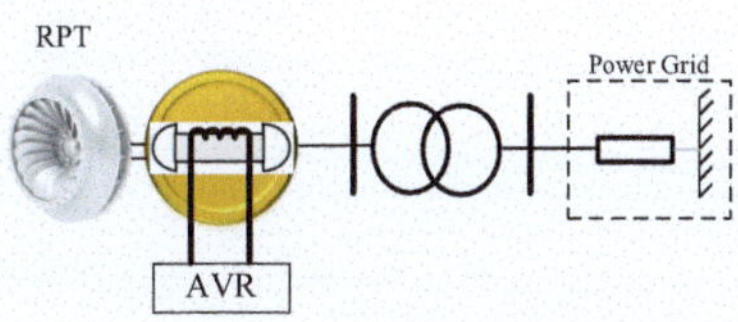

Fig. 1 FS-PSPP configuration

Solving this issue, designers of PSPPs employed the hydraulic system with two configurations, which are explained later; 1—a RPT (usually Francis type), and 2—a separate pump and turbine, known as a ternary system. It should be mentioned that the use of RPT in PSPPs is most common nowadays.

1.1.2 Ternary PSPP

In Switzerland, the Hongrin-Leman PSPP is an example of a horizontal-shaft ternary unit. ANDRITZ firm erected two new 120 MW ternary Pelton generating units in 2017. As seen in Fig. 2, a ternary arrangement links two distinct pumps and turbines to a motor-generator. The turbine can be either a Francis or a Pelton design, depending on the hydraulic parameters of the location. Note that Pelton turbines often being utilized for very high head installations. Unlike RPT applications, where the direction of rotation must be reversed when switching from pumping to production mode and vice versa, ternary sets do not require the rotational direction to be reversed in both operational modes.

On the other hand, in the design of RPTs, the pumping operation is often prioritized; hence, the total cycle efficiency of RPTs is smaller than that of two independent pump turbines. In comparison, both the pump and turbine may be adjusted in ternary sets, resulting in increased hydraulic efficiency and a shorter time required to transition between modes. The following components may be used to transition between turbine and pump operation: a clutch that may be operated at a standstill, a beginning turbine, or a synchronizing torque converter. Due to the fact that the pump and turbine can run concurrently, the hydraulic flow is more complicated than in a standard pumped storage unit. Additionally, the plant's complete power spectrum is available by utilizing the hydraulic short circuit principle. In other words, the operation of ternary systems in hydraulic short circuits enables the pumping power to be regulated. While the transition time for ternary units is substantially quicker, on the order of 0.5–1 min, the transition time for an RPT in the other direction, from pumping to producing mode, is between 1.5 and 5 min. This chapter focuses on the variable-speed functioning of reversible pump-turbine systems, which will be discussed in detail in the following sections.

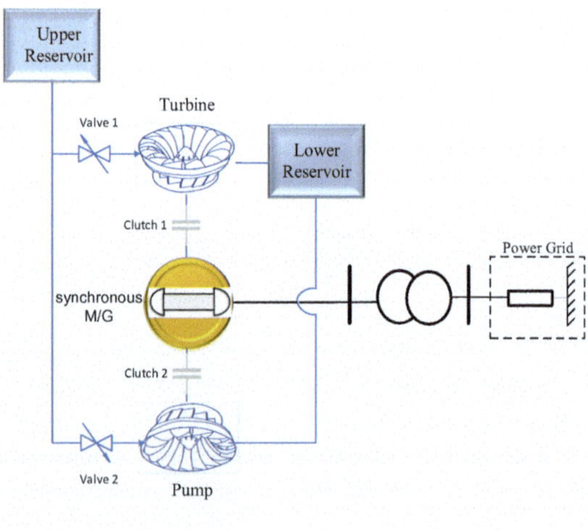

Fig. 2 Ternary PSPP configuration

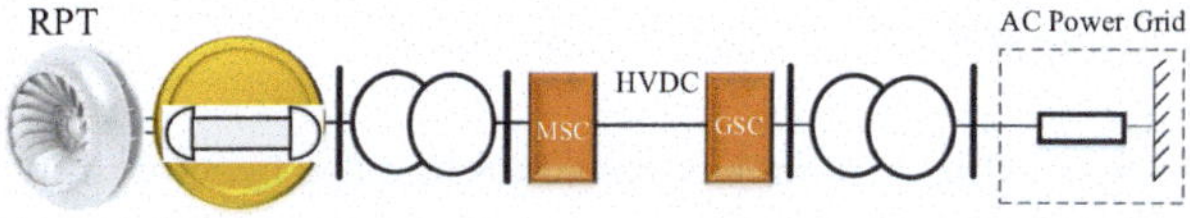

Fig. 3 SM-based VS-PSPP used in HVDC configuration

1.2 Diverse Solutions for Variable-Speed PSPPs

Different topologies can vary the speed of the state-of-the-art PSPP. Three strategies may be used for medium or large-size PSPP to operate at variable speeds, which are explained in the following.

1.2.1 HVDC

The synchronous machine, used in conventional applications, connecting to a high voltage direct current (HVDC) link is one of these solutions, as shown in Fig. 3. With this strategy, the range of speed variation may be within ±25% around the synchronous speed. This solution is attractive for power plants far away from the power grid [18]. Two other schemes are available in short distance or AC transmission usage. The unique feature of this solution is that the generator can be directly connected to the sending terminal of the HVDC. These plants can operate optimum turbine efficiency points according to hydraulic conditions and adjust absorbed power in pumping operational mode by changing the rotor speed synchronous of the machine.

1.2.2 SM Equipped with a Full-Converter

The second solution uses the synchronous machine which is connected to the grid through a full-scale converter, as shown in Fig. 4. Although this scheme has been installed in Grimsel 2 by ABB in Switzerland [19], it is not widely applied to other projects. It is noted that this method may be cost-effective for microhydropower sites that use a permanent magnet synchronous machine (PMSM) [4].

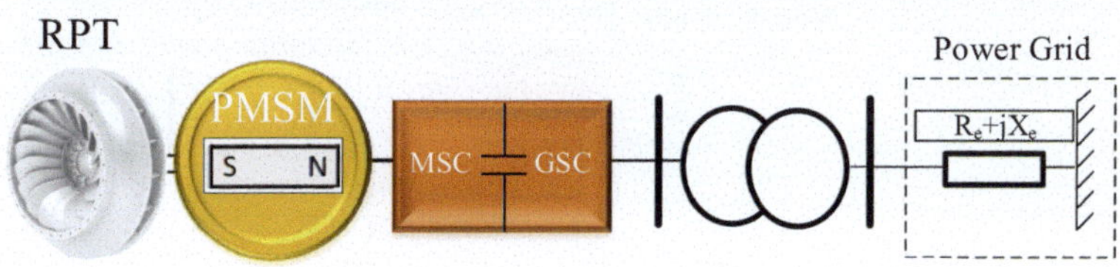

Fig. 4 SM-based VS-PSPP equipped with full-scale back-to-back converter

1.2.3 DFIM Equipped with a Partial Converter

A cyclo-converter is a direct frequency converter that has been used in early DFIM-based PSPP in the world. In cyclo-converters, as shown in Fig. 5a, unlike modern back-to-back converters that convert voltage with a middle DC link, the voltage is not changed in any other form. However, advanced frequency converters are known as back-to-back converters whose power rating is a partial scale of the DFIM rated power are used, as shown in Fig. 5b. This converter is located between the rotor and stator terminal that can transfer the mechanical power through both the stator and rotor. The list provided in Table 3 shows the increase in the tendency of various countries during recent years for the DFIM-based PSPPs installation in the world. More details for designing controllers of converters will discuss in Sects. 4 and 5 of this chapter.

1.3 Historic DFIM-Based PSPPs Installation in the World

The first-worldwide DFIM-based PSPP, installed in Japan during the late 1980s and into the mid-1990s., employs the cyclo-converter, while the first one in Europe was commissioned in 2004 (Goldisthal in Germany). This practical strategy has been recently used in many projects worldwide using state-of-the-art power electronic converters, i.e., back-to-back voltage source converters, such as Avce in Slovenia or Linthal in Switzerland [20]. The interesting point is that nor is it important whether the PSPP is FS or VS, both of them are popular according to the site's features. In other words, although DFIM-based PSPPs are chosen in areas where the water head variation is considerable, a designer may select each type of PSPP, i.e., either FS or VS, considering the other configuration of a site. In the early eighties, high-power converters and thyristors allowed for the design of medium voltage VS machines.

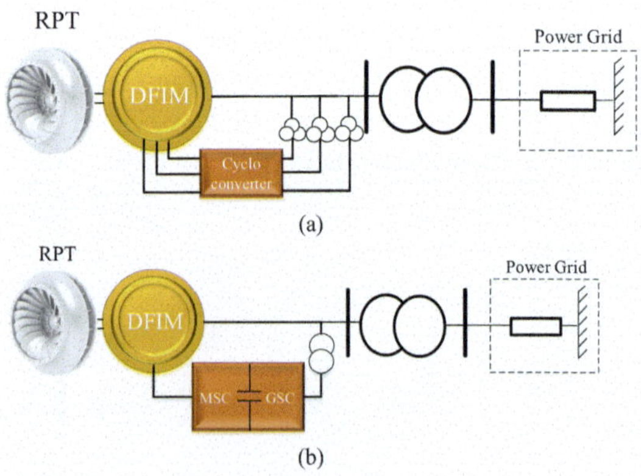

Fig. 5 DFIM-based VS-PSPP equipped **a** with partial-scale cyclo-converter, **b** with partial-scale back-to-back converter

Table 3 Large-size VS-PSPPs installed in different locations during last years

Unit	Country	Upgrade/new	HPT (MW)	MG (MVA)	Power electronic		Speed range (rpm)	Rated speed (rpm)	Man.[b]	Year
					Type [a]	Size (MVA)				
Narude	Japan	New	22	22	1	3.83	190–210 (±5%)	200	H	1987
Yagisawa	Japan	New	2 × 82	2 × 85	1	25.8	130–156 (+4/−13%)	150	T	1990
Takami	Japan	New	105	140	2	24.3	209–254		M	1992
Ohkawachi	Japan	New	392	395	1	72	330–390 (±8.3%)	360	H	1993
Shiobara	Japan	New	330	360	1	51.1	356–394 (±8%)		T	1995
Okukiyotsu	Japan	New	340	345	2	31.5	408–450 (±5%)		T	1996
Yanbaru	Japan	New	30	31.5	2	3.96	423–477 (±5%)		T	1999
Goldisthal	Germany	New	2 × 265	2 × 340	1	2 × 57.6	300–347 (−10/+4%)	333	V	2003
Avce	Slovenia	New	185	195	3	10	576–636 (±4%)	600	ABB,M	2008
Omarugawa III	Japan	New	2 × 330 2 × 330	2 × 350 2 × 370	3 1	2 × 27.4	576–624 (±4%)	600	M H	2008 2010
Grimsel II[c]	Switzerland	Upgrade	80	100	4	2 × 43.3	690–765 (−10/+2%)	750	ABB	2010
Okutataragi	Kansai Electric Power Co	Upgrade	2 × 320	–	–	–	–	300	H	2014
Kyogoku	Japan		228	230	2		–		T	2015
Linthal	Switzerland	New	4 × 250	4 × 280	3	–	470–530 (±6%)	500	A	2015
Frades II	Portugal	New	2 × 390	2 × 440	3	–	350–381 (+2%, −7%)	364	Vo, S	2015

(continued)

Table 3 (continued)

Unit	Country	Upgrade/new	HPT (MW)	MG (MVA)	Power electronic		Speed range (rpm)	Rated speed (rpm)	Man.[b]	Year
					Type [a]	Size (MVA)				
Vende Nova III	Portugal		2 × 380	2 × 420	–	45	–	–	Vo, S	2015
Le cheylas	France	Upgrade	1 × 250	–	–	–	(±7%)	300		
Tehri	India	New	4 × 250	4 × 278	3	40	−10.73% to + 8.33%	230.77	GE, T	2016
Kazunogawa	Japan	New	2 × 475	2 × 475	2	25	480–520	–	T	2017
Kyogoku	Japan	New		230					T	
Iowa Hill	USA	New	3 × 133	3 × 150	2	10	481–547 (±6.5%)	514.3	–	2018
Nant de Drance	Switzerland	New	6 × 157	6 × 175	3	–	399–459 (±7%)	428.6	A	2019
Kozjak	Slovenia	New	2 × 220	2 × 230	–	–	–	600	V	–
Fengning II	China	New	2 × 300	2 × 330	–	–	–	–	GE	2021

[a] Power Electronic Type: 1—Cyclo-converter, 2—GTO-back-to-back (VSI), 3—IGCT/IGET-back-to-back (VSI), 4—full converter (VSI)
[b] Manufacturer: H: HITACHI, T: TOSHIBA, M: MITSUBISHI, V: VATECH, Vo: VOITH, A: ALSTOM, GE: General Electric, S: Siemens
[c] Only pumping mode operation

In the late eighties, the developments of VS drive were in hydropower using cyclo-converter technology. The first VS machines, in commercial operation, used the cyclo-converter technology with thyristors or Gate Turn Off (GTO) thyristors to provide the three-phase alternating current field excitation required on the rotor. The latest technological progress introduced multi-level voltage source inverter (MLVSI) excitation and Pulse Width Modulation (PWM) techniques.

The VSI excitation benefits from using a two-winding transformer on the input compared to a three-winding transformer with the cyclo-converter. To develop these new techniques, diverse manufacturers have developed excitation systems with several other semiconductor devices. Some use the thyristor technology with the more recent Insulated Gate Commutated Thyristor (IGCT), or Insulated Gate Bipolar Transistor (IGBT), and those who favor transistor technology with the more recent Injection Enhanced Gate Transistor (IEGT). The ability to operate at the partial load results in savings in water to be used for a later generation. The list provided in Table 3 shows the increase in the tendency of various countries during recent years for the installation of VS-PSPPs.

1.4 Range of Rotor Speed Adjustment in VS-PSPPs

A variable-speed PSPP can operate in the following modes; generate mode, motor mode, synchronous condenser mode in turbine direction, synchronous condenser mode in pumping direction, synchronous condenser mode without motor-generator running (only with the operation of GSC as a STATCOM), a start-up in pump mode (stator winding is short-circuited), and electrical braking mode. However, a designer of a new VS PSPP or upgraded FS-PSPP is required to adjust the rotor speed to either control active power consumed in pumping mode or improve the efficiency of the plant during partial load operation in generating mode. Therefore, in this section, these two significant contributions of VS-PSPPs compared to FS-PSPPs are more elucidated.

1.4.1 Frequency Regulation in Pumping Mode

For appreciating the importance of frequency regulation in pumping mode, a real upgraded case is introduced at Grimsel PSPP in the following. Grimsel PSPP has four units that two of which have been converted from fixed to variable-speed operation, ranging from 680 rpm, i.e., 10% lower than synchronous speed ($n_s = 750$ rpm) to 765 rpm, which is +2% higher). Since a FS unit can only operate at full power in pumping operation, the only way to adjust the amount of consumed power is to regulate the number of pumps employed. It is crucial that designers install a system for VS operation, but they also have to keep unchanged the electrical machine from 40 or so years ago. Thus, they could cope with this challenge by employing a full-scale back-to-back converter. As shown in Fig. 6, the frequency of MSC and the

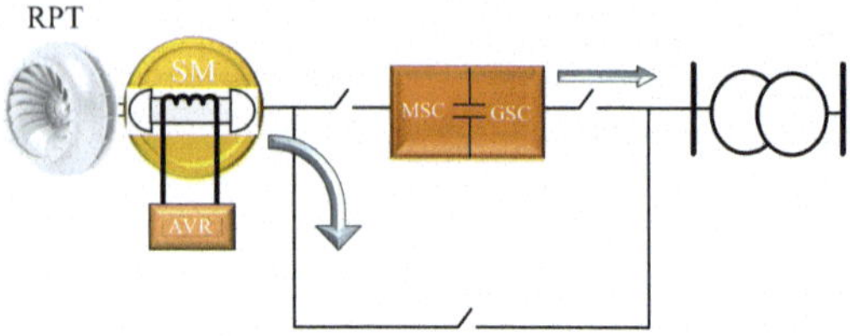

Fig. 6 Schematic of VS-PSPP used in Grimsal 2

rotational speed can be adjusted, which causes to vary absorbed power. It should be mentioned, with this solution, there is no need for SM to be replaced with a new machine. For instance, in this case, the SM now operates between 600 and 765 rpm in pumping mode. Also, due to the high losses of the full-scale converter, it should be bypassed, and PSPP runs at a fixed speed, as shown in Fig. 6.

By installing a new idea for PSPPs rather than common DFIM-based VS plants, the plant's power can vary continuously between 60 and 100 MW. Note that the lower limit is defined by the risk of cavitation in the pumping operation. This case shows a specific example of VS-PSPPs for the aim of frequency regulation in pumping mode, which will be described in the following. A PSPP operates in consuming operational mode for hours and can pump extra water into the upper reservoir to control the grid frequency. According to affinity law, i.e., $P_{\mathrm{new}} = P_{\mathrm{old}}\left(\frac{N_{\mathrm{new}}}{N_{\mathrm{old}}}\right)^3$, the power consumed by the machine is proportional to the cube of the pump speed. Therefore, the machine's power will also be constant because of the constant speed in a FS unit. Therefore, the only way to control the power consumption is to use the unit power off-grid method. It is also possible to control the amount of power input in small-pump motors; however, it is impossible to use this method in large units due to energy loss and mechanical damage. A VS-PSPP operates within a range limited by the minimum and maximum speed of the pump, full power, the minimum and maximum head of the system, the reverse flow limit in the high head (limited by the low pump discharge), and the cavitation limit in the low head (limited by the high pump discharge).

The pump's operating point is always obtained from the intersection of the characteristic curve of the system in which it is installed and the characteristic curve of the pump itself, as shown in Fig. 7. The intersection point of the system curve and the pump curve are always constant for a given working head if the rotational speed of the machine does not change. However, the pump curve can be adjusted by changing the rotation speed of the motor so that the motor pump operates at our desired operating point and the motor pump can control the power consumption.

1.4.2 Efficiency Improvement in Generating Mode

This part discusses the impact of a VS operation on PSPP efficiency in generator mode, especially in partial load conditions. In the generator mode, the Hill diagram of a RPT is usually extracted by the manufacturer and performed by several tests

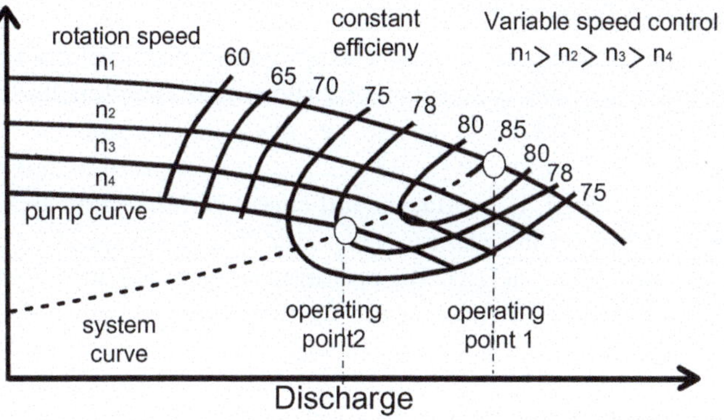

Fig. 7 The operating point of a variable-speed pump

conducted at the plant on the RPT and provided to the user. As stated, the Hill diagram for a pump-turbine Francis by the pump-turbine manufacturer shows at different work points that these graphs are based on the physical properties of a pump-turbine or briefly on a parameter called a specific speed set. Specific speeds in hydraulic turbines are usually determined using the equation [21].

$$^*\Omega = 0.0189 \frac{nQ^{1/2}}{h^{3/4}} \tag{2}$$

where n is the rotational speed in (rpm), Q is the flow rate in (m^3/sec), and h is the head in (m). As shown in Fig. 8, various hydro PT can be classified versus the specific speed.

Hydro Turbine is one of the most critical pieces of the hydropower plant's equipment. The turbine drive wheel and the generator rotor are usually coaxial, so the torque required to rotate the rotor axis is created by turning the turbine axis. It is well known that there are two hydro turbine types; 1—Impulse turbines such as Pelton type, with a specific speed less than 0.2 installed for units that have low flow rate and high head, and 2—reaction turbines such as Francis and Kaplan types. Francis turbines being radial flow, have specific speeds in a wide range from 0.2 to 2. Francis turbines can be in three groups; fast speed, medium speed, and low speed. Francis

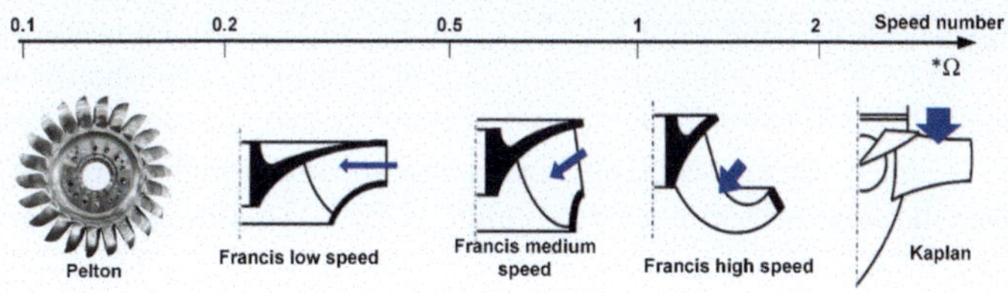

Fig. 8 Hydro-turbines classified by specific speed $^*\Omega$

turbines are used in many situations because of their speed number, which is consistent with geographical locations' values. The last type of reaction turbine, axial flow, is used for high flow and low head situations. The turbine is a propeller if the blades are fixed; otherwise, it is a Kaplan turbine, i.e., blades are variable.

The efficiency hill chart for a typical Francis turbine is given in unit values N_{11}–Q_{11} as shown in Fig. 9b–d. However, the hill curves for this study are used in unit values N_{11}–P_{11} [22]. The unit values, unit power (P_{11}), unit speed (N_{11}), and unit discharge (Q_{11}), are defined in [22]. The operating range of the turbine valve is 50 to 100%. It is usually not used in practice for the percentage of openings of the input gate less than 50% due to mechanical problems caused to the turbine. According to Fig. 9a, the maximum turbine efficiency is about 90%, which occurs at the nominal operating point of the turbine and about 67% gate opening. Vertical line 2 shows the FS and rated head operation when different gate openings change the output power. Since the inverse relationship between the unit speed and the head, if the plant head increases from the rated one and speed maintains without change, the turbine operates in the area between the two vertical lines 1 and 2, i.e., area A, shown in Fig. 9a. When the output power is at the maximum point (gate opening is 100%), efficiency improvement could only be achieved by about 4% (80–84%) by increasing the speed by 10% over the synchronous speed. Efficiency must be improved by about 7% (80–87%) when the gate opening decreases from 100 to 90% and the speed increases by 10%. Also, if the plant head decreases from the rated one and speed is maintained without change, the turbine operates in the area between the two vertical lines 2 and 3, i.e., area B, shown in Fig. 3. In this operation area when the output power is at minimum point (gate opening is 50%) efficiency of the turbine is about 80% which could be improved up to 85% by reducing the speed by 10% under the synchronous speed of the machine. Furthermore, efficiency may be reached 89% when the gate opening is optimized to 60%.

For deduction from the above explanations, the maximum efficiency improvement for a VS-PSPP can be achieved by decreasing the speed when both the head and output power of the unit are low in the turbine mode. In other words, the minimum speed of a VS-PSPP is determined by the cost analysis of the efficiency improvement in the turbine mode at partial load. It should be noted that the power variation range in pumping mode is another parameter that impacts the required speed variation range. Some more detailed explanations are given in the following for various Hill diagrams shown in Fig. 9b–d.

The efficiency diagram shows that the efficiency varies from zero to 95% when the discharge Q varies from zero to 0.5, as shown in Fig. 9b. Notably, efficiency increases by up to 90% by a flow rate of 0.25. Since turbine performance is usually not performed at very low loads due to cavitation vortex rope issues, the Pleton-type Turbine with a characteristic speed and a very low specific speed has high efficiency of 95%. Over the load performance, the range is 50%. Therefore, there is no need to install variable remote equipment to improve plant efficiency. On the other hand, it can be said that the targeting of these units is usually to control the frequency and change the active power in pump mode. Another important thing to mention

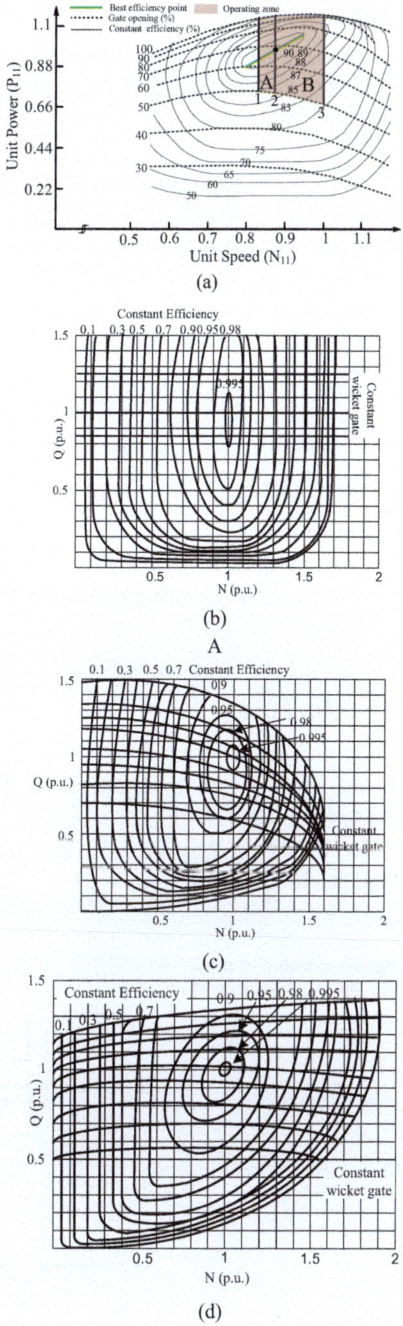

Fig. 9 The hill chart curves with Different specific Speed [21] **a** Francis RPT, **b** Pelton turbine $^*\Omega = 0.04$, **c** Francis turbine $^*\Omega = 0.25$, **d** Francis turbine $^*\Omega = 0.75$

in this chart is the runaway speed in this unit. For equal discharge p.u1 and η zero efficiencies, the escape velocity value is 1.75 p.u.

As shown in Fig. 9c, d. it should be noteworthy to examine the above analysis for lower specific speed values. Significant differences in these curves can be found. For example, a Francis turbine with a flow rate of 0.55 will have a maximum efficiency of 90% and a flow rate maximum of 0.62 with a 95% maximum efficiency. In comparison, a Francis turbine with a flow rate of 0.6 and 0.75 will have a maximum efficiency of 90 and 95%. This indicates that the Francis turbines will have lower efficiency at lower loads and above 50% with an increasing specific speed, thus requiring the use of variable-speed units and a greater variation range than synchronous speed. Also, the amount of runaway speed increases in Francis turbines with increasing specific speed.

2 Modeling

This part presents a mathematical model of various hydraulic and electrical sections to design a vector controller for a VS-PSPP.

2.1 *Turbine*

Figure 10 shows the components model, including penstock, servomotor, control, and turbine (IEEE nonlinear model).

The following equations will determine the water column as a non-elastic penstock characteristic [11].

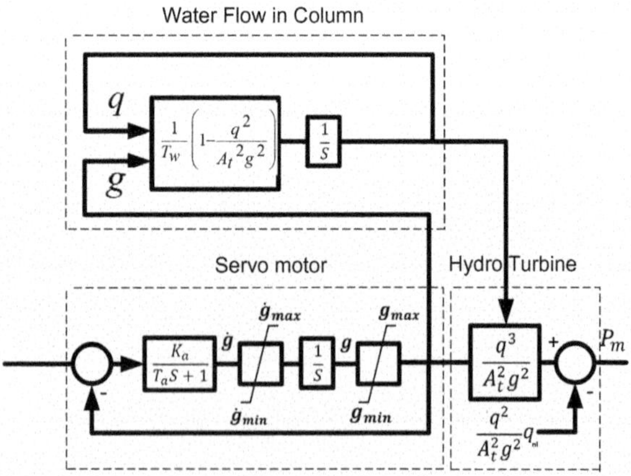

Fig. 10 Nonlinear model of non-electrical Section of the unit in turbine mode

$$\dot{q} = \frac{1}{T_w}(h_s - h_d - h_f) \tag{3}$$

where h_s is the static head of water equal to 1, h_d is water head at the turbine inlet, h_f is the head loss due to the friction, T_w is starting time of water which is defined as the time required for water at nominal head h_n so that it would be reached from rest mode to nominal speed U_n. Also, h_s, h_d, h_f and T_w are obtained as the following.

$$\begin{aligned} h_d &= \left(\frac{q}{g}\right)^2 \\ h_f &= f_p Q^2 \\ h_s &= 1 \\ T_w &= \frac{Lq_n}{9.81Ah_n} \end{aligned} \tag{4}$$

where f_p is the friction coefficient of head loss, other parameters are also defined as follows:

$$A_t = \frac{1}{(G_n - G_{nl})} \; G = A_t g \; T_w = \frac{Lq_n}{9.81Ah_n}$$

Additionally, mechanical power produced by the turbine can be defined as Eq. (5):

$$P_m = \eta_T P_{rT} h_d (q - q_{nl}) \tag{5}$$

where P_{rT} is a constant coefficient to convert per-unit value from the base of the turbine to the machine, which is defined as follows.

$$P_{rT} = \frac{P_{nT}(\text{MW})}{S_{nM}(\text{MVA})} \tag{6}$$

where P_{nT} is the nominal power of the turbine and S_{nM} is the apparent nominal power of the machine.

2.2 Pump

The pump's operating point is always obtained from the intersection of the characteristic curve of the system in which it is installed (Eq. (7)) and the characteristic curve of the pump itself (Eq. (8)), as shown in Fig. 11a.

$$h_d = a_2 q^2 + a_1 \omega_r |q| + a_0 \omega_r \tag{7}$$

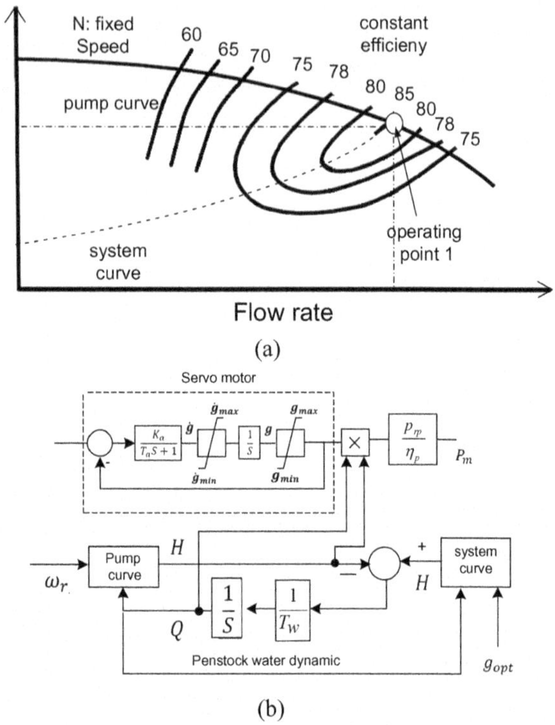

Fig. 11 Operating point of a pump, **b** nonlinear model of non-electrical section of the unit in pumping operational mode

$$h_d = h_s + \left(f_e + f_g\right)q^2 \tag{8}$$

where a_0, a_1, and a_2 are constant coefficients that are different based on the pump type, and they are used to estimate the pump curve. f_e and f_g are head loss friction and friction loss caused by the wicket gates. Also, h_s is the static head, which is equal to 1. In the end, the mechanical power used by the pump can be shown through Eq. (8).

$$P_m = \frac{p_{rp}}{\eta_p} h_d q = \frac{p_{rp}}{\eta_p}\left(a_2 q^3 + a_1 \omega_r q|q| + a_0 q \omega_r^2\right) \tag{9}$$

where P_{rp} is a constant coefficient to convert per-unit value from the turbine base to the machine, which is defined as follows.

$$p_{rp} = \frac{P_{np}(\text{MW})}{S_{nM}(\text{MVA})}$$

where P_{np} is the nominal power of the turbine, and S_{nM} is the apparent nominal power of the machine.

2.3 DFIM

Two three-phase windings are available in a wound rotor induction machine: stator winding connects to grid voltage, and a voltage applied by MSC feeds rotor winding. For convenience and to get rid of three-phase equations in the time domain, these equations would commonly be presented in a *dq* reference frame rotating at an angular speed of ω. Using this conversion, Eq. (10) could be defined as follows in the per-unit system for motor direction, i.e., the stator and rotor currents are of positive direction into the machine.

$$\begin{aligned} V_{sdq} &= R_s i_{sdq} + j\omega\varphi_{sqd} + \frac{1}{\omega_b}\frac{d\varphi_{sdq}}{dt} \\ V_{rdq} &= R_r i_{rdq} + j\omega s\varphi_{sqd} + \frac{1}{\omega_b}\frac{d\varphi_{rdq}}{dt} \end{aligned} \tag{10}$$

It should be noticed that there are different conversions for the transformation of three-phase values to reference frames; a rotational reference frame rotating with synchronous speed is more valuable and prevalent in induction machines. In this case, as far as the reference frame is rotating with synchronous speed, the stator variables will be related to the value of synchronous speed, and rotor variables are linked to slip frequency.

$$\begin{cases} \varphi_{ds} = L_s i_{ds} + L_m i_{dr} \\ \varphi_{qs} = L_s i_{qs} + L_m i_{qr} \end{cases} \quad \begin{cases} \varphi_{dr} = L_r i_{dr} + L_m i_{ds} \\ \varphi_{qr} = L_r i_{qr} + L_m i_{qs} \end{cases} \tag{11}$$

Hereinafter, two motor and generator operational modes could be separated from each other to model the active and reactive power of the machine explained in the following.

2.3.1 Generator Mode

Given * as a conjugate form of a complex value, in generating mode, electrical torque T_{e_g} and apparent power S_{s_g} are defined as Eq. (12). Also, active power P_{s_g} and reactive power Q_{s_g} can be written as Eq. (13).

$$\begin{aligned} T_{e\_g} &= \tfrac{L_m}{L_s}\mathrm{Im}\left(\varphi_s i_r^*\right) = \tfrac{L_m}{L_s}\left(-\varphi_{ds} i_{qr} + \varphi_{qs} i_{dr}\right) \\ S_{s\_g} &= -V_s I_s^* = -\left(v_{ds} + jv_{qs}\right)\left(i_{ds} + ji_{qs}\right)^* \end{aligned} \tag{12}$$

$$P_{s\_g} = -\mathrm{Re}\left(V_s I_s^*\right) = -v_{ds} i_{ds} - v_{qs} i_{qs}$$

$$Q_{s\_g} = -\operatorname{Im}(V_s I_s^*) = v_{ds} i_{qs} - v_{qs} i_{ds} \tag{13}$$

2.3.2 Motor Mode

Equations (12) and (13) are modified for motor mode as follows.

$$\begin{aligned} S_{s\_m} &= V_s I_s^* = (v_{ds} + j v_{qs})(i_{ds} + j i_{qs})^* \\ T_{e\_m} &= -\tfrac{L_m}{L_s} \operatorname{Im}(\varphi_s i_r^*) = \tfrac{L_m}{L_s} (\varphi_{ds} i_{qr} - \varphi_{qs} i_{dr}) \end{aligned} \tag{14}$$

$$\begin{aligned} P_{s\_m} &= \operatorname{Re}(V_s I_s^*) = v_{ds} i_{ds} + v_{qs} i_{qs} \\ Q_{s\_m} &= \operatorname{Im}(V_s I_s^*) = -v_{ds} i_{qs} + v_{qs} i_{ds} \end{aligned} \tag{15}$$

2.4 *Excitation Transformer and DC-Link Capacitor*

To model the GSC section and usually due to the existence of inductance in the GSC path (filter inductance in WTs or leakage inductance of excitation transformer in PSPP), dynamics of the current variable passing through inductances are used for modeling of converters. Note that the dynamic of converters, MSC and GSC, are ignored when designing the controller. Writing down Kirchhoff's Voltage Law between converter voltage and grid voltage, the current passing through the transformer is achieved. Transforming the equation to *dq* frame, the following equations result. On the other hand, DC-link capacitor voltage dynamics also result from the active power balance between MSC and GSC. It should be mentioned that the direction of chosen current from the grid to the transformer is supposed to be positive.

$$\begin{aligned} \frac{L_{tr}}{\omega_b} \frac{di_{dg}}{dt} &= -R_{tr} i_{dg} + L_{tr} i_{qg} - v_{dg} + v_{ds} \\ \frac{L_{tr}}{\omega_b} \frac{di_{qg}}{dt} &= -R_{tr} i_{qg} - L_{tr} i_{dg} - v_{qg} \\ C \frac{dV_{dc}}{dt} &= -\frac{P_r}{V_{dc}} + \frac{P_g}{V_{dc}} \end{aligned} \tag{16}$$

where P_r and P_g are active power of rotor and grid side, respectively, defined as follows.

$$P_r = v_{dr} i_{dr} + v_{qr} i_{qr}, \quad P_g = v_{dg} i_{dg} + v_{qg} i_{qg}$$

3 Designing Vector Controller

The controller of a DFIM-based VS-PSPP is shown in Fig. 12. All parts of this controller for both generating and motor operational modes are explored in this part.

3.1 *Generator Mode*

The DFIM-based VS-PSPP controller is divided into three parts in generator mode: 1—MSC controller; 2—GSC controller; and 3—Hydro RPT controller. The controller of the turbine, known as the governor, in generator mode is responsible for the generator's speed control in a VS-PSPP. Moreover, MSC controls the active power generated by the unit contributing to grid frequency control and stator voltage regulating reactive power. The GSC controller also is responsible for maintaining DC-link capacitor voltage and reactive power exchanged from this converter.

3.1.1 MSC Controller

Using the stator voltage-oriented method, the stator voltage can be assumed in the direction of d-axis of the reference frame, that is, $v_{ds} = vs$ and $v_{qs} = 0$. On the other hand, using that and regardless of the stator resistance, it can be concluded that $\varphi_{qs} = \varphi_s$ and $\varphi_{ds} = 0$. Therefore, Eqs. (10) and (11) can be rewritten if the derivative of stator flux variables in dq axes are equal to zero in steady-state conditions.

$$v_{dr} = R_r i_{dr} - s\frac{L_m}{L_s}\varphi_{qs} - sL'_r i_{qr} + \frac{L'_r}{\omega_b}\frac{di_{dr}}{dt}$$

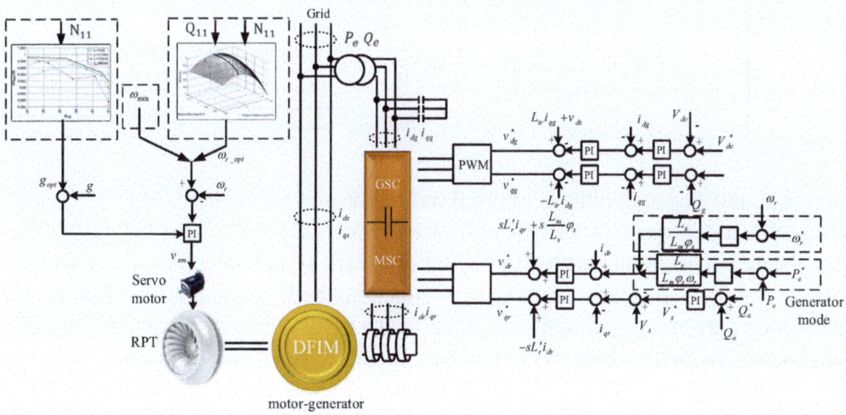

Fig. 12 An overview of the electrical equipment used in a VS-PSPP unit and its controller

$$v_{qr} = R_r i_{qr} + sL'_r i_{dr} + \frac{L'_r}{\omega_b}\frac{di_{qr}}{dt} \tag{17}$$

where P_{e_g} and Q_{s_g} can be calculated as follows.

$$\begin{aligned} P_{e\_g} &= \omega_r \frac{L_m}{L_s}\varphi_s i_{dr} \\ Q_{s\_g} &= -\frac{V_s^2}{L_s} - \frac{L_m}{L_s}V_s i_{qr} \end{aligned} \tag{18}$$

According to Eqs. (17) and (18), it is obvious that two cascades control loops can be used to control active and reactive power by *dq* axes voltages from MSC. That is:

$$P_{e\_g} \propto i_{dr} \rightarrow i_{dr} \propto v_{dr}$$

$$Q_{s\_g} \propto i_{qr} \rightarrow i_{qr} \propto v_{qr}$$

Regarding two cascade controller loops, the former or inner loop controls rotor current on *dq* axes, which their input is reference rotor current, and their output is rotor voltages. Usually, in references, this loop is called fast loop. The latter or outer control loop is called the slow loop, with active and reactive power input (stator voltage) and reference rotor current as of its output.

To adjust proportional and integral (PI) control coefficients, it should start from the inner current loop of the *dq* axes. Equation (17) can be stated as Eq. (19) considering derivative of *dq* axes stator flux caused by sudden changes in grid voltage and severe disturbances such as electrical faults.

$$\begin{cases} v_{rd} = \left(R_r + \frac{L'_r}{\omega_b}\frac{d}{dt}\right)i_{rd} - sL'_r i_{rq} \underbrace{- s\frac{L_m}{L_s}\varphi_{sq} + \frac{L_m}{\omega_b L_s}\frac{d\varphi_{sd}}{dt}}_{E'_{rd}} \\ v_{rq} = \left(R_r + \frac{L'_r}{\omega_b}\frac{d}{dt}\right)i_{rq} + sL'_r i_{rd} + \underbrace{\frac{L_m}{\omega_b L_s}\frac{d\varphi_{sq}}{dt}}_{E'_{rq}} \end{cases} \tag{19}$$

where E'_{rd} and E'_{rq} are voltages induced on the rotor winding, including the steady-state term (first part of Eq. (19)) and transient term (second part of Eq. (19)). The steady-state part is compensated by the feedforward signal added to the controller; however, its transient part resulting from quick stator flux changes remains in the closed-loop current. Additionally, $-sL'_r i_{rq}$ and $sL'_r i_{rd}$ the cross-coupling terms are compensated in *dq* axes controllers to decouple the *dq* rotor current controller loops.

$$E'_{rd} = -s\frac{L_m}{L_s}\varphi_{sq} + \frac{L_m}{\omega_b L_s}\frac{d\varphi_{sd}}{dt}$$

$$E'_{rq} = 0 + \frac{L_m}{\omega_b L_s}\frac{d\varphi_{sq}}{dt} \tag{20}$$

Summation of voltage induced on the rotor winding in steady-state and coupling in *dq* axes are called disturbances on *d*-axis (d_d) and *q*-axis (d_q), as shown in Eq. 22.

$$\left(\frac{L'_r}{\omega_b}S + R_r\right)i_{rd} = v_{rd} + \underbrace{sL'_r i_{rq} + s\frac{L_m}{L_s}\varphi_{sq}}_{d_d}$$

$$\left(\frac{L'_r}{\omega_b}S + R_r\right)i_{rq} = v_{rq} - \underbrace{sL'_r i_{rd}}_{d_q} \tag{21}$$

Compensating all disturbances defined in *dq* axes, i.e., d_d and d_q, transfer functions i_{dr}/v_{dr} and i_{qr}/v_{qr} are obtained as a linear function defined in the following.

$$\frac{i_{dr}}{v_{dr}} = \frac{i_{qr}}{v_{qr}} = \frac{1}{\frac{L'_r}{\omega_b}S + R_r} \tag{22}$$

Considering uncompensated voltage induced on the *dq* axes rotor winding in the transient state, i.e., e'_{rd} and e'_{rq} terms, controller of current loops for *dq* axes can be drawn as shown in Fig. 13.

The linear approximation method can be used to tune PI coefficients in the current loop. As it is known from linear control and classic forms of designing control, zero of the controller is selected to cancel the pole of the plant. Therefore, considering the open-loop bandwidth of the rotor current loop β_{ol-i}, Eq. (23) can be computed.

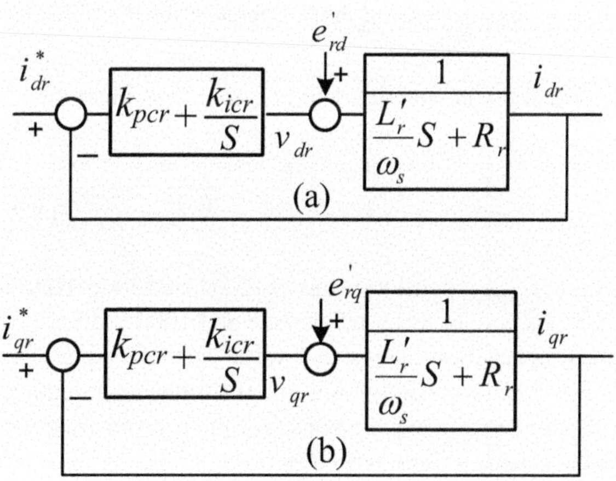

Fig. 13 Inner control loop for MSC

$$\frac{k_{icr}}{k_{pcr}} = \frac{\omega_b R_r}{L_r^{'}} = \beta_{ol-i} \tag{23}$$

By substituting Eq. 23 in the closed-loop transfer function of the rotor current, as computed in Eq. 24, closed-loop bandwidth β_{cl-i} is equal to $\frac{\omega_b k_{pcr}}{L_r^{'}}$.

$$\frac{i_{dr}}{i_{dr}^*} = \frac{i_{qr}}{i_{qr}^*} = \frac{\beta_{cl-i}}{S + \beta_{cl-i}} \tag{24}$$

Usually, bandwidth is selected based on two points in practical systems and for implementation. 1—Higher bandwidth results in a shorter response time (controlled variable better follows reference signal); 2—selecting high bandwidth does not result in noises and harmonic frequencies being filtered by the system. Also, more expensive types of equipment are required to implement a high bandwidth system. Thus, closed-loop system bandwidth is usually selected 10–20% of the converter's switching frequency f_{sw}. In the present study, converters' switching frequency and closed-loop bandwidth are assumed to be 10 kHz and 1000 rad/sec, respectively. Accordingly, control coefficients are computed; other assumed parameters are available in the appendix.

$$\begin{aligned} \sigma &= 1 - \frac{L_m^2}{L_r L_s} = 0.121 \rightarrow L_r^{'} = L_r \sigma = 0.52, \\ \omega_b &= 2\pi f = 377 \text{ red}/sec \end{aligned} \tag{25}$$

$$\begin{cases} k_{pcr} = \dfrac{L_r^{'} \beta_{cl}}{\omega_b} = 1.37 \\ k_{icr} = k_{pcr} \beta_{ol-i} = k_{pcr} \dfrac{\omega_b R_r}{L_r^{'}} = 1.99 \end{cases} \tag{26}$$

The transfer function from i_{dq} to $e_{rdq}^{'}$ is given by

$$\frac{i_{rd}}{e_{rd}^{'}} = \frac{i_{rq}}{e_{rq}^{'}} = \frac{S}{\frac{L_r^{'}}{\omega_b}(S + \beta_{ol-i})(S + \beta_{cl-i})} \tag{27}$$

where β_{ol-i} open-loop and β_{cl-i} closed-loop bandwidths can be calculated by previous equations as 1.5 rad/sec and 1000 rad/sec (10% of f_{sw}), respectively. This transfer function shows that rotor current response on *dq* axes depends on open and closed-loop bandwidths and the transient reactance value of rotor winding ($L_r^{'}$). Figure 14 shows step response for the transfer function of induced transient voltage in the rotor winding as well as rotor current. As it is observed, the less would be the bandwidth of rotor closed-loop current, the less would be the effect of disturbance (induced voltage on the rotor winding) on rotor current.

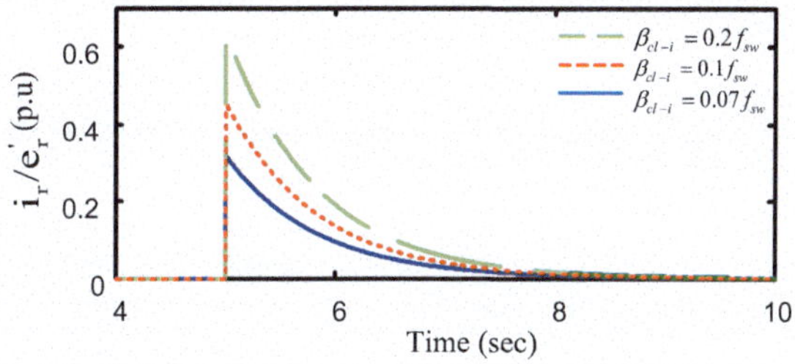

Fig. 14 Behavior of rotor flow against sudden disturbance from e_r'

Upon design performed, the transfer function of closed-loop current $\frac{\beta_{cl-i}}{S+\beta_{cl-i}}$ can be approximated as similar to a low-pass filter with a passband of β_{cl-i}. So, active and reactive power control loops can be shown in Fig. 15. Inner loops of current act as a low-pass filter in controlling active power and stator voltage, and for the appropriate performance of these types of loops, the closed-loop bandwidth would be selected as 10% of the filter's passband. Hence, the bandwidth of active power closed-loop and stator voltage and outer loops must be 10% of the closed-loop bandwidth of the inner current loop, i.e., 100 rad/sec [22] (see Fig. 16).

Again, by pole cancelation of the system by appropriate selection of zero in PI, the control coefficients are calculated using the active power closed-loop bandwidth (β_{cl-p}) and the voltage-closed-loop bandwidth (β_{cl-v}). It should be noted that the closed-loop bandwidth of the outer loop should be about 10% of the bandwidth of the inner loop [22], i.e., 100 rad/sec. Accordingly, Eq. (28) can be written to determine the control coefficients.

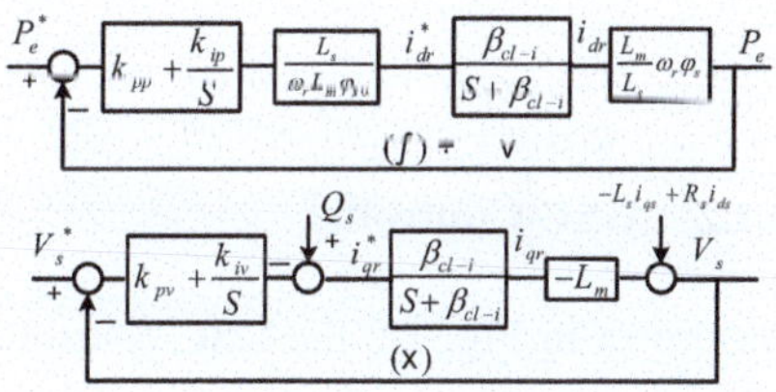

Fig. 15 Simplified outer control loop MSC

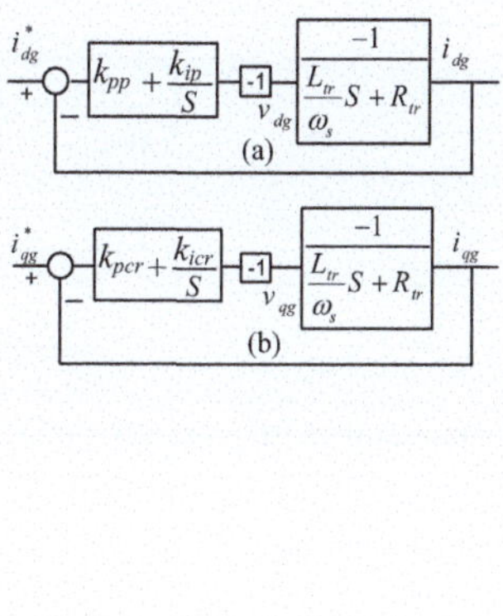

Fig. 16 Inner control loop for network-side converter

$$\begin{cases} \dfrac{k_{ip}}{k_{pp}} = \beta_{cl-i} \\ k_{pp}\dfrac{L_m}{L_r}\omega_{r0}\varphi_{s0}\beta_{cl-i} = \beta_{cl-p} \end{cases} \tag{28}$$

According to the equations obtained, k_{pp} and k_{pi} are designed roughly 0.1 and 100, respectively. The same steps would be repeated for the voltage control loop.

$$\begin{cases} \dfrac{k_{iv}}{k_{pv}} = \beta_{cl-i} \\ k_{pv}\beta_{cl-i}L_m = \beta_{cl-v} \end{cases} \tag{29}$$

The above equations show k_{pp} and k_{pi} are roughly 0.025 and 25, respectively.

Finally, the reference signal of dq axes rotor voltage and rotor current can be defined as follows.

$$\begin{aligned} v_{dr}^{*} &= k_{pcr}\left(i_{dr}^{*} - i_{dr}\right) + k_{icr}\int\left(i_{dr}^{*} - i_{dr}\right) - sL_r' i_{qr} - s\frac{L_m}{L_s}\varphi_s \\ v_{qr}^{*} &= k_{pcr}\left(i_{qr}^{*} - i_{qr}\right) + k_{icr}\int\left(i_{qr}^{*} - i_{qr}\right) + sL_r' i_{dr} \end{aligned} \tag{30}$$

$$\begin{aligned} i_{dr\_g}^{*} &= \frac{L_s}{\omega_r L_m \varphi_s}\left[k_{pp}\left(P_e^{*} - P_e\right) + k_{ip}\int\left(P_e^{*} - P_e\right)\right] \\ i_{qr\_g}^{*} &= -k_{pv}\left(V_s^{*} - V_s\right) - k_{iv}\int\left(V_s^{*} - V_s\right) - Q_s \end{aligned} \tag{31}$$

3.1.2 GSC Controller

As shown in Fig. 12, the GSC controller has two cascade control loops similar to MSC to control DC-ink voltage and reactive power through dq axes voltages. So, Eq. (16) can be rearranged in Laplace space as follows.

$$\begin{aligned} \left(\frac{L_{tr}}{\omega_b}S + R_{tr}\right)i_{dg} &= -v_{dg} + \underbrace{L_{tr}i_{qg} + v_{ds}}_{d_d} \\ \left(\frac{L_{tr}}{\omega_b}S + R_{tr}\right)i_{qg} &= -v_{qg} - \underbrace{L_{tr}i_{dg}}_{d_q} \end{aligned} \tag{32}$$

Through full compensation of steady-state disturbances shown in Eq. (32), i.e., d_d and d_q, transfer functions of i_{dtr}/v_{dtr} and i_{qtr}/v_{qtr} are obtained as a linear function in the following:

$$\frac{i_{dtr}}{v_{dtr}} = \frac{i_{qtr}}{v_{qtr}} = \frac{-1}{\frac{L_{tr}}{\omega_b} S + R_{tr}} \tag{33}$$

Applying the method used to design MSC loops controller, control coefficients are designed for GSC loops. As stated before, the design is primarily started from the current control loop.

$$\begin{cases} k_{pcr} = \dfrac{L_{tr} \beta_{cl-i}}{\omega_b} \\ k_{icr} = k_{pcr} \beta_{ol-i} = k_{pcr} \dfrac{\omega_b R_{tr}}{L_{tr}} \end{cases} \tag{34}$$

Selecting the switching frequency of GSC, $L_{tr,}$ and R_{tr} as 10 kHz, 0.15 p.u., and 0.005 p.u., respectively; then, controller coefficients k_{pcr} and k_{icr} are equal to 0.4 and 5, respectively.

Ignoring active and reactive power losses in GSC, active and reactive power transmitted through GSC can be stated by Eq. (35).

$$C\frac{dV_{dc}}{dt} = \frac{V_s i_{dg}}{V_{dc}} - \frac{P_r}{V_{dc}} \rightarrow 0.5C\frac{dV_{dc}^2}{dt} = V_s i_{dg} - P_r \tag{35}$$

where

$P_{tr} = V_s i_{dtr}, Q_{tr} = V_s i_{qtr}$

$P_r = v_{dr} i_{dr} + v_{qr} i_{qr}$

To simplify and reduce the order of transfer function obtained, the inner loop's pole is ignored, and the transfer function is rewritten.

$$\frac{V_{dc}}{V_{dc}^*} = \frac{V_s\left(k_{pvdc} S + k_{ivdc}\right)}{V_{dc0} C S^2 + V_s\left(k_{pvdc} S + k_{ivdc}\right)} = \frac{V_s}{C V_{dc0}} \frac{\left(k_{pvdc} S + k_{ivdc}\right)}{S^2 + \frac{V_s k_{pvdc}}{C V_{dc0}} S + \frac{V_s k_{ivdc}}{C V_{dc0}}} \tag{36}$$

$$\begin{cases} \xi = \dfrac{V_s k_{pvdc}}{2 C V_{dc0} \omega_n} \\ \omega_n = \sqrt{\dfrac{V_s k_{ivdc}}{C V_{dc0}}} \end{cases} \rightarrow \begin{cases} k_{pvdc} = \dfrac{2\xi \omega_n C V_{dc0}}{V_s} \approx 8.4 \\ k_{ivdc} = \dfrac{C V_{dc0} \omega_n^2}{V_s} \approx 600 \end{cases} \tag{37}$$

Note that ω_n frequency should be 10% of the bandwidth of rotor closed-loop current (inner loop: 1000 rad/sec). Also, V_{dc0} and C is assumed approximately 0.2 and 0.3, respectively. Closed-loop transfer function dc-link to rotor power from Fig. 17a can be obtained as (see Fig. 18)

$$\frac{V_{dc}}{(-P_r/V_{dc0})} = \frac{S}{\frac{L_r'}{\omega_b}(S + \beta_{ol-i})(S + \beta_{cl-i})} \tag{38}$$

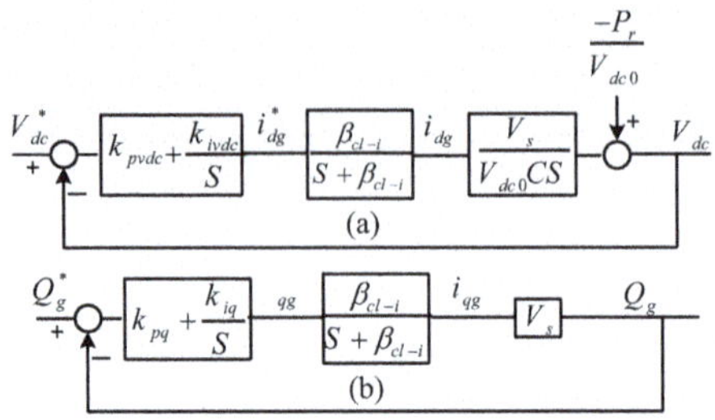

Fig. 17 Simplified outer control loop for GSC

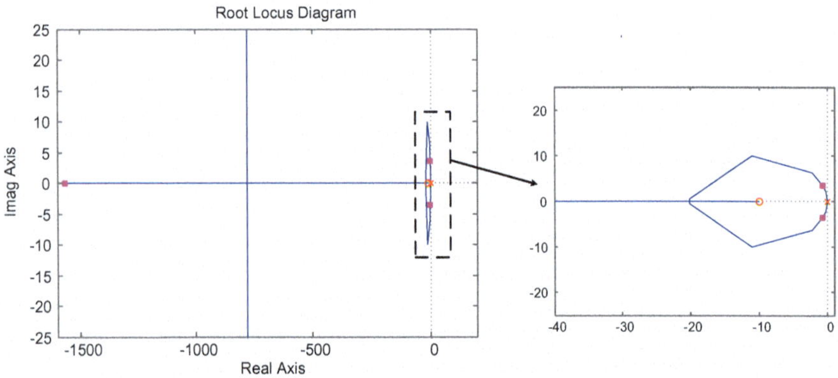

Fig. 18 Root Locus block diagram for zero and poles of the closed-loop controller in GSC

Analyzing the transfer function shown above, the following points can be concluded.

- Reduction of the inner current loop bandwidth is followed by reduction of outer bandwidth, increasing disturbances' effects of MSC on DC link.
- If the bandwidth of the inner loop of the current remains fixed, reduction of outer loop bandwidth increase disturbance's effect on the output.

Moreover, the closed-loop transfer function of reactive power is obtained as follows:

$$\frac{Q_g}{Q_g^*} = \frac{V_s\left(k_{pq}S + k_{iq}\right)\beta_{cl-i}}{S^2 + \left(\beta_{cl-i} + V_s k_{pq}\beta_{cl-i}\right)S + V_s k_{iq}\beta_{cl-i}} \tag{39}$$

As stated, ω_n frequency has to be 10% of the bandwidth of the closed-loop rotor current (inner loop: 250 rad/sec). According to the above equations, k_{pq} and k_{iq} will be 0.1 and 100, respectively.

$$\begin{cases} \dfrac{k_{iq}}{k_{pq}} = \beta_{cl-i} \\ k_{pq} V_s \beta_{cl-i} = \beta_{cl-q} \end{cases} \tag{40}$$

Finally, the reference signal of dq axes grid voltage and grid current can be defined as follows.

$$\begin{aligned} v_{dg}^{*} &= -k_{pcg}\left(i_{dg}^{*} - i_{dg}\right) - k_{icg}\int \left(i_{dg}^{*} - i_{dg}\right) + L_{tr} i_{qg} + v_{ds} \\ v_{qg}^{*} &= -k_{pcg}\left(i_{qg}^{*} - i_{qg}\right) - k_{icg}\int \left(i_{qg}^{*} - i_{qg}\right) - L_{tr} i_{dg} \end{aligned} \tag{41}$$

$$\begin{aligned} i_{dg}^{*} &= k_{pvdc}\left(V_{dc}^{*} - V_{dc}\right) + k_{ivdc}\int \left(V_{dc}^{*} - V_{dc}\right) \\ i_{qg}^{*} &= k_{ppf}\left(Q_{g}^{*} - Q_{g}\right) + k_{ipf}\int \left(Q_{g}^{*} - Q_{g}\right) \end{aligned} \tag{42}$$

3.1.3 Turbine Controller

Here, the turbine controller or governor is discussed. Figure 19 shows various nonlinear parts for the turbine, a servomotor for changing the turbine valve, a non-elastic system, and a high length penstocks [23].

Existing loops are simplified to design the controller, as shown in Fig. 20a. The servomotor's loop has one pole in the open loop with a high frequency; thus, it can be ignored in the mechanical loop of the turbine. So, the transfer function of the

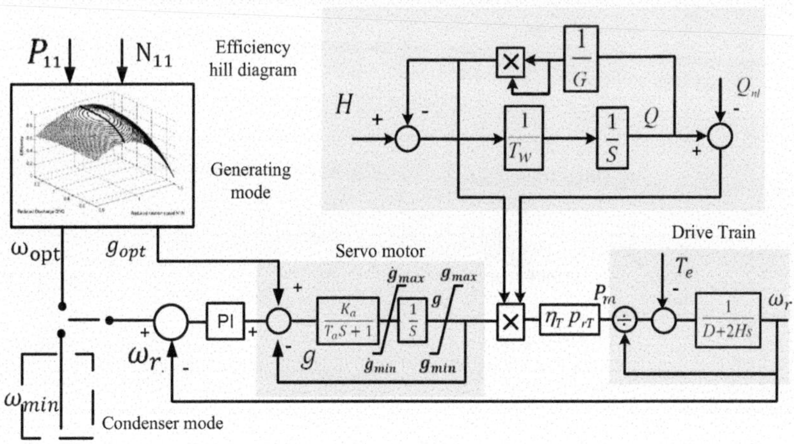

Fig. 19 Block diagram of hydro RPT in turbine mode and condenser mode

closed-loop in the servomotor's circuit can be written as $\frac{k_a}{k_a+S}$. A linearized model of a RPT is shown in [24] through transfer function as $\frac{1-T_wS}{1+0.5T_wS}$ in which $T_w = 2.67$ for this study. Figure 20b shows root locus diagram the rotor speed loop's pole-zeros. It can be found that the system has a zero which is located in the right-half plane; that is, hydro RPT in turbine mode is a non-minimum phase system.

The following points can be drawn from Fig. 20

(1) The pole resulting from servomotor has little effect on overall system response due to its high frequency; hence, it can be eliminated.
(2) The turbine pole is moving toward zero in PI via increasing proportional gain k_{pw} in the closed loop. The turbine pole can be removed by selecting zero of PI equal to 0.1 of the pole of the turbine.
(3) Remained system will include a zero (turbine) and two poles (turbine shaft and PI), and an appropriate response could be obtained by specifying a second-order system.
(4) As it is observed, since there is a non-minimum phase system, too much increase in k_{pw}, a right-half pole will appear, and the system will become unstable. So, attention has to be paid to its design.
(5) Studying system closed-loop model shows that the system is stable against $0 < k_{pw} < 0$ $0.4 < k_{iw} < 4$, with values chosen, respectively. Therefore, 2 and 0.2 in

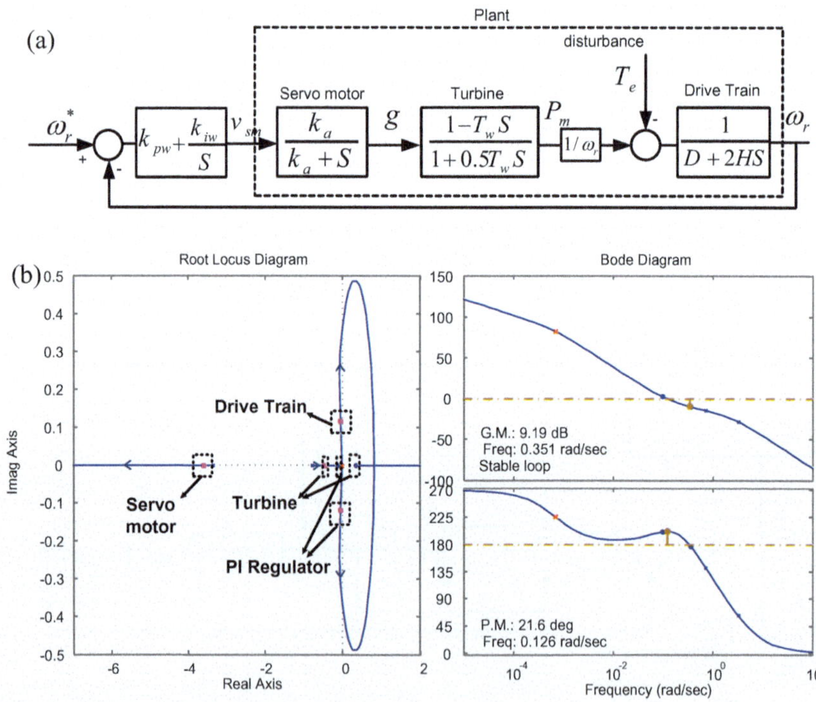

Fig. 20 Hydro RPT in turbine mode **a** simplified block diagram **b** root locus diagram

this study.

$$\dot{\omega}_{r\_T,P} = \pm\frac{1}{2H}(T_m - T_e) \tag{43}$$

$$v_{sm\_T} = k_{ps}\left(\omega_r^* - \omega_r\right) + k_{is}\int\left(\omega_r^* - \omega_r\right) + G^* \tag{44}$$

3.2 Motor Mode

Contrary to generator mode, the VS-PSPP controller is divided into two parts as shown in Fig. 21: MSC controller and GSC controller. The MSC controls the motor's rotation speed, and power consumed (see Fig. 21). In other words, the MSC can handle the active power used by the motor by controlling motor rotation speed so that grid frequency control would be assisted. Also, reactive power/stator voltage control is done by another loop of the MSC. On the other hand, controller of GSC like generator mode is responsible for voltage control of DC-link capacitor and reactive power exchanged from this converter.

3.2.1 MSC Controller

Firstly, relationships between objectives and control variables can be identified from modeling performed in the previous section.

$$T_{e\_m} = -\frac{L_m}{L_s}\varphi_s i_{dr} \rightarrow v_{dr} = R_r i_{dr} - s\frac{L_m}{L_s}\varphi_{qs} - sL_r^{'} i_{qr} + \frac{L_r^{'}}{\omega_b}\frac{di_{dr}}{dt}$$

$$Q_{s\_g} = \frac{V_s^2}{L_s} + \frac{L_m}{L_s}V_s i_{qr} \rightarrow v_{qr} = R_r i_{qr} + sL_r^{'} i_{dr} + \frac{L_r^{'}}{\omega_b}\frac{di_{qr}}{dt}$$

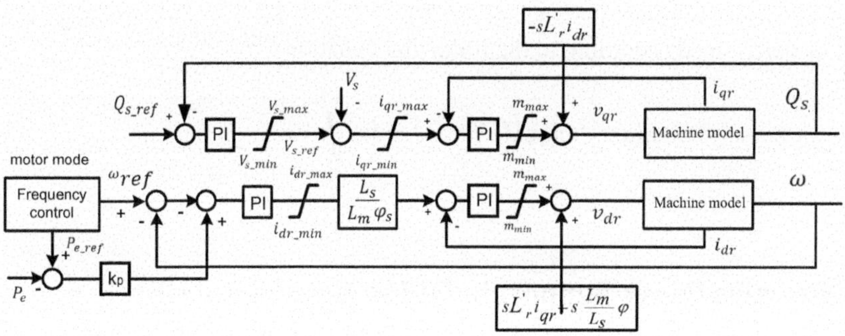

Fig. 21 MSC control schematic in motor mode

$$\begin{aligned} &\omega_r \propto T_{e\_m} \propto i_{dr} \rightarrow i_{dr} \propto v_{dr} \\ &Q_{s\_m} \propto i_{qr} \rightarrow i_{qr} \propto v_{qr} \end{aligned} \tag{45}$$

Based on Eq. (45), it could be suggested that to control rotor speed and reactive power through *dq* axes voltages from the MSC, two cascades control loops are used. The first or inner loop controls rotor current in *dq* axes, with their input being reference rotor current, usually called a fast loop. The second or outer control loop (slow loop) has control variables for that rotor speed/reactive power (stator voltage).

Therefore, the following equations compute the reference value of the rotor current and voltage on the *dq* axes.

$$\begin{aligned} v_{dr}^* &= k_{pcr}\left(i_{dr}^* - i_{dr}\right) + k_{icr}\int\left(i_{dr}^* - i_{dr}\right) - sL_r' i_{qr} - s\frac{L_m}{L_s}\varphi_s \\ v_{qr}^* &= k_{pcr}\left(i_{qr}^* - i_{qr}\right) + k_{icr}\int\left(i_{qr}^* - i_{qr}\right) + sL_r' i_{dr} \end{aligned} \tag{46}$$

$$\begin{aligned} i_{dr\_m}^* &= \frac{-L_s}{L_m\varphi_s}\left[k_{pw}\left(\omega_r^* - \omega_r\right) + k_{iw}\int\left(\omega_r^* - \omega_r\right)\right] + k_p\left(P_e^* - P_e\right) \\ i_{qr\_m}^* &= -k_{pv}\left(V_s^* - V_s\right) - k_{iv}\int\left(V_s^* - V_s\right) + Q_s \end{aligned} \tag{47}$$

3.2.2 GSC Controller

GSC controllers are not different in motor and generator modes. So, Eqs. (41) and (42) are also established here.

3.2.3 Pump Controller

The pump input gate in motor mode is specified based on maximum efficiency and has no control over other parameters (see Fig. 22). So, the servomotor voltage value will be determined according to the following equation.

$$v_{sm\_P} = k_{pg}\left(G^* - G\right) + k_{ig}\int\left(G^* - G\right) \tag{48}$$

4 Reactive Power Support in DFIM-Based VS-PSPP

The reactive power of a plant needs to be controlled not only in the condenser mode, but also in generator and motor operational modes within the P-Q curve capability of

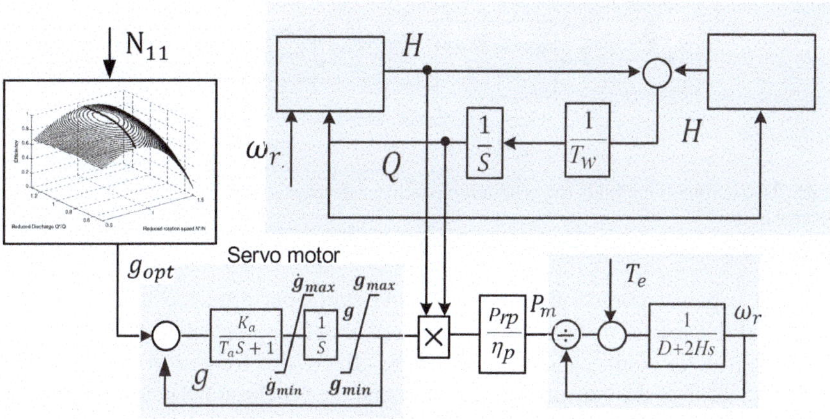

Fig. 22 Diagram of the hydraulic part in pump mode

the machine. Therefore, due to the importance of reactive power control for PSPPs, reactive power support in DFIM-based VS-PSPPs is discussed in this section.

As explained in the last part, the reactive power of a DFIM-based unit is produced or consumed in two ways; 1—the stator of the DFIM, 2—GSC. In general, these two paths can be designed based on either voltage control or power factor correction aims. It is practically impossible to meet these two aims concurrently. In conclusion, in a small number of WTs connected to the distribution network, automatic voltage control can be a better option than power factor control, while in large WTs and VS-PSPP connected to the transmission network, it is necessary to inject additional reactive power into the grid to maintain the stability of the network, especially in fault conditions. Therefore, it will be more appropriate to use power factor control in VS-PSPPs.

Now it can be discussed the components that produce reactive power. In Fig. 23, the reactive reference power calculated by the circuit is reconnected to the main controller in the equipment that is to generates the reactive power. In general, reactive power generation in a DFIM can be done in two ways. 1—stator, which of course, is fed and controlled through the MSC. 2—GSC. a brief description of the methods of generating reactive power through each of the mentioned sections is given. These methods, which are found in the application of WT, are divided into the following three categories.

1. The most cost-effective method at first glance can be considered by simultaneously using the GSC and the stator to produce reactive power. This reduces the cost of designing the rated reactive power between the two parts. Of course, it should be noted that in this solution, the rotation of the reactive current between the stator and GSC is one of the main problems in this control method that can be avoided by choosing the appropriate strategy.

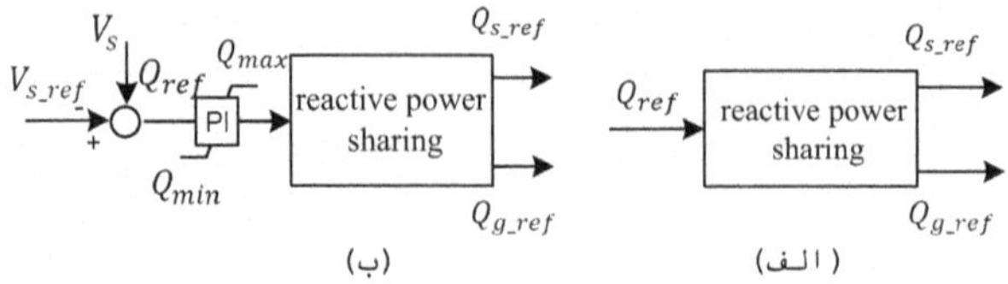

Fig. 23 Two common methods for generating reactive reference power in **a** Power factor control **b** Voltage control

2. Reactive power control through GSC and stator performance in unity power factor is the second method that can be used to exchange reactive power with the power grid. Of course, this method can be effective as long as the active power passing through the converters is low (around the synchronous point). Of course, the operation around the synchronous point also has thermal limitation for the switches due to the switching frequency reduction. Therefore, for the production of reactive power by the GSC, it is necessary to pay an additional cost for the new size of the converter in exchange for the production of reactive power.
3. The third method can be considered the opposite of the second method. In this method, all the reactive power exchanged with the network is done through the stator, and GSC always works in the unity power factor.

To decide which of the above methods is suitable for use in a VS-PSPP, it is essential to note that the active power is usually between full load and partial load is changing. Also, in the condenser mode, the reference speed is set to the minimum acceptable speed of a unit so that the rotational losses of the mechanical equipment are minimized. Therefore, the first and third types can be the most suitable method for generating reactive power in a DFIM-based VS-PSPP. Although using the DFIM stator and the TSC concurrently can be a cost-effective strategy to control reactive power, uncoordinated control of the reactive power can be caused by the reactive current circulation between the TSC and the machine stator [25].

4.1 P-Q Machine Capability Curve

The P-Q capability curve is needed for designing the plant's controller and the grid's power flow. Therefore, it is necessary to consider the P-Q capability curves of DFIM-based PSPPs in this part. Unlike salient pole SM, which adjusts the output voltage and the reactive power of the machine only through the excitation system, DFIMs can control the reactive power via the stator winding, which is controlled indirectly by the MSC, and the TSC.

The apparent power of the stator for an induction machine is obtained by Eq. 47.

$$|S_s| = \left|V_s I_s^*\right| = P_s + jQ_s \tag{49}$$

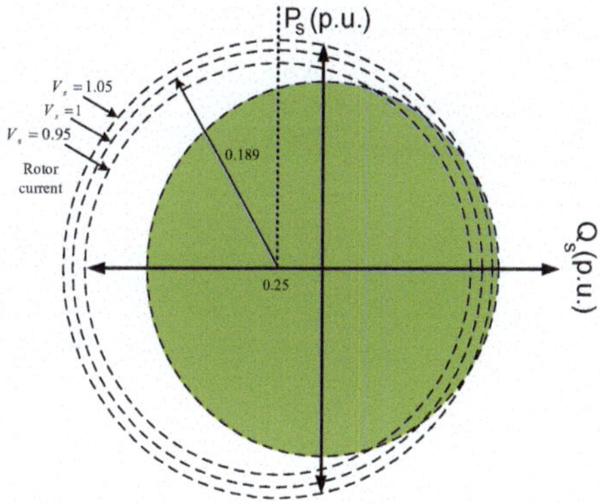

Fig. 24 Capacity curve based on the rated current of the rotor and for different values of stator voltage

Therefore, the stator's active and reactive power can be considered a circle based on the nominal current of the stator with the center of origin and radius $V_s I_s$ as Fig. 24.

$$P_s^2 + Q_s^2 = |V_s I_s|^2 \tag{50}$$

In addition to the stator current, the rotor current will not be able to produce more than its nominal value. Therefore, we try again to arrange the apparent power relationship of the stator based on the rotor current. In this case, we will have Eq. (51) considering $\vec{I}_r = \vec{I}_s + \vec{I}_m$.

$$|S_s| = |V_s(I_r - I_m)| = P_s + jQ_s \tag{51}$$

Now, if we consider the reactive power consumed by the magnetizing current $Q_m \approx V_s I_m$ as follows:

$$\vec{S}_s = P_s + j(Q_s + Q_m) = \vec{V}_s \vec{I}_r \tag{52}$$

Therefore, we can write

$$P_s^2 + (Q_s + Q_m)^2 = |V_s I_r|^2 \tag{53}$$

Taking into account the nominal values for rotor and stator current and stator voltage, Fig. 25 can be drawn for generating mode.

Finally, using the values designed in Tables 4, 5, and 6, the capability curve of active-reactive power for generator, motor, and condenser modes is obtained as shown in Fig. 26.

Consequently, the following notes can be mentioned for generating and motor modes.

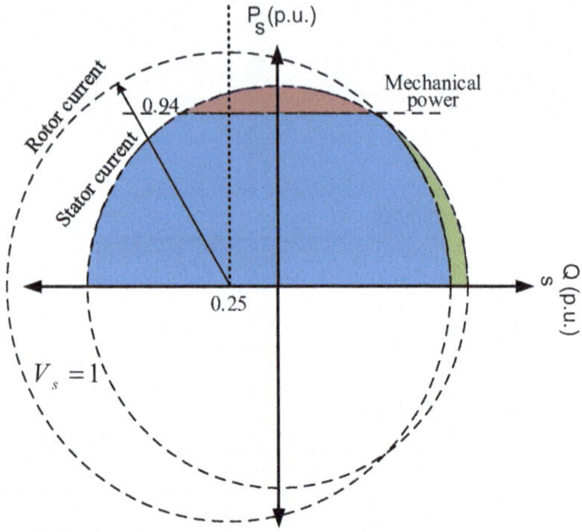

Fig. 25 Machine stator capability curve based on the rated current of the stator and rotor

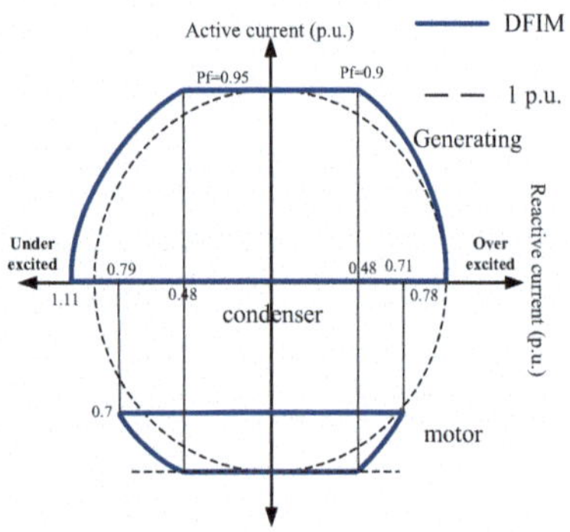

Fig. 26 Total P-Q ability curve of two-way induction machine in three modes of generator, motor and condenser

(1) ***Generator mode***: For a DFIM-based PSPP, the over-excited operation is restricted to the rated current of the MSC, i.e., the rotor winding current. It should be noted that DFIM-based PSPP can control the reactive power via both the stator winding and the GSC. On the other hand, the total apparent current of the GSC is designed based on vector summation of the active and reactive current. It means that the capability of exchanging reactive power of the GSC depends on the rotor speed operating point, i.e., output real power of the PSPP. Note that if the reactive power is needed at full active power production, the GSC must be oversized. Otherwise, when the little active power is transferred through the rotor, i.e., sub-synchronous operating region, the GSC can exchange further reactive power with the grid.
(2) ***Motor mode***: the active power consumption of the DFIM-based PSPP in pumping mode can vary within a specific range, which is determined by the

rotor speed variation range. Therefore, considering the limitations similar to the generator mode, exchanging the reactive power of the DFIM-based PSPP as the motor can derive, as shown in Fig. 26.

5 Conclusion

This chapter presents a model of a DFIM-based VS-PSPP in generator, motor, and condenser operating modes. Modeling for the electric machine was performed by considering the stator voltage-oriented method. Also, the control loops of each part were simplified, and the control coefficients were tuned and proposed. Finally, PQ capability curve of DFIM for both operational modes was plotted and analyzed. In generator and condenser mode under excitation, the asynchronous machine will absorb almost two times the reactive power compared to the synchronous machine because the core is heated in the end winding of the machine. Also, The speed variation range of DFIM can be a trade-off problem between the cost of the system, required efficiency improvement, and the active power variation range in pumping mode.

Appendix

See Tables 4, 5, and 6.

Table 4 Nominal values electrical parts of PSPP

Variable	P_n	S_n	Ip_n	Iq_n	I_n	V_n
Unit	MW	MVA	kA	kA	kA	kv
Stator winding	320.6	361.08	10.28	5.33	11.58	18
MSC	25.23	30.67	7.15	3.29 + 1.79	8.68	2.04
GSC $Pf = 0.9$	22.4	24.93	6.34	3.08	7.05	2.04
GSC $Pf = 1$	22.4	22.4	6.34	0	6.34	2.04

Table 5 DFIM parameter

L_s	4.26	V_{dcn}	kV 4.54	C	F 0.3
L_r	4.272	S_n	MVA 381.1	s	7%
R_s	0.00174	S_{ns}	MVA 361.07	H	7.7 s
R_r	0.002	$S_{n\text{-}RSC}$	MVA 49.25	f	60 Hz
L_m	4	$S_{n\text{-}GSC}$	26.93	P	16

Table 6 Pump-turbine data

D_{ref}	m 4.41
P_n	343 MW
N_n	450 rpm
Q_n	54.38 m^3/sec
H_n	709 m

References

1. Donalek P (1995) Application of adjustable-speed machines in conventional and pumped-storage hydro projects. Final report
2. Padoan AC et al (2010) Dynamical behavior comparison between variable speed and synchronous machines with PSS. IEEE Trans Power Syst 25(3):1555–1565. https://doi.org/10.1109/TPWRS.2009.2039586
3. Pannatier Y, Kawkabani B, Nicolet C, Simond J-J, Schwery A, Allenbach P (2009) Investigation of control strategies for variable-speed pump-turbine units by using a simplified model of the converters. IEEE Trans Ind Electron 57(9):3039–3049
4. Belhadji L, Bacha S, Munteanu I, Rumeau A, Roye D (2013) Adaptive MPPT applied to variable-speed microhydropower plant. IEEE Trans Energy Convers 28(1):34–43. https://doi.org/10.1109/TEC.2012.2220776
5. Valavi M, Nysveen A (2016) Variable-speed operation of hydropower plants: past, present, and future. In: Proceedings of the 2016 22nd international conference on electrical machines, ICEM 2016, vol PP, no October, pp 640–646. https://doi.org/10.1109/ICELMACH.2016.7732593
6. Azbe V, Mihalic R (2017) Transient stability of a large doubly-fed induction machine in a pumped-storage plant. Electr Power Syst Res 142:29–35
7. Joseph A, Desingu K, Semwal RR, Chelliah TR, Ieee M, Khare D (2017) Dynamic performance of pumping mode of 250 MW variable speed hydro-generating unit subjected to power and control circuit faults, vol 1, no c, pp 1–11
8. Joseph A, Selvaraj R, Chelliah TR, Sarma SVA (2018) Starting and braking of a large variable speed hydro-generating unit subjected to converter and sensor faults. IEEE Trans Ind Appl 9994(i):1–11
9. Joseph A, Chelliah TR (2018) A review of power electronic converters for variable speed pumped storage plants: configurations, operational challenges, and future scopes. IEEE J Emerg Sel Top Power Electron 6(1):103–119
10. Yang W, Yang J (2019) Advantage of variable-speed pumped storage plants for mitigating wind power variations: integrated modelling and performance assessment. Appl Energy 237:720–732
11. Bidgoli MA, Yang W, Ahmadian A (2020) DFIM versus synchronous machine for variable speed pumped storage hydropower plants: a comparative evaluation of technical performance. Renew Energy 159:72–86
12. Xu B, Li H, Campana PE, Hredzak B, Chen D (2020) Dynamic regulation reliability of a pumped-storage power generating system: Effects of wind power injection. Energy Convers Manag 222:113226
13. Vasudevan KR, Ramachandaramurthy VK, Venugopal G, Ekanayake JB, Tiong SK (2021) Variable speed pumped hydro storage: a review of converters, controls and energy management strategies. Renew Sustain Energy Rev 135. https://doi.org/10.1016/j.rser.2020.110156
14. Bidgoli MA, Ganjali D, Yang W, Atrian S (2021) Small signal stability improvement of pumped storage hydropower using wide area signal considering wind farm. In Wide area power systems stability, protection, and security. Springer, pp 249–271
15. Bidgoli MA, Atrian S, Yang W, Gonzalez-Longatt FM (2021) Transient stability assessment of power system incorporating DFIM-based pumped storage hydropower and wind farm. In:

Modelling and simulation of power electronic converter dominated power systems in power factory. Springer, pp 131–152
16. MWH Americas (2009) Technical analysis of pumped storage and integration with wind power in the pacific Northwest final report U.S. army corps of engineers northwest division hydroelectric design center
17. Liang J, Harley RG (2010) Pumped storage hydro-plant models for system transient and long-term dynamic studies. IEEE PES Gen Meet PES 2010:1–8. https://doi.org/10.1109/PES.2010.5589330
18. Kerkman RJ, Lipo TA, Newman WG, Thirkell JE (1980) An inquiry into adjustable speed operation of a pumped hydro plant part 1—machine design and performance. IEEE Trans Power Appar Syst. PAS-99(5):1828–1837. https://doi.org/10.1109/TPAS.1980.319773
19. Schlunegger H, Thöni A (2013) 100 MW full-size converter in the Grimsel 2 pumped-storage plant. Hydro
20. Auber S (Jul 2012) Power on tap from variable speed pumped water storage scheme. Energize
21. Kjølle A (2001) Mechanical equipment. Waterpower Lab NTNU
22. Rahimi M, Parniani M (2010) Dynamic behavior analysis of doubly-fed induction generator wind turbines-the influence of rotor and speed controller parameters. Int J Electr Power Energy Syst 32(5):464–477. https://doi.org/10.1016/j.ijepes.2009.09.017
23. Alizadeh Bidgoli M, Yang W, Ahmadian A (2020) DFIM versus synchronous machine for variable speed pumped storage hydropower plants: a comparative evaluation of technical performance. Renew Energy 159:72–86. https://doi.org/10.1016/j.renene.2020.05.163
24. Lung JK, Lu Y, Hung WL, Kao WS (2007) Modeling and dynamic simulations of doubly fed adjustable-speed pumped storage units. IEEE Trans Energy Convers 22(2):250–258. https://doi.org/10.1109/tec.2006.875481
25. Plants D, Kayıkc M (2007) Reactive power control strategies for. IEEE Trans Energy Convers 22(2):389–396

Chapter 7
Integrated Voltage and Frequency Controllers (IVFCs) for Pico Hydro Off-Grid Power Generation

Veramalla Rajagopal and Bhim Singh

Renewable energy sources are receiving a lot of attention as a result of the exhaust of fossil fuels and the increase in carbon dioxide emissions. Wind energy and small hydro are popular clean energy sources for supplying electricity in inaccessible and remote areas where the power of the grid is unavailable. As a result, small generating plants are being built in remote areas where there is a lot of small hydro potential that can be used to generate electricity. An asynchronous generator is more robust, brushless, and smaller in size than a conventional synchronous generator [1–6]. As a result of these benefits, the uncontrolled hydro asynchronous generator is a viable option for remote power generation. The reactive power is supplied by a capacitor bank connected across its terminals. The rating of the capacitor bank is chosen to produce the rated voltage when driven at the rated speed [7–12]. These generators have inherent frequency and voltage regulation issues in small hydropower generation. Along with the frequency and voltage regulation issues, it also has poor power quality. As a result, a VFC with an appropriate control strategy has required this system [13–35].

For isolated hydro and wind power applications, these controllers have been extensively researched [13–35]. In the primary stage, load controllers (electronic) do not have an unbroken control and are realized in distinct steps [13]. The following controller stage employs fully controlled thyristor bridges with a supplementary dump load. The total active power consumed by the loads and the dump loadset to

V. Rajagopal (✉)
Department of Electrical and Electronics Engineering, Kakatiya Institute of Technology and Science, Warangal 506015, India
e-mail: vrg.eee@kitsw.ac.in

B. Singh
Department of Electrical Engineering, Indian Institute of Technology, Delhi, New Delhi-110016, India
e-mail: bsingh@ee.iitd.ac.in

A. K. Giri et al. (eds.), *Power Quality: Infrastructures and Control*,
Studies in Infrastructure and Control,
https://doi.org/10.1007/978-981-19-7956-9_7

equal the turbine power [14]. Later electronic load controllers have used continuous control, but they have power quality issues and distort generated voltage and generator currents, reducing asynchronous generator life and limiting the ability to feed classified consumer loads [15–35]. To control the frequency and voltage, the projected integrated improved power quality controllers employ an IGBT chopper with dump load and VSC (IGBT based) [22]. To control the frequency, the active power is controlled and the control strategy is the heart of these systems. The control algorithm (based on unit templates) has been reported for estimating reference source currents for controlling the voltage and frequency of isolated asynchronous generators for small hydropower generation. However, when a unit template-based control algorithm is applied to the hardware implementation of voltage and frequency controllers, the voltage of the DC link is found to be very sensitive to load perturbations, causing stress on the components of the VSC [27].

As a result, an enhanced control strategy for VFC in generators driven by a hydro turbine is necessary. Various control strategies, however, have been proposed, including synchronous reference frame strategy [36–40], instantaneous symmetrical component theory strategy [41–44], instantaneous reactive power theory strategy [45–49], I$\cos\varphi$ strategy [50], synchronous detection strategy [51–53], power balance strategy [54], single-phase theory strategy[55], and least mean square based or Adaline control strategy [56–58]. In the proposed research work, these old and new control strategies are applied to VFCs to feed consumer loads. Furthermore, in such isolated small hydropower generation, low-cost commercially available machines can be used as IAG systems. The eight IGBT-based VSC is used for neutral current compensation and enhancement in power quality [33]. Furthermore, numerous configurations in transformer [57–61] have been reported in the previously reported work for neutral current compensation such as star/hexagon, star/polygon, star/delta, T-connected and star-zigzag transformer.

Many reported voltage and frequency controllers in the literature are required to operate the power converters at optimum voltage. In light of this requirement, isolated VFC topologies, particularly for 03-phase 04-wire systems with a range of transformer configurations such as star-delta [57], zigzag-star [58, 59], T-T connected [60], star-hexagon [61], and star-polygon connections, may be investigated.

Integrated and decoupled voltage and frequency controllers are designed and developed for pico-hydro power applications [62, 63]. Shunt, series, and shunt-series active power filters are used for the mitigation of power quality problems [64]. The control algorithm based on multi-objective is applied for combined hydro-solar PV power generation [65]. Control algorithm based on adaptive sliding for isolated 1-phase micro-grid with wind-hydro-solar PV-based generation [66]. The control is based on a reweighted ZA-LMS algorithm for wind-hydro power generation rural applications [67]. A improved power quality is achieved by using a resilient FLL control for a hydro-solar generation [68]. The ELD-OSG-based control is used for battery-based ELC to control voltage and frequency for small hydro applications [69]. In light of these gaps in technology, improved power quality-based VFCs are designed and developed in a variety of configurations in this research work, utilizing a number of new control algorithms.

1 Scope of Work

In view of existing controllers of IAG, the main objectives of the proposed research work are aimed to investigate three-phase, four-wire, and three-wire, three-phase IVFC for power generation with Pico-hydro for IAG.

The detailed objectives of the proposed research work are given here.

2 Review of 3-Phase 4-Wire IVFC for Pico Hydro Power Generation

IVFC is employed to regulate frequency and voltage in an integrated manner for isolated asynchronous generators in pico-hydro plants supplying four-wire, three-phase consumer loads. This work deals with various IVFCs, which are designed, modeled, and simulated for asynchronous generators in the off-grid mode in power generation for pico-hydro. These IVFCs use star/delta, T-connected, zigzag/star, star/polygon, star/hexagon, and zigzag/single-phase transformers for optimizing the DC bus voltage and converting 3-wire system to 4-wire system by connecting the neutral of the load to that of the transformer. Various control algorithms are proposed in this work, which is more appropriate for isolated generating systems. The prototype of IVFC for 4-wire 3-phase system is developed and tested in the laboratory environment with various control algorithms. Simulation results of some VFCs are validated by experimental results to demonstrate their satisfactory performance in IAG systems.

2.1 Design of 3-Leg 4-Wire IVFC

Figures 1, 2, 3, 4 and 5 show the proposed 3-leg 4-wire IVFCs using a zigzag-star, star-delta, star-hexagon, star-polygon, or T-T-connected transformers for the IAG system. Besides each of these have an asynchronous generator, bank of capacitors, interfacing inductors, DC link capacitors, a VSCand an auxiliary load. The components' design has been described in detail in the present section.

2.1.1 Electrical System Design

The system is a combination of the commercially available asynchronous machine of 3.7 kW rating with a 4.4kVAR capacitor for rated voltage generation.

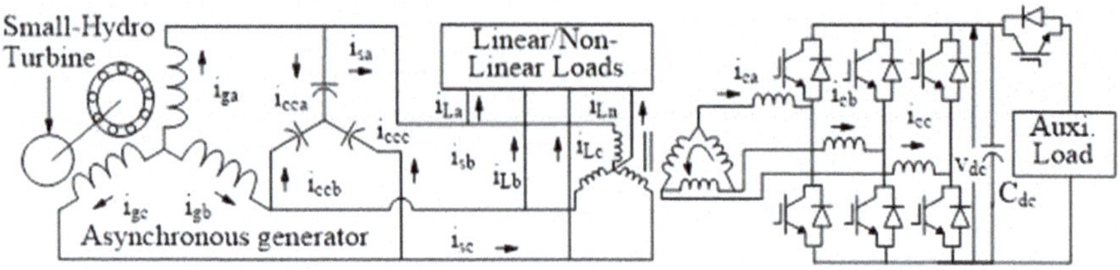

Fig. 1 Star-delta transformer-based IVFC

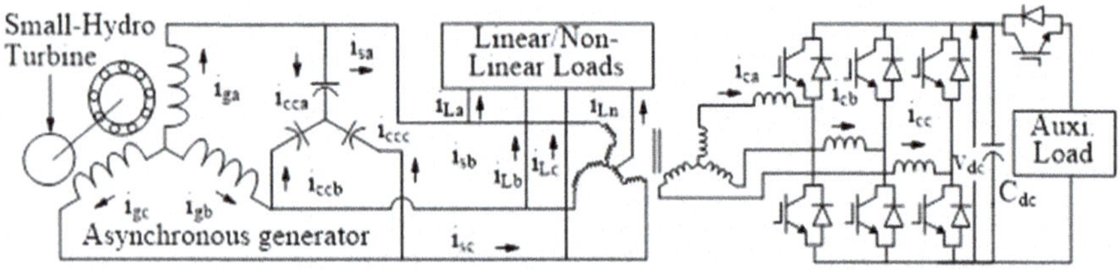

Fig. 2 Zigzag-star transformer-based IVFC

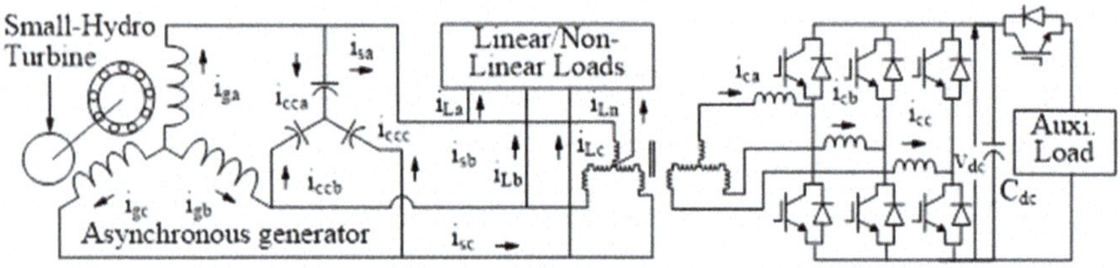

Fig. 3 T-T transformer-based IVFC

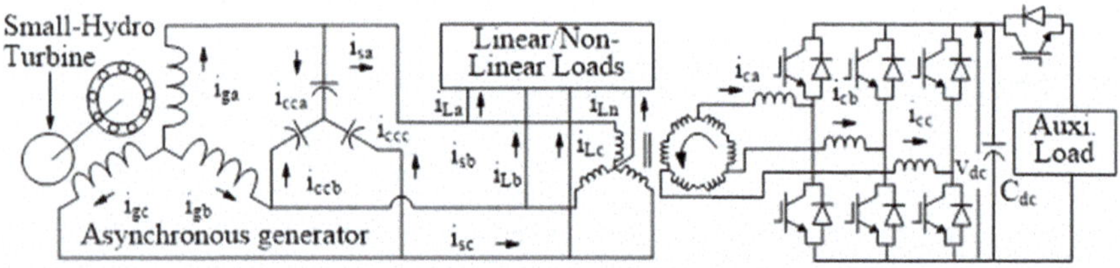

Fig. 4 Star-hexagon transformer-based IVFC

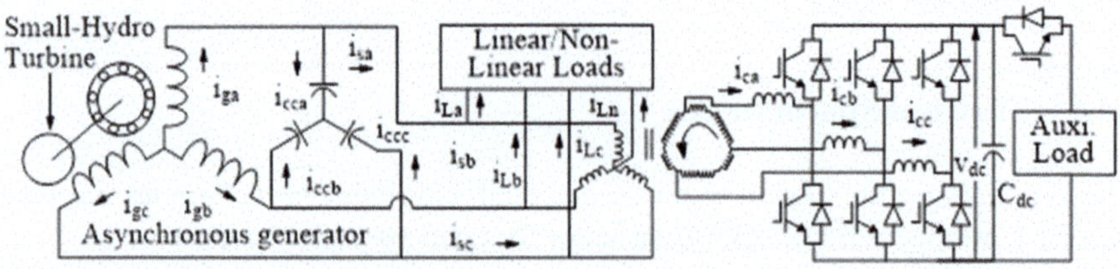

Fig. 5 Star-polygon transformer-based IVFC

Determining Voltage of DC Bus

The smallest value of voltage to be maintained at the VSC DC bus must be superior to double the upper value of the voltage of phase of the IAG system to facilitate successful PWM control. If the modulation index is 'm' and IVFC voltage (output) is given by $V_{LL,}$ the voltage of the DC bus can be obtained using the following equation:

$$V_{dc} = \frac{V_{LL} * 2\sqrt{2}}{m * \sqrt{3}} \tag{1}$$

Normally m = 1 while the calculated value of V_{dc} is 217 V for $V_{LL} = 132.8$ V and chosen value is $V_{dc} = 300$ V.

Design of AC Inductor

The magnitude of the inductor (L_f) on the AC side is dependent on $i_{Lripple\ (p\text{–}p)}$ (ripple current), f_s (frequency of switching), and V_{dc} (voltage of DC bus). This can be obtained from the following equation:

$$\frac{\left(\sqrt{3}\right)V_{dc} * m_a}{\left(12 * i_{Lripple((p-p)} * a_L * f_s\right)} \tag{2}$$

where a_L = overloading factor, $i_{Lripple\ (p\text{–}p)} = 5\%$ of full load current, $m = 1, f_s =$ 10 kHz, $a = 1.2$, $V_{dc} = 300$ V, the L_f (calculated value) is 4.5 mH and it is refined to 4mH for this investigation.

DC Bus Capacitor Design

The C_{dc} (capacitor-DC bus) on VSC is dependent on the amount of energy (instantaneous) during the momentary state and is available to the VSC. This can be deduced as

$$\left(\frac{C_{dc}}{2}\right)\left[V_{dr^2} - V_{dc\,\min^2}\right] = 3V_{ph}(a_L I)t \tag{3}$$

where V_{dr} = DC voltage (reference), V_{dcmin} = DC bus voltage (minimum), a_L (overloading factor), phase voltage (V_{ph}), I (current of phase), and t = time for recovery of DC bus.

Considering, $V_{dr} = 300$ V, and $V_{ph} = 76.7$ V (secondary), 132.8 V (primary) and $V_{dcmin} = 296$ V, $a_L = 1.2$, t = 350 μs, the C_{dc} calculated value is 1304 μF and chosen value as 1650 μF.

2.1.2 Design of Three-Phase Four-Wire IVFC

The VFC employs IGBTs as switches. Under dynamic condition, V_{sw} (voltage rating) of the IGBTsare is obtained as

$$V_{sw} = (V_d + V_{dc}) \tag{4}$$

where V_d is 10% of V_{dc} under load perturbations.

The rating of the switch voltage is as

$V_{SW} = 10\%$ of $300 + 300 = 330$ V.

Let 'I' is the secondary rated line current of 3-leg VSC of the transformer (zigzag-star) with crest value given by $I_{s(peak)} = \sqrt{2}I$. Considering 1.2 as a safety factor, the maximum current that can be allowed in the device be able to be computed using

$$I_{sw} = 1.2\left(i_{Lripple(p-p)} + I_{s(peak)}\right) = 1.2(5\% \text{ of crest value}) = 28.5\text{A} \tag{5}$$

The voltage and current ratings are decided on the voltage of DC bus and generator active power, which is fed into the load. The chopper current rating is given as, $I_{SW} = \frac{P_{gen}}{V_{dc}} = 15$ A (taking safety factor of 1.2); $V_{SW} = 600$ V.

Based on the above calculations, the desired ratings of the IGBT switches can be chosen as 600 V and 30 A.

The auxiliary resistance = R_d is chosen as per the power rating of the IAG and the voltage of the DC link and is given as

$$R_d = \left(\frac{1}{P_{gen}}\right) * (V_{dc})^2 = 24\Omega \tag{6}$$

Various components selected and designed values of 3-leg IVFCs are given in Table 1.

Table 1 Computation of components of 3-leg VSC-based IVFC

S. No	Component	Expression (IVFC)	Computed value	Selected value
1	L_{an}, L_{bn}, L_{cn}	$\frac{(\sqrt{3})V_{dc}*m_a}{(12*i_{Lripple((p-p)}*a_L*f_s)}$	4.5 mH	4 mH
2	V_{dc}	$\geq V_{dc} = \frac{V_{LL}*2\sqrt{2}}{m*\sqrt{3}}$	217 V	300 V
3	C_{dc}	$\left(\frac{C_{dc}}{2}\right)\left[V_{dr^2} - V_{dc\,min^2}\right] = 3V_{ph}(a_L I)t$	1304 μF	1650 μF
4	V_{SW}	$V_{sw} = (V_d + V_{dc})$	330 V	600 V
5	I_{SW}	$I_{sw} = 1.2\left(i_{Lripple(p-p)} + I_{s(peak)}\right)$	28.5A	30A
6	R_d	$R_d = \left(\frac{1}{P_{gen}}\right) * (V_{dc})^2$	24Ω	24Ω
7	I_{ch}	$\frac{V_{dc}}{R_d}$	15A	30A

2.2 Configuration and Operating Principle

System configuration is shown in Fig. 6 for the IAG system for pico-hydro systems. The major components of the proposed IVFC system are IGBT-based chopper, IGBT-based VSC (3-leg), auxiliary load, VSC DC link capacitor. The IVFC is applied for optimizing the voltage of DC bus and compensation of neutral current using a transformer (zigzag-star)connected through the filtering inductors at PCC (point of common coupling). The reactive power requirement for the system is given by a 3-phase bank of capacitors and the input of the system is kept steady with the changing consumer loads. When the consumer loads are smaller than the IAG power, the IVFCtransfers the excessive active power to the auxiliary load, which in turn controls the frequency.

The requirement of reactive power of the system is provided by IVFC which maintains a constant voltage. By this, both the frequency and voltage of IAG remain unaffected by the load consumer perturbations.

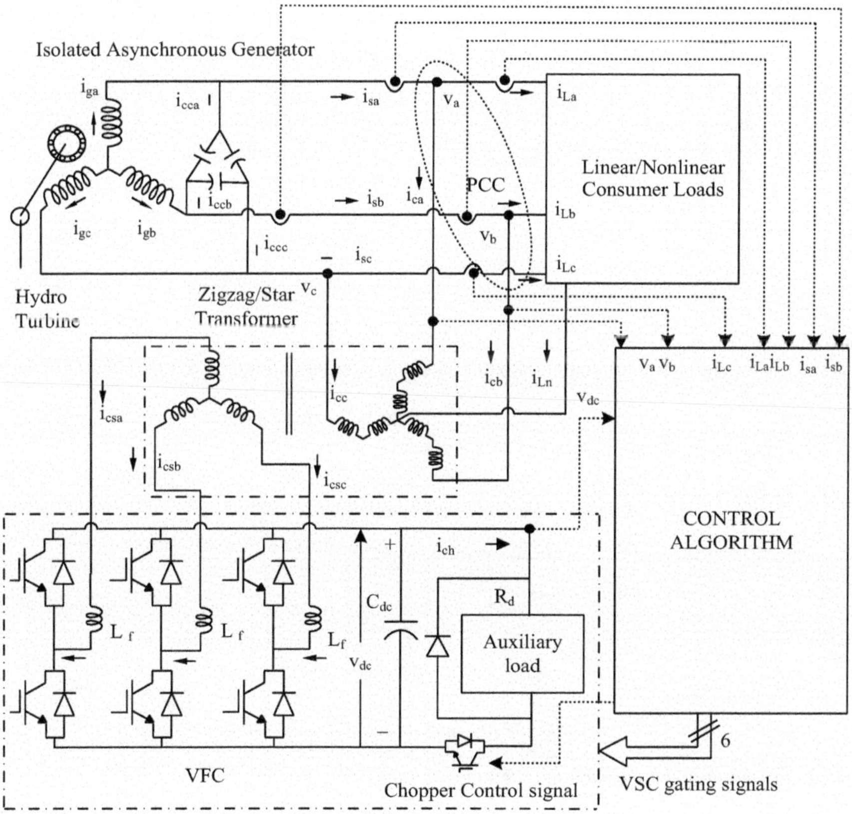

Fig. 6 Schematic diagram IAG with IVFC

2.3 Pi Controller-Based Algorithm

Figure 7 depicts the proposed control scheme for the IVFC to standardize the frequency and voltage of IAG. The governing equations of the control algorithm proposed for maintaining the frequency and voltage constant are specified in the following sections.

2.3.1 In-Phase Source Currents

In-phase currents (reference) of the source are projected from the PI controller of DC bus (I_{smd}). The voltage error of DC bus is as follows:

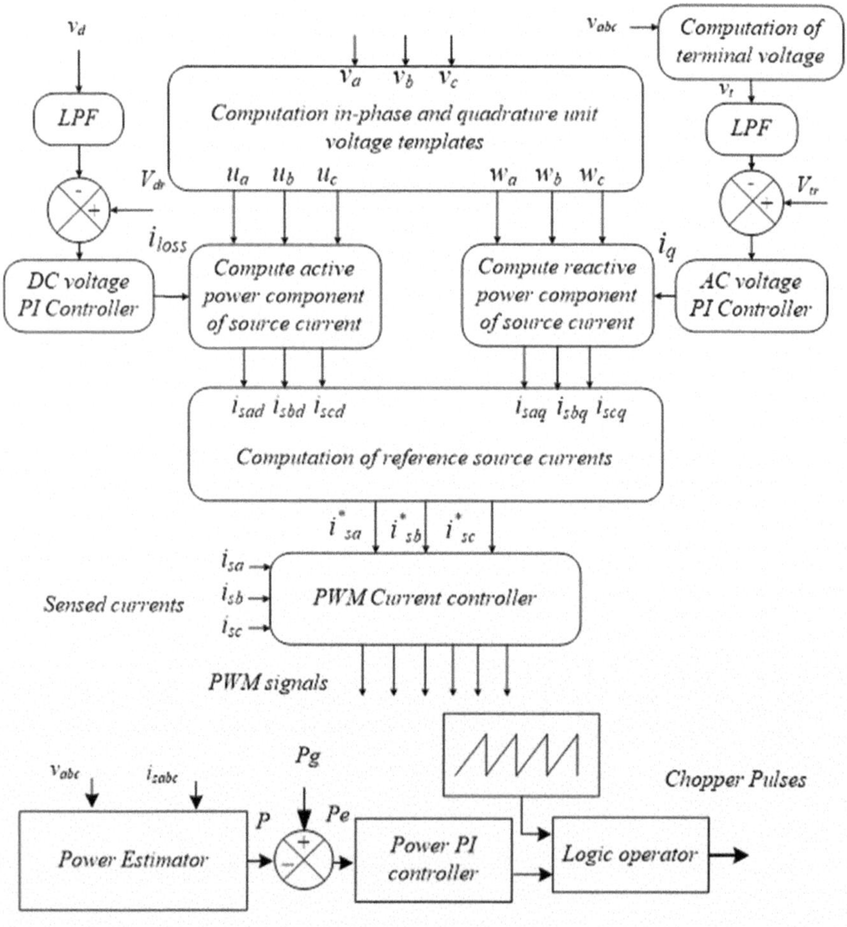

Fig. 7 PI-based control algorithm for IVFC

$$V_{de(n)} = V_{dr(n)} - V_{dc(n)} \quad (7)$$

where $V_{dr\,(n)}$ is the voltage of DC bus (reference) and $V_{dc(n)}$ is the voltage of DC link (sensed) of VSC.

The DC PI controller equation is given by

$$I_{smd(n)} = I_{smd(n-1)} + K_{id} V_{dce(n)} + K_{pd}\left\{V_{de(n)} - V_{de(n-1)}\right\} \quad (8)$$

where source current active component ($I_{smd(n)}$), K_{id} and K_{pd} being the controller gain constants.

where i_{smd} = active source current.

The amplitude of the three-phase sinusoidal voltages at the IAG terminals (v_a, v_b, and v_c) is calculated as

$$V_t = \left\{\frac{2}{3}\left(V_c^2 + V_b^2 + V_a^2\right)\right\}^{\frac{1}{2}} \quad (9)$$

The v_a, v_b, and v_c (unit vectors) for in-phase reference source current components can be calculated using

$$\begin{bmatrix} u_a \\ u_b \\ u_c \end{bmatrix} = \frac{1}{V_t}\begin{bmatrix} v_a \\ v_b \\ v_c \end{bmatrix} \quad (10)$$

$$\begin{bmatrix} i^*_{sad} \\ i^*_{sbd} \\ i^*_{scd} \end{bmatrix} = I_{smd}\begin{bmatrix} u_a \\ u_b \\ u_c \end{bmatrix} \quad (11)$$

2.3.2 Source Quadrature Currents

The error in the voltage (terminal) V_{er} is given as

$$V_{er(n)} = V_{tr(n)} - V_{t(n)} \quad (12)$$

where $V_{tr(n)}$ and $V_{t(n)}$ are the reference and sensed terminal voltage values at nth sampling instant.

The PI controller's output $I_{smq(n)}$ for retaining the terminal voltage at nth sampling instant is given as

$$I_{smq(n)} = K_{id} * V_{er(n)} + K_{pa} * \left\{V_{er(n)} - V_{er(n-1)}\right\} + I_{smq(n-1)} \quad (13)$$

with K_{pa} = proportional gain and K_{ia} = integral gain of PI controller.

$V_{te(n-1)}$ and $V_{er(n)}$ are the errors (voltage) in $(n-1)th$ and nth instant.

$I_{smq(n-1)}$ = current (quadrature component) at $(n-1)th$ instant.

The quadrature components are estimated as

$$\begin{bmatrix} i^*_{saq} \\ i^*_{sbq} \\ i^*_{scq} \end{bmatrix} = I_{smq} \begin{bmatrix} w_a \\ w_b \\ w_c \end{bmatrix} \tag{14}$$

These w_a, w_b, and w_c (unit vectors) may be derived as

$$\begin{bmatrix} w_a \\ w_b \\ w_c \end{bmatrix} = \begin{bmatrix} 0 & -\frac{1}{\sqrt{3}} & \frac{1}{\sqrt{3}} \\ \frac{\sqrt{3}}{2} & \frac{1}{2\sqrt{3}} & -\frac{1}{2\sqrt{3}} \\ -\frac{\sqrt{3}}{2} & \frac{1}{2\sqrt{3}} & -\frac{1}{2\sqrt{3}} \end{bmatrix} \begin{bmatrix} u_a \\ u_b \\ u_c \end{bmatrix} \tag{15}$$

2.3.3 Reference Source Currents

Total reference source currents are calculated as

$$\begin{bmatrix} i^*_{sa} \\ i^*_{sb} \\ i^*_{sc} \end{bmatrix} = \begin{bmatrix} i^*_{sap} \\ i^*_{sbp} \\ i^*_{scp} \end{bmatrix} + \begin{bmatrix} i^*_{saq} \\ i^*_{sbq} \\ i^*_{scq} \end{bmatrix} \tag{16}$$

These sensed currents (i_{sa}, i_{sb}, and i_{sc}) are compared with currents ($i*_{sa}$, $i*_{sb}$, and $i*_{sc}$) and errors are compared and amplified with a triangular wave of 10 kHz frequency and signals are given to switches of IVFC.

2.3.4 Chopper PWM Controller

The power error of the IAG is defined as

$$\mathrm{P}_{\mathrm{e(n)}} = \mathrm{P}_{\mathrm{g(n)}} - \mathrm{P}_{\mathrm{(n)}} \tag{17}$$

where "P" is the sensed load power and "P_g" is the rated generator power.

The instantaneous active value of fundamental power (P) is computed as

$$P = \{(v_a * i_{sa} + v_b * i_{sb} + v_c * i_{sc})\} \tag{18}$$

The PI controller (power) at sampling instant (nth) is as

$$v_{c(n)} = K_{pp}\{P_{e(n)} - P_{e(n-1)}\} + K_{ip}P_{e(n)} + V_{c(n-1)} \tag{19}$$

This PI controller's (power) output $V_{c(n)}$ is compared with a 3.3 kHz triangular carrier wave to produce gating pulses for the IGBT of the chopper of the IVFC.

2.4 *Performance of IVFC at Linear Loads*

Performance in the steady state of the 3-wire 3-phase IVFC under the application of balanced/unbalanced linear loads using simulation in MATLAB is revealed in Fig. 8. A 3.5 kW load is connected to the system. The load is made unbalanced by disconnecting one of the phases at 5.4 s. This results in IVFC diverting to an auxiliary load the excess active power. With the detached load connected back at 5.6 s, the IAG feeds power back to the consumer load. In the duration lasting from 5.4 to 5.6 s, IVFC injects marginally higher currents into the PCC to see balanced source currents. During the unequal conditions, the DC bus voltage is regulated to 400 V. It has been observed currents at the source remain balanced during the consumer load variations while the combined power of auxiliary load and consumer load equals the generated power with the frequency and terminal voltage remaining constant.

Figure 9 illustrates the experiment results in the steady state of the IAG system with 3-phase 4-wire IVFC under the application of balanced linear consumer loads. Figure 9a–c shows the generator currents with terminal voltage.

Figure 9d–f displays the capacitor currents with voltage at the terminals, while Fig. 9g–i displays source currents with terminal voltage and Fig. 9j–l is for 3-phase balanced linear load currents. Figure 9m–o shows the IVFC currents with PCC terminal voltage. Figure 9p–r shows the power consumed by the auxiliary, power of load, and generated power. Figure 9s illustrates the voltage at the DC bus of VSC of IVFC with the chopper current. The generator and source currents have been found to be balanced and harmonic free. The IVFC injects balanced currents at the PCC because the consumer load currents are balanced. The generator is delivering a power of 3.29 kW with a consumer load consuming 2.06 kW and the remaining power of 1.09 kW is diverted to an auxiliary load of IVFC. The voltage at the DC bus of VSC of IVFC is maintained within the limits of 400 V with a chopper current of 4.82A. During this load condition, the harmonic spectra and the terminal voltage THD and generator current THD are both within IEEE-519 std. limits.

Figure 10 depicts the steady-state experiment results of IVFC in IAG system with the removal of single-phase load (linear). Figure 10a–l depicts the terminal voltage in relation to 3-phase generator currents, capacitor currents, source currents, and unbalanced linear load currents. Figure 10m–o displays the IVFC currents with terminal voltage. Figure 10p displays the chopper current with voltage at the DC bus. Figure 10q displays the chopper current with terminal voltage and Fig. 10r displays the generated power of the IAG system. It has been seen that when the "a" phase of the consumer load is detached, the IVFC consumes higher current compared to the previous balanced load condition.

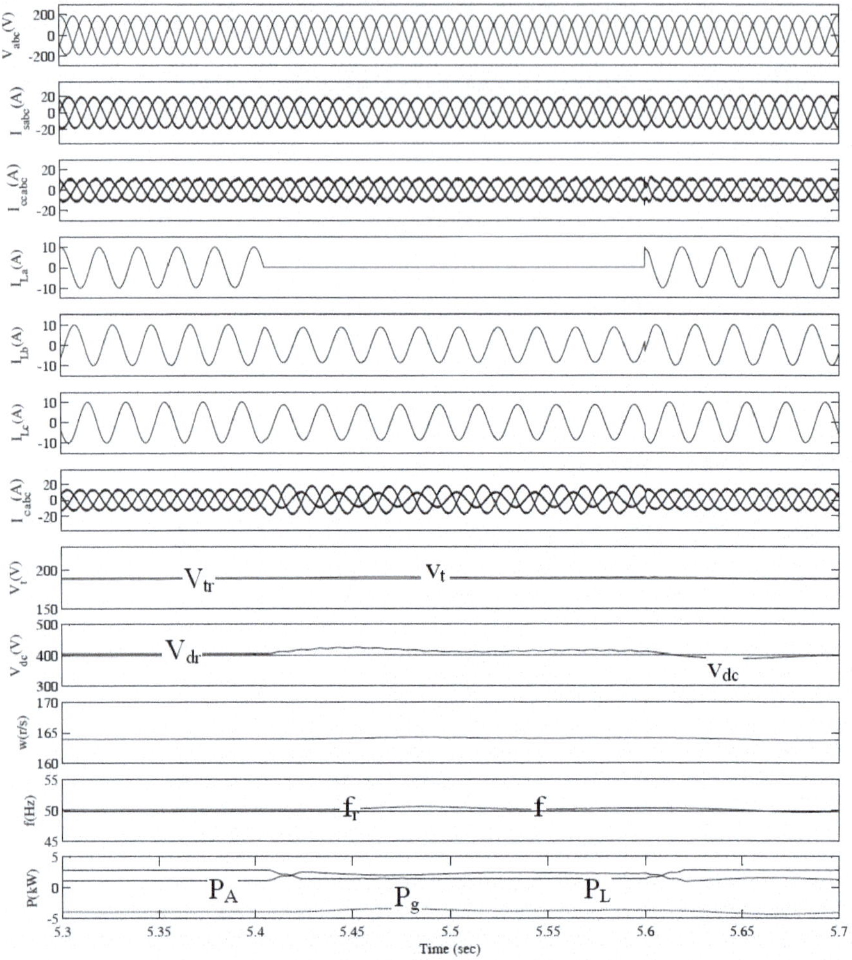

Fig. 8 Performance (Dynamic) with linear loads

The IVFC phase 'a' injects slightly elevated current into the system compared to the other two phases of IVFC currents, thus maintaining a balanced source and generator currents. The voltage at the DC bus is maintained constant within the limits of 400 V. The chopper current has increased from 4.62A shown in Fig. 10s to 6.29A shown in Fig. 10p This shows the increase in consumption of auxiliary load active power.

The delivered power by the IAG system is 3.25 kW, which is nearly the same as the previous balanced condition and thus maintains the frequency constant.

Figure 11 displays the experimental results at unbalanced load conditions. The balanced currents of the generator, currents of source, currents of load, and currents of IVFC are revealed in Fig. 11a–d, correspondingly. Figure 11e displays the IVFC

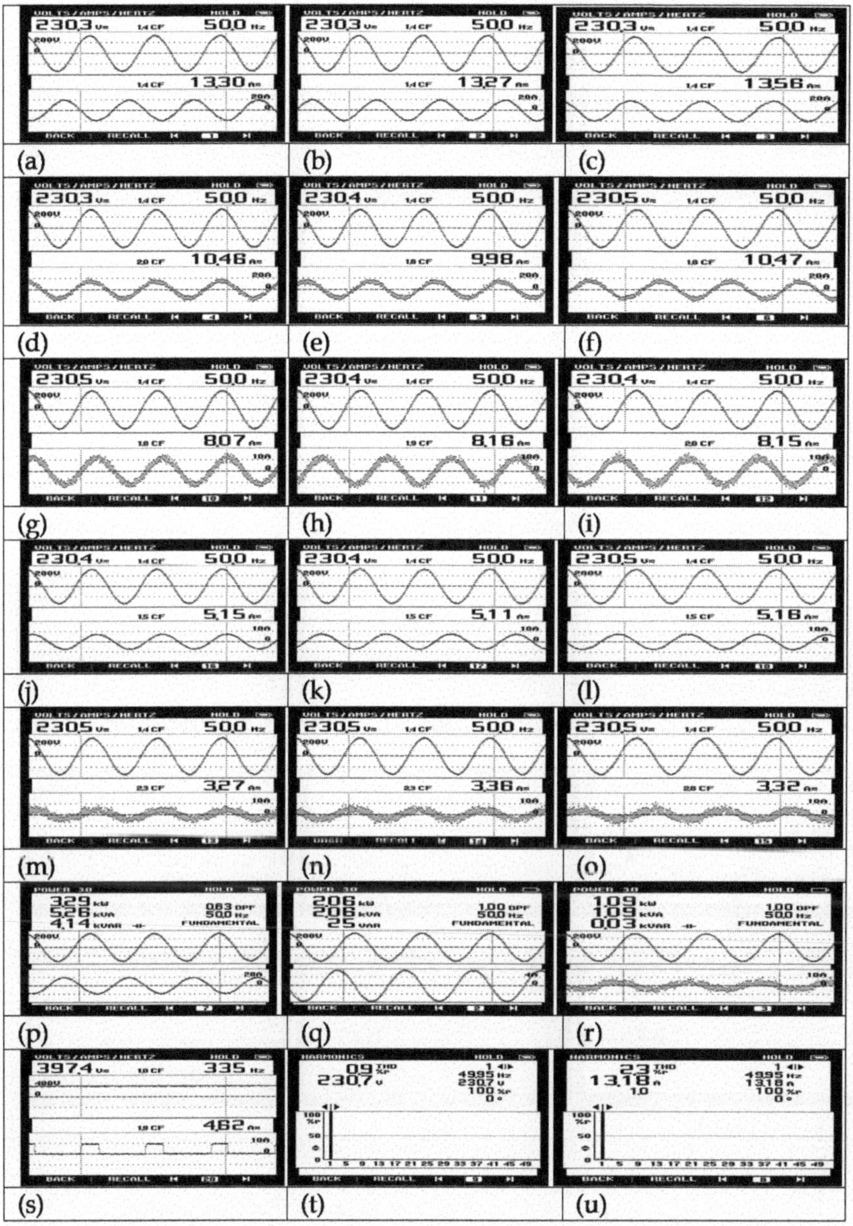

Fig. 9 Performance of induction generator with IVFC, **a** i_{ga}, **b** i_{gb} **c** i_{gc}, **d** i_{cca}, **e** i_{ccb}, **f** i_{ccc}, **g** i_{sa}, **h** i_{sb}, **i** i_{sc}, **j** i_{La}, **k** i_{Lb} **l** i_{Lc}, **m** i_{ca}, **n** i_{cb}, **o** i_{cc}, **p** power of the source (p_g), **q** load power (p_L), **r** auxiliary load power (P_A), **s** v_{dc} with i_{ch}, **t** THD with Harmonic spectrum of (v_{ab}), **u** THD with Harmonic spectrum of i_{ga}

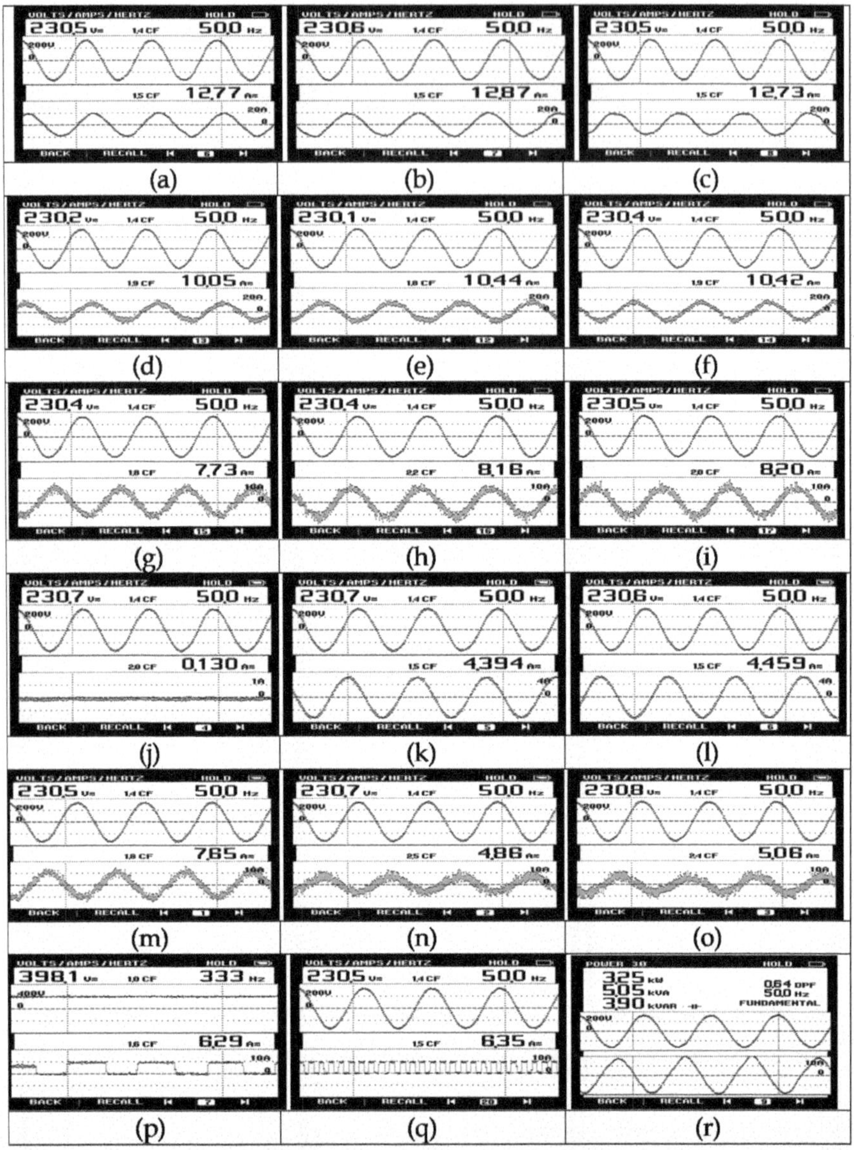

Fig. 10 Performance of IVFC, **a** i_{ga}, **b** i_{gb}, **c** i_{gc}, **d** i_{cca}, **e** i_{ccb}, **f** i_{ccc}, **g** i_{sa}, **h** i_{sb}, **i** i_{sc}, **j** i_{La}, **k** i_{Lb}, **l** i_{Lc}, **m** i_{ca}, **n** i_{cb}, **o** i_{cc}, **p** v_{dc} with i_{ch}, **q** v_{ab} with i_{ch}, **r** power of the source (p_g)

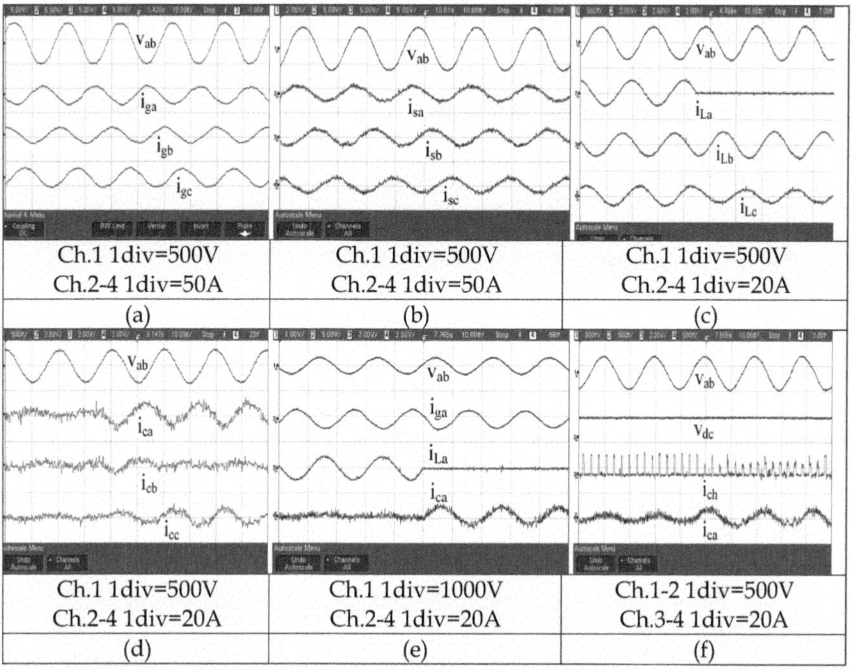

Fig. 11 Experiment results (dynamic) of IAG system with linear load removal

current, load current, generator current, and terminal voltage. Figure 11f displays the IVFC current, DC chopper current, DC bus voltage, and terminal voltage.

It has been seen that the source currents and generator currents are balanced, when the loads of consumer are unbalanced and balanced. When load 'a' phase is removed the IVFC is injecting higher current at PCC to maintain a balanced generator and source currents. The current of the chopper is greater than before when the single-phase load is detached, confirming the consumption of extra active power in the IVFC auxiliary load. The VSC DC bus voltage is held constant during this operation.

Figure 12 depicts the dynamic performance of an IAG system with IVFC (single-phase linear load applied back). Figure 12a–d depicts three-phase balanced currents of the generator, currents of source, currents of load, and currents of IVFC, respectively. The terminal voltage, current of the generator, current of load, and current of IVFC are depicted in Fig. 12e. The terminal voltage, the voltage at DC bus, current of DC chopper, and current of IVFC are shown in Fig. 12f.

It is seen that during applying the single-phase load after removal the generator currents and currents of the source remain constant and free from harmonics. The IVFC current of the chopper reduces and settles to the previous value showing the reduction of auxiliary load power consumption. Hence the frequency and terminal voltage are sustained constantly.

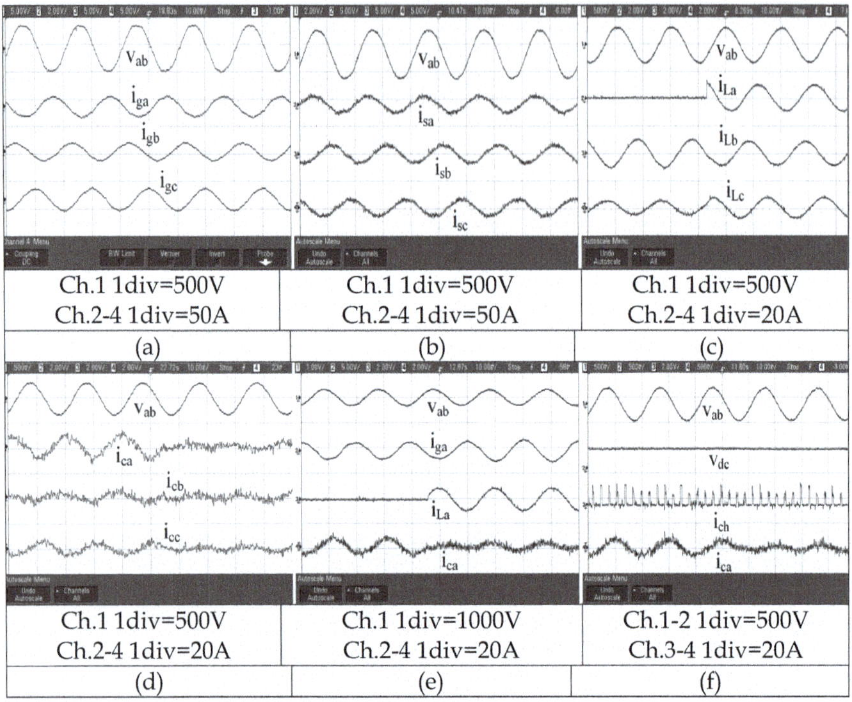

Fig. 12 Dynamic results with single-phase linear load application

2.5 *Performance of IVFC for IAG Under Non-linear Loads*

The effectiveness of a presented IAG scheme with IVFC for frequency and voltage control, as well as power quality enhancement, is investigated when it is feeding non-linear user loads. An IVFC prototype is made up of a three-phase asynchronous generator with a 230 V, 3.7 kW output and an IVFC that is experienced for non-linear loads for frequency and voltage regulation. The simulation and test results are depicted in Figs. 13, 14, 15, 16 and 17.

Figure 13 depicts the performance of an IVFC on the IAG system when unbalanced/balanced non-linear consumer loads are applied using MATLAB/Simulink simulations. A three-phase non-linear load was applied to the IAG at 5.3 s. The IVFC does not use any active power. At 5.4 s, one phase load is activated, and IVFC feeds the excess active power between the generator and consumer load power to the IVFC auxiliary load. At 5.6 s, the removed phase load is applied back and the IVFC consumes less auxiliary load power and load consumes more power. During the non-linear balanced/unbalanced load operation, the currents at the source are harmonic free and balanced. The voltage at the DC bus during this operation is maintained constant with the limits. The summation of auxiliary load and consumer load power

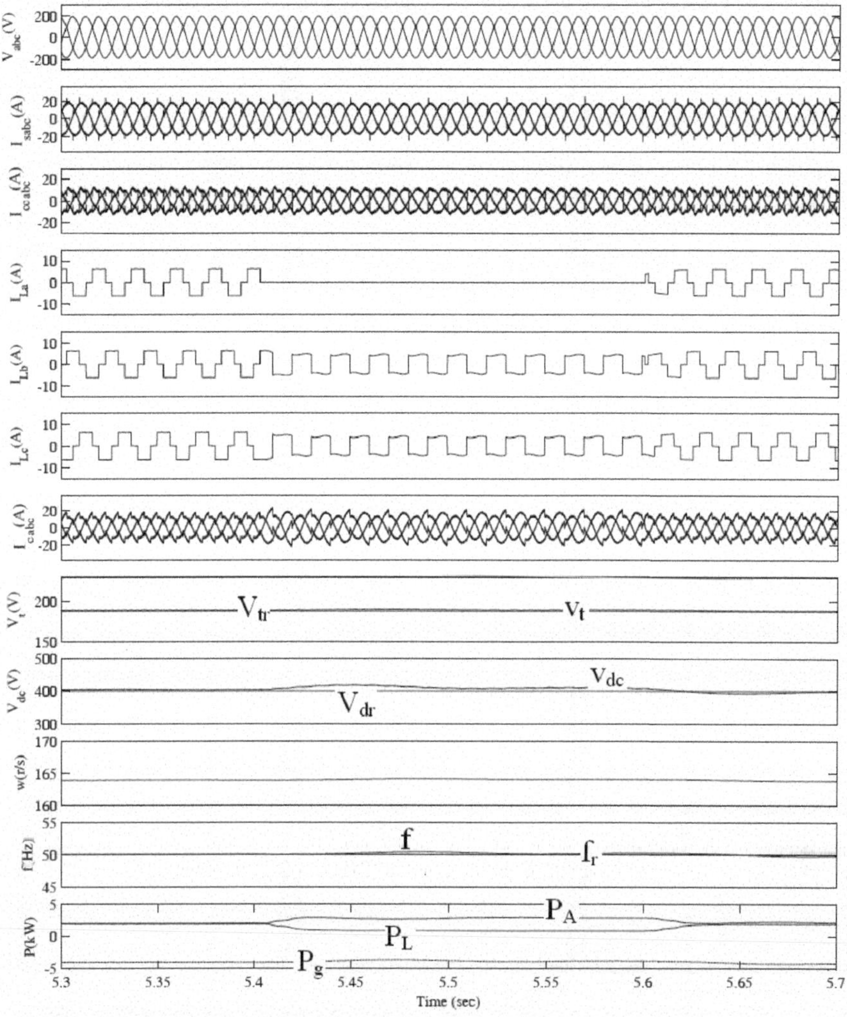

Fig. 13 Dynamic performance feeding non-linear loads

is equivalent to the delivered power by IAG, which is constant. Hence the terminal voltage and frequency are maintained constant.

The experimental results of the IAG system with IVFC feeding balanced non-linear loads are shown in Fig. 14. Figure 14a–c displays the generator currents the terminal voltage. Figure 14d–f displays the capacitor currents with terminal voltage. Figure 14g–i displays the source currents with terminal voltage. Figure 14j–l displays the load currents with terminal voltage. Figure 14m–o displays the IVFC currents with terminal voltage. Figure 14p displays the reactive and active power of the generator. Figure 14q displays the chopper current with terminal voltage. Figure 14r shows the chopper current with voltage at the DC bus.

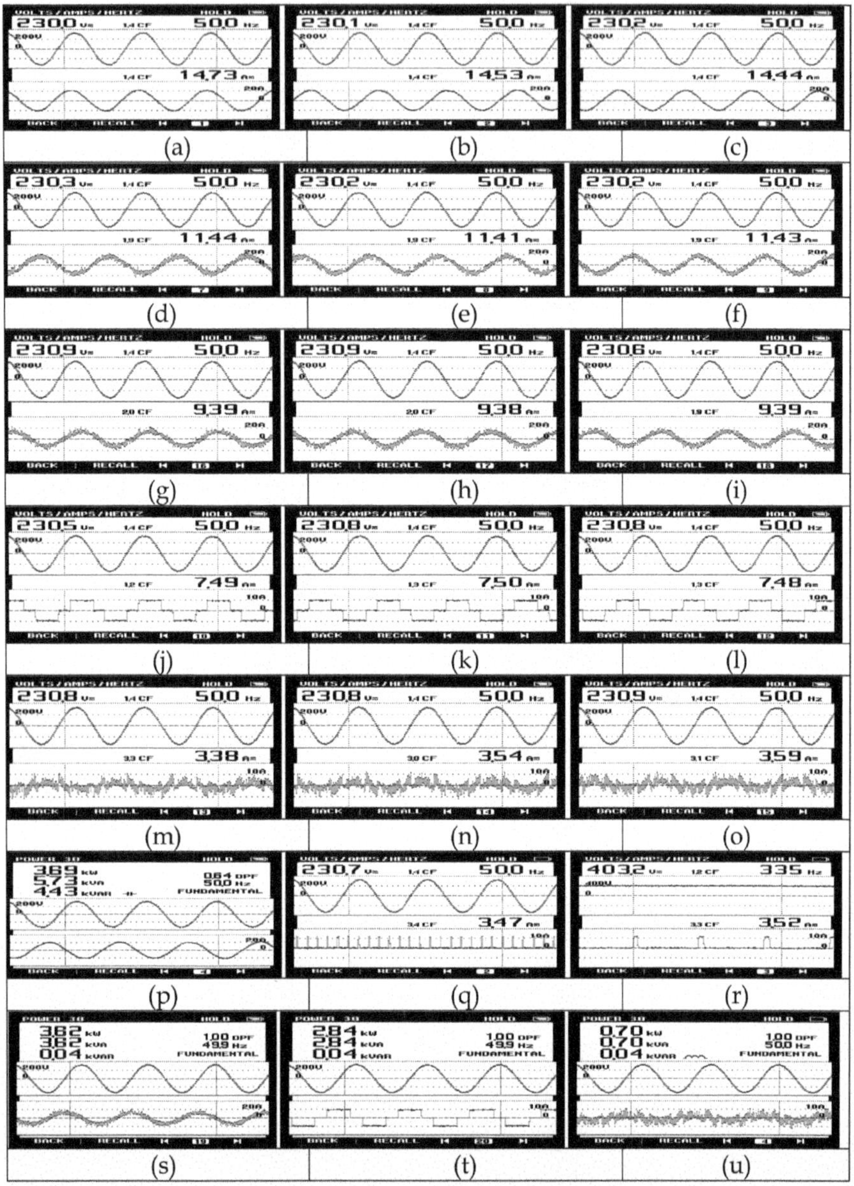

Fig. 14 Performance of induction generator with IVFC, **a** i_{ga}, **b** i_{gb}, **c** i_{gc}, **d** i_{cca}, **e** i_{ccb}, **f** i_{ccc}, **g** i_{sa}, **h** i_{sb}, **i** i_{sc}, **j** i_{La}, **k** i_{Lb}, **l** i_{Lc}, **m** i_{ca}, **n** i_{cb}, **o** i_{cc}, **p** generator power (p_g), **q** v_{ab} with i_{ch}, **r** v_{dc} with i_{ch}, **s** source power (P_s), **t** load power (p_L), **u** auxiliary load power (P_A)

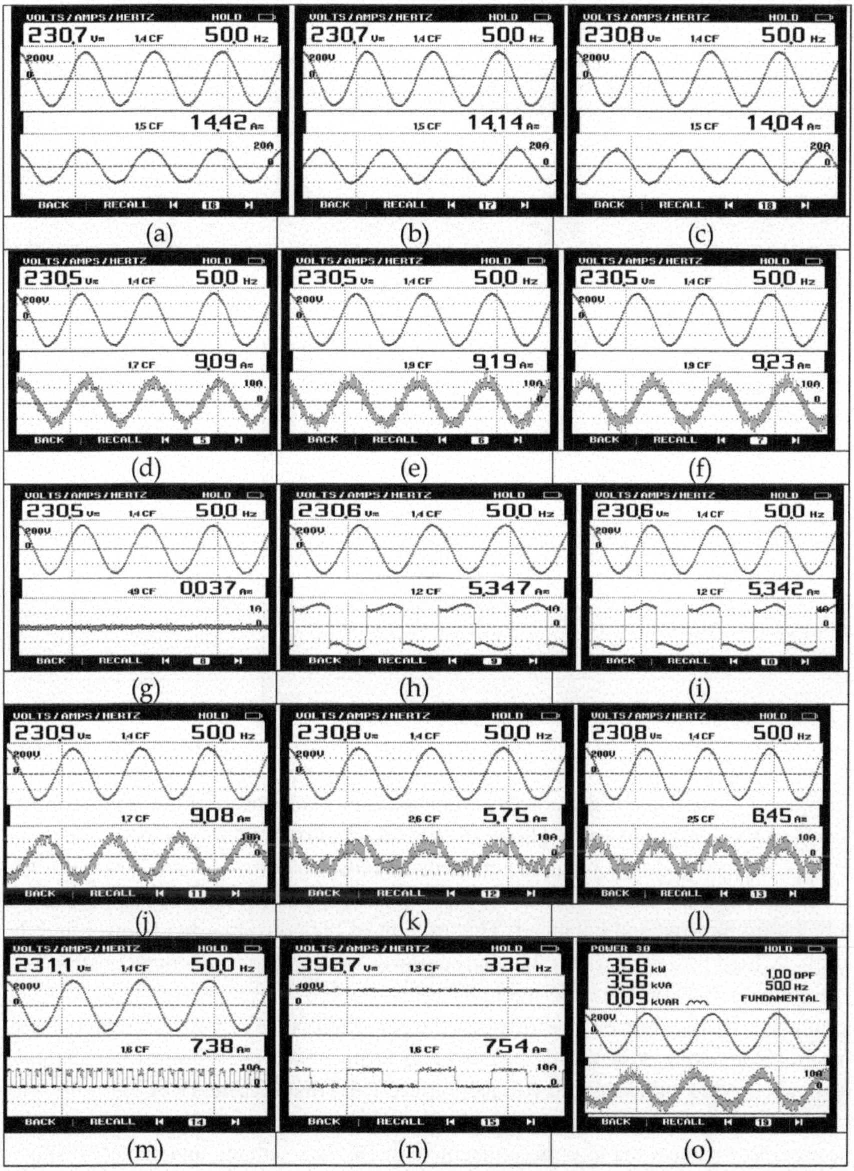

Fig. 15 Performance of induction generator with VFC, **a** i_{ga}, **b** i_{gb}, **c** i_{gc}, **d** i_{sa}, **e** i_{sb}, **f** i_{sc}, **g** i_{La}, **h** i_{Lb}, **i** i_{Lc}, **j** i_{ca}, **k** i_{cb}, **l** i_{cc}, **m** i_{ch}, **n** v_{dc}, with i_{ch}, **s** source power (P_s)

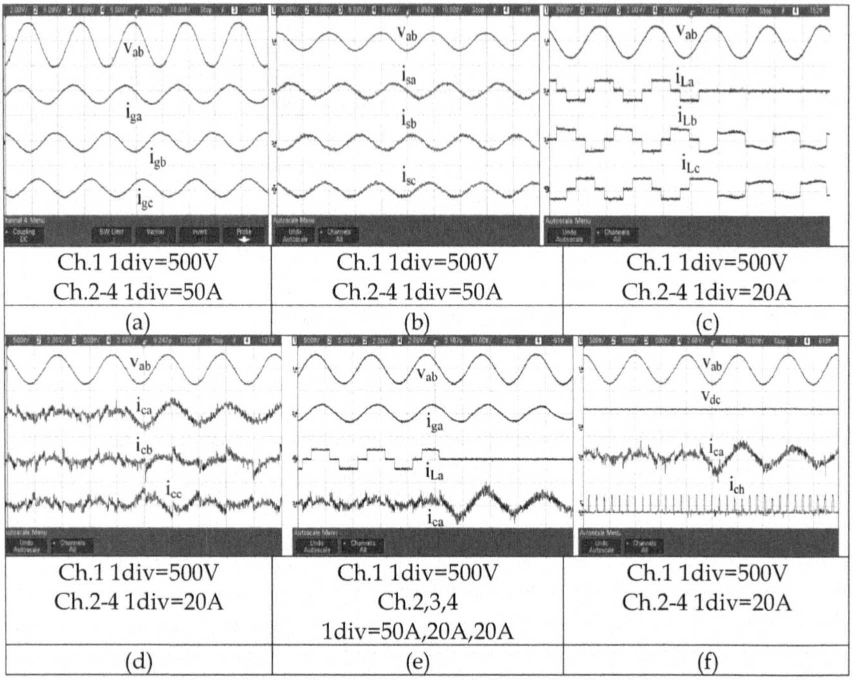

Fig. 16 Dynamic results with removal of single-phase load

Figure 14s–u displays the power of the auxiliary load, power of the generator, and power of the consumer load. The non-linear load currents are balanced; hence the IVFC is injecting almost balanced currents at PCC.

Even when the load currents are highly non-linear, the generator currents are found to be harmonic free and balanced. The amount of the consumer load power and the auxiliary load power provides the source power to keep the frequency constant.

Figure 15 displays the experimental results of the IAG system with IVFC feeding unbalanced loads. Figure 15a–c displays the generator currents with terminal voltage. Figure 15d–f displays the source currents with terminal voltage. Figure 15g–i displays the load currents with terminal voltage. Figure 15j–l displays the IVFC currents with terminal voltage. Figure 15m displays the chopper current with terminal voltage. Figure 15n displays the chopper current with DC bus voltage and Fig. 15o shows the power of the source. It has been seen that during load operation in unbalanced mode, the IVFC is injecting an elevated current compared to the other two phases, and also the IVFC currents are drawing higher currents compared to the previous values during the unbalanced load operation, which is displayed in Fig. 15j–l. The generator is delivering almost the same power as shown in Fig. 15o. The chopper current has increased from the previous value of 3.52A shown in Fig. 14r to 7.54A shown in Fig. 15n depicts the consumption of additional active power into the IVFC auxiliary load in order to keep the frequency constant.

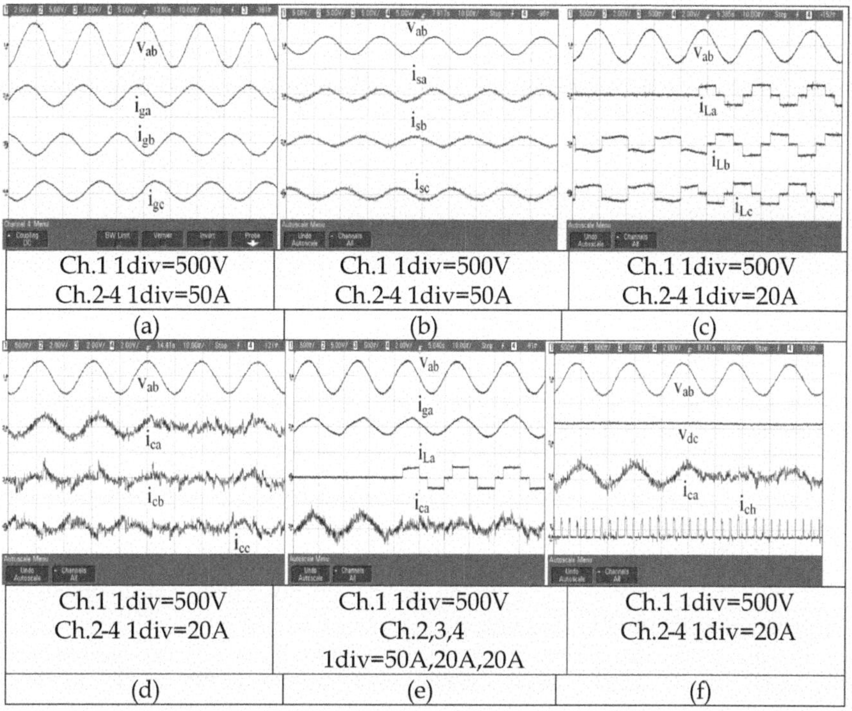

Fig. 17 Dynamic experimental results with removal of single-phase load

Experimental results of the system with IVFC under the removal of single-phase non linear load are displayed in Fig. 16. The currents of the generator, currents of the source, currents of the load, and currents of the IVFC are displayed in Fig. 16a–d in that order. Figure 16e displays the terminal voltage, current of the generator, current of the load, and current of the IVFC. Figure 16f displays the terminal voltage, voltage of the dc bus, current of the IVFC, and current of DC chopper. During the removal of single-phase load, the IVFC is injecting elevated current in 'a' phase to keep up the balanced currents at the generator and source.

The overall IVFC currents and chopper current of IVFC are also increased from their previous values. This demonstrates the consumption of additional active power by the IVFC auxiliary load to maintain the frequency constant.

Experimental results of IVFC for the system under the application of single-phase non-linear load after removal are revealed in Fig. 17. Figure 17a–d depicts currents of the generator, currents of the source, currents of the load, and currents of the IVFC, respectively. Figure 17e illustrates the terminal voltage, currents of the generator, currents of the load, and currents of the IVFC. Figure 17f displays the terminal voltage, voltage at the DC bus, current of the IVFC, and current of the DC chopper.

During the single-phase non-linear load application, the IVFC current, which has been injecting elevated current into the system reduces to its previous balanced non-linear load operation value. The chopper current and IVFC currents are also reduced to maintain the constant active power on the IAG system, thus maintaining the frequency constant. During the load removal and application, the terminal voltage is regulated while improving the power quality.

2.6 *Power Quality Performance of IVFC for IAG*

Figure 18 shows the power quality performance of IVFC-IAG system supplying non-linear loads with MATLAB-based simulations and test results. The harmonic spectra of generator voltage and its waveform are shown in Fig. 18a, g, with THDs of 1.14 percent and 0.9 percent, respectively. Figure 18b, h shows the harmonic spectrum of the generator current and its waveform with a THD of 1.64 and 2.0%, respectively. Figure 18c, i depicts the harmonic spectrum of load current and its waveform, with THDs of 30.77 and 28.3%, respectively. Figure 18d, j depicts the harmonic spectrum of the capacitor current and its waveform, which have a THD of 13.27 and 8.2%, respectively. The harmonic spectrum of the source current and its waveform THD of 3.73 and 5.5% are shown in Fig. 18e, k, respectively. Figure 18f, l depicts the harmonic spectrum of IVFC current and its waveform, which have THDs of 15.14 and 46.6%, respectively. When the load current is highly non-linear and contains lower order harmonics, the generator voltage and current have THD of less than 5%, meeting the IEEE-519 standard. Table 2 compares simulations and test results with IAG feeding non-linear loads by tabulating the THDs of all the voltages and currents in the IAG system.

3 Conclusions

A standalone small-hydro-based three-phase asynchronous generator of 3.7 kW 230 V has been implemented, and its performance has been evaluated using a single-phase p-q theory-based IVFC control algorithm. The unit template control algorithm is used on a zigzag/star-based IVFC for an IAG system powered by a small-scale hydro turbine. The performance of an IAG system under various loading conditions (balanced/unbalanced linear/non-linear) has been demonstrated, as it has the improved power quality of the IAG system's generator currents. The THDs of the terminal voltage and generator current are found to be well within the IEEE-519 standard limit of 5%. The IVFC has been found to perform satisfactorily under a variety of loading conditions, including voltage and frequency control, load balancing, and harmonics elimination of three-phase four-wire loads. The isolated zigzag/star transformer connections have effectively attenuated the neutral current and zero-sequence harmonic currents on the source side, lowering the IVFC's VSC rating.

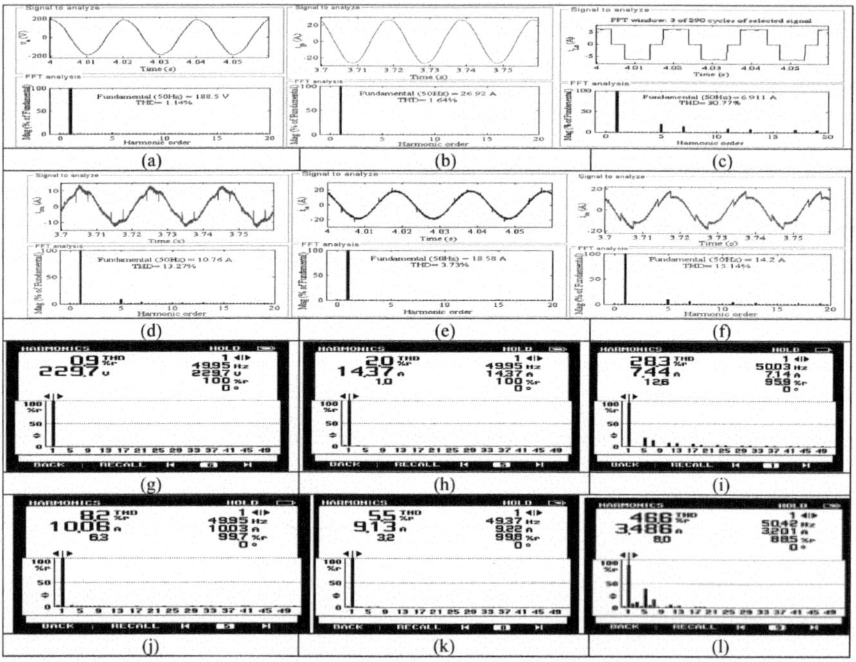

Fig. 18 Power Quality Performance of IVFC for IAG feeding non-linear loads

Table 2 THD summary of voltage and currents

Generator voltage THD	Generator current THD	Load current THD
0.69%	1.39%	0.69%
1.4%	2.6%	1.1%
Capacitor current THD	Source current THD	IVFC current THD
6.36%	2.99%	4.41%
6.6%	6.7%	13.9%

4 Appendices

A. Parameters of Three-Phase Induction MachineY-connected 230 V 3.7 kW, 50 Hz, 4-Pole

$X_m = 24.08114\ \Omega$, $R_s = 0.3939\ \Omega$, $R_r = 0.4791\ \Omega$, $X_{lr} = 0.6335\Omega$, $X_{ls} = 0.7898\ \Omega$,

B. Parameters of VFC Hardware

SKM100GB128DN: IGBT Module 3-leg.

SKM100GB128DN: Diode Module 3-leg.

SKHI22B Driver: Chopper and VSCGate driver.

LEMCT100S: Current Sensors Hall Effect.

LEMCV3-1500: Voltage sensors Hall Effect.

1650 μF/800 V: Capacitance of VSCDC bus.

C. VFC Processor

Processor: TMS320F240 slave DSP and Controller board dSPACE DS1104R&D with.

D. Consumer Loads

(1) Linear Loads: Three-phase resistive loads.

Non-linear Loads:

(2) Diode-bridge three-phase rectifier with dc link R-L load.

Single-phase (03 Nos) diode bridge rectifiers with R-L at its dc link for three-phase four-wire application.

References

1. Simoes M G, Farret FA (2004) Renewable energy systems. CRC Press, Florida
2. Boldea I (2006) The electric generators handbook, variable speed generators. CRC, Taylor & Francis
3. Lai LL, Chan TF (2007) Distributed generation -induction and permanent magnet generators. John Wiley and Sons Ltd, Chichester, UK
4. Gupta S (2004) Analysis and development of load controller and static compensator for self excited induction generator based autonomous generating system. Ph. D. Thesis, Dept of Electrical Engineering, IIT Delhi, India
5. Perumal V (2006) Generalized impedance controller for a self excited induction generator. PhD thesis, Dept of Electrical Engineering, IIT Delhi, India
6. Kasal GK (2009) Analysis, design and development of voltage and frequency controllers for asynchronous generators in isolated power generation. Ph.D. Thesis, Dept of Electrical Engineering, Indian Institute of Technology, Delhi, India
7. Malik NH, Al-Bahrani AH (1990) Influence of terminal capacitor on the performance characteristics of self-excited induction generator. IEE Proc Generat Trans Distrib 137(2):168–173
8. Chan TF (1993) Capacitance requirements of self-excited induction generators. IEEE Trans. Energy Convers EC-8(2):304–311
9. Murthy SS, Singh B (1996) Capacitive VAR controllers for induction generators for autonomous power generation. In Proceedings of international conference on power electronics, drives and energy systems for industrial growth. New Delhi, pp 679–686
10. Wang L, Lee CH (1997) A novel analysis on the performance of an isolated self-excited induction generator. IEEE Trans Energy Convers EC-12(2):109–117
11. Chakraborty C, Bhadra SN, Chattopadhyay AK (1998) Excitation requirements for stand alone three-phase induction generator. IEEE Trans Energy Convers 13(4):358–365

12. Eltamaly AM (2002) New formula to determine the minimum capacitance required for self-excited induction generator. Proc IEEE PESC 1(23–27):106–110
13. Elder JM, Boys JT, Woodward JL (1985) Integral cycle control of stand-alone generators. IEE Proc 132(2):57–66
14. Profumo F, Colombo B, Mocci F (1989) Frequency controller for induction generators in stand-by mini hydro power plants source. In: Proc IEEE Electrical Machines and Drives. pp 256–260
15. Bim E, Szajner J, Burian Y (1989) Voltage compensation of an induction generator with long-shunt connection. IEEE Trans Energy Convers 4(3):526–529
16. Mishra RK, Singh B, Vasantha K (1992) Voltage regulator for an isolated self-excited induction generator. J Electr Power Syst Res 24(1)
17. Sutanto D, Grantham C, Rahman F (1993) A regulated self-excited induction generator for use in a remote area power supply. In: 1993 Sixth international conference on electrical machines and drives (Conf. Publ. No. 376). pp 234–239
18. Murthy SS, Jose R, Singh B (1998) Experience in the development of micro hvdel grid independent power generation scheme using induction generators for Indian condition. In: Proceedings of IEEE TENCON'98. IEEE region 10 international conference on global connectivity in energy, computer, communication and control (Cat. No. 98CH36229). vol. 2, pp 461–465
19. Bonert R, Rajakaruna S (1998) Self-excited induction generator with excellent voltage and frequency control. IEE Proc Gen Trans Distrib 145(1):33–39
20. Singh B, Sridhar L, Jha CS (1999) Improvements in the performance in the self excited generator using series compensator. IEE Proc Gen Trans Distrib 146(6):602–608
21. Murthy SS, Jose R, Singh B, Singh BP (2000) Development of standalone micro hydel power generation system using self excited induction generator (SEIG) and pump operated as turbine. In: Proceedings National Power Systems Conference. Bangalore, India, pp 3–8
22. Singh B, Murthy SS, Gupta S (2003) An improved electronic load controller for self-excited induction generator in micro-Hydel applications. In: IECON'03. 29th annual conference of the IEEE industrial electronics society (IEEE Cat. No. 03CH37468). (vol. 3, pp 2741–2746
23. Chan TF, Nigim KA, Lai LL (2004) Voltage and frequency control of self excited slip ring induction generators. IEEE Trans Energy Convers 19(1):81–87
24. Singh B, Murthy SS, Gupta S (2006) A voltage and frequency controller for self excited induction generators. J Electr Power Comp Syst 34(2):141–157
25. Murthy SS, Singh B, Gupta S, Kulkarni A, Sivarajan R (2006) Water, Water..... Anywhere field experience on a novel pico-hydro system to supply power for remote locations. IEEE Indus Appl Magaz 65–75
26. Chatterjee JK, Perumal BV, Gopu NR (2006) Analysis of operation of a self-excited induction generator with generalized impedance controller. IEEE Trans Energy Convers 22(2):307–315
27. Singh B, Kasal GK (2006) Voltage and frequency controller for isolated asynchronous generators feeding 3-phase 4-wire loads. In: 2006 IEEE international conference on industrial technology. Pp 2773–2778
28. Ramirez JM, Emmanuel TM (2007) An electronic load controller for self-excited induction generators. In IEEE power engineering society general meeting. pp 1–8
29. Ramirez JM, Emmanuel MT (2007) An electronic load controller for the self-excited induction generator. IEEE Trans Energy Convers 22(2):546–548
30. Tamrakar I, Shilpkar LB, Fernandes BG, Nilsen R (2007) Voltage and frequency control of parallel operated synchronous generator and induction generator with STATCOM in micro hydro scheme. IET Gen Transm Distrib 1(5):743–750
31. Kasal GK, Singh B (2008) Decoupled voltage and frequency controller for isolated asynchronous generators feeding three-phase four-wire loads. IEEE Trans Power Delivery 23(2):966–973
32. Singh B, Kasal GK, Chandra A, Kamal-Al-Haddad (2008) ,An independent active and reactive power control of an isolated asynchronous generator in 3-phase 4-wire applications. In: 2008 IEEE power electronics specialists conference. pp 2057–2063

33. Singh B, Kasal GK, Chandra A, Kamal-Al-Haddad (2008) Voltage and frequency controller for an autonomous micro hydro generating system. In: 2008 IEEE power and energy society general meeting-conversion and delivery of electrical energy in the 21st Century. pp 1–9
34. Ion CP, Marinescu C (2010) Control of parallel operating micro hydro power plants. In: Proceedings of the 2010 12th international conference on optimization of electrical and electronic equipment. Basov, pp 1204–1209
35. Chauhan YK, Jain SK, Singh B (2010) A prospective on voltage regulation of self-excited induction generators for industry applications. IEEE Trans Indus Appl 46(2):720–730
36. Bhattacharya S, Divan D (1995) Design and implementation of a hybrid series active filter system. Proc Power Electron Specialists Conf PESC '95 1:189–195
37. Cheng P-T, Bhattacharya S, Divan DM (1996) Hybrid solutions for improving passive filter performance in high power applications. In: Proceedings of the applied power electronics conference, APEC '96. (vol. 2, pp.911–917)
38. Bhattacharya S, Cheng P-T, Divan DM (1997) Hybrid solutions for improving passive filter performance in high power applications. IEEE Trans Industry Applications 33(3):732–747
39. Newman MJ, Zmood DN, Holmes DG (2002) Stationary frame harmonic reference generation for active filter systems. IEEE Trans Indus Appl 38(6):1591–1599
40. Da Silva CH, Pereira RR, Da Silva LEB, Lambert-Torres G, Bose BK, Ahn SU (2010) A digital PLL scheme for three-phase system using modified synchronous reference frame. IEEE Trans Indus Electron 57(11):3814–3821
41. Mishra MK, Karthikeyan K (2005) A novel three-phase active power filter control algorithm with unbalanced and distorted supply voltages. Proc TENCON Conf 2005(21–24):1–7
42. Rao UK, Mishra MK, Ghosh A (2008) Control strategies for load compensation using instantaneous symmetrical component theory under different supply voltages. IEEE Trans Power Delivery 23(4):2310–2317
43. Lohia P, Mishra MK, Karthikeya K, Vasudevan K (2008) A minimally switched control algorithm for three-phase four-leg VSI topology to compensate unbalanced and nonlinear load. IEEE Trans Power Electronics 23(4):1935–1944
44. Mishra MK, Ghosh A, Joshi A, Suryawanshi HM (2007) A novel method of load compensation under unbalanced and distorted voltages. IEEE Trans Power Delivery 22(1):288–295
45. Akagi H, Kanazawa Y, Nabae A (1984) Instantaneous reactive power compensators comprising switching devices without energy storage components. IEEE Trans Indus Appl IA-20(3):625–630
46. Fujita H, Akagi H (1998) The unified power quality conditioner: the integration of series and shunt-active filters. IEEE Trans Power Electron 13(2):315–322
47. Akagi H, Ogasawara S, Kim H (1999) The theory of instantaneous power in three-phase four-wire systems: a comprehensive approach. In: Conference Record of the 1999 IEEE Industry Applications Conference. Thirty-Forth IAS Annual Meeting (Cat. No. 99CH36370). (vol.1, pp. 431–439)
48. Akagi H (2005) Active harmonic filters. Proc IEEE 93(12):2128–2141
49. Akagi H, Watanabe EH, Aredes M (2007) Instantaneous power theory and applications to power conditioning. John Wiley & Sons, New Jersey
50. Bhuvaneswari G, Nair MG (2008) Design, simulation, and analog circuit implementation of a three-phase shunt active filter using the icosφ algorithm. IEEE Trans Power Delivery 23(2):1222–1235
51. Jung Y-G, Kim W-Y, Lim Y-C, Yang S-H, Harashima F (2003) The algorithm of expanded current synchronous detection for active power filters considering three-phase unbalanced power system. IEEE Trans Indus Electron 50(5):1000–1006
52. Chen CL, Lin CE, Huang CL (1994) An active filter for unbalanced three-phase system using synchronous detection method. In: Proceedings of 1994 Power Electronics Specialist Conference-PESC'94. (vol. 2, pp 1451–1455)
53. Syed MK, Sanker Ram BV (2008) Active power filtering in asymmetric power system by modified synchronous detection algorithm. In : Proceedings of the TENCON 2008-2008 IEEE Region 10 Conference. pp 1–4

54. Singh BN, Singh B, Chandra A, Al-Haddad K (2005) Design and digital implementation of active filter with power balance theory. IEE Proc Electric Power Appl 152(5):1149–1160
55. Haque MT (2002) Single-phase PQ theory. In: Proceedings of the IEEE Power Electronics Specialists Conference. (vol. 4, pp 1815–1820)
56. Singh B, Solanki J, Verma V (2005) Neural network based control of reduced rating DSTATCOM. Proc IEEE INDICON 2005(11–13):516–520
57. Singh B, Jayaprakash P, Kothari DP (2008) Three-phase four-wire DSTATCOM with H-bridge VSC and star/delta transformer for power quality improvement. In: *Proceedings of the 2008 annual IEEE India conference*. (vol. 2, pp 412–417)
58. Khera PP (1990) Application of zigzag transformers for reducing harmonics in the neutral conductor of low voltage distribution system. In: Conference record of the 1990 IEEE industry applications society annual meeting. vol. 2, p 1092
59. Jou H-L, Wu J-C, Wu K-D, Chiang W-J, Chen Y-H (2005) Analysis of zig-zag transformer applying in the three-phase four-wire distribution power system. IEEE Trans Power Delivery 20(2):1168–1173
60. Cogbill BA, Hetrick JA (1968) Analysis of T-T connections of two single-phase transformers. IEEE Trans Power Appar Syst PAS-87(2):388–394
61. Singh B, Jayaprakash P, Kothari DP (2008) Star/hexagon transformer based three-phase four-wire DSTATCOM for power quality improvement. Int J Emerg Electr Power Syst 9(6):1–18
62. Rajagopal V (2012) Design and development of voltage and frequencycontrollers for asynchronous generators in isolated smallhydro power generation. PhD Thesis, Supervised by Bhim Singh
63. Chilipi R (2014) Design and development of VFD based voltage and frequency controllers for a SEIG in micro-hydro power generation. PhD Thesis, Supervised by Bhim Singh and SS Murthy
64. Singh B, Chandra A, Al-Haddad K (2015) Power quality problems and mitigation techniques. John Wiley & Sons Ltd, United Kingdom
65. Mahendra Reddy K, Singh B (2018) Multi-objective control algorithm for small hydro and SPV generation-based dual mode reconfigurable system. IEEE Trans Smart Grid 9(5):4942–4952
66. Kalla UK, Singh B, Murthy SS, Jain C, Kant K (2018) Adaptive sliding mode control of standalone single-phase microgrid using hydro, wind and solar PV array-based generation. IEEE Trans Smart Grid 9(6):6806–6814
67. Pathak G, Singh B, Panigrahi BK (2019) Wind–hydro microgrid and its control for rural energy system. IEEE Trans Indus Appl 55(3):3037–3045
68. Sharma R, Singh B (2021) A resilient FLL control for solar-hydro generation based utility interactive microgrid with enhanced power quality. IEEE Trans Indus Appl 57(6):5716–5725
69. Tyagi S, Singh B, Das S (2022) ELD-OSG control of a battery-based electronic load controller for a small hydro energy conversion system. IEEE Trans Indus Appl 58(3):3142–3152

Chapter 8
Control of Hybrid Off-Microgrid Configurations Based on Renewable and Non-renewable Energy Sources

Miloud Rezkallah, Hussein Ibrahim, Ambrish Chandra, Fadoul, Abdelhamid Hamadi, and Abdeldjebar Hazzab

Abstract This chapter examines the control and design of hybrid off-grid configurations using renewable and non-renewable energy sources (R&NR-ESs). Details relating to power quality, voltage, frequency regulation, and synchronization between off-grid configuration elements are provided. Analyzing and testing a hybrid off-grid configuration based on a solar PV array (SPA) and a variable speed doubly fed induction generator (DFIG) is done to better understand this technology. A rotor side converter (RSC) is designed with improved vector control to vary the speed of the diesel engine accordingly to load power demand, generated power from SPA, and battery charge level. To achieve maximum power point tracking (MPPT) from SPA, an improved Perturb and Observe algorithm (IP&O) is used. Improved oriented stator flux control is designed to maintain constant the PCC voltage and frequency, as well as improve the power quality (PQ) at the point of common coupling (PCC) for load-side converters (LSC). An algorithm based on a proportional–integral controller that prevents the BES from overcharging is employed. Simulations are conducted using MATLAB/Simulink to evaluate the performance of selected hybrid off-grid configurations and their control strategies.

M. Rezkallah (✉) · H. Ibrahim
CR2Ie, Sept-Îles, QC, Canada
e-mail: miloud.rezkallah@itmi.ca; miloud.rezkallah@etsmtl.ca

M. Rezkallah · A. Chandra · Fadoul
École de Technologie supérieure (ETS), Montréal, Canada
e-mail: ambrish.chandra@etsmtl.ca

Fadoul
e-mail: souleyman.fadoul.1@ens.etsmtl.ca

A. Hamadi
GSE Solution, Montréal, Canada
e-mail: abdelhamid.hamadi@gsesolutions.ca

A. Hazzab
University of Bechar, Algeria and ETS, Béchar, Algeria
e-mail: hazzab.abdeldjebar@etsmtl.ca; hazzab.abdeldjebar@univ-bechar.dz

A. K. Giri et al. (eds.), *Power Quality: Infrastructures and Control*,
Studies in Infrastructure and Control,
https://doi.org/10.1007/978-981-19-7956-9_8

Keywords Off-microgrid configuration · Renewable and non-renewable energy sources · Variable speed diesel generator · DFIG · Control system · Power quality improvement

1 Introduction

In several remote areas, especially for those that rely exclusively on diesel generators for their electricity needs, hybrid off-microgrid systems consisting of renewable energy sources (RESs) such as solar photovoltaic arrays (SPAs), wind turbines (WTs), micro-hydro-powers (MHPs), etc., have become a cost-effective and sustainable alternative energy source. Generally, MHPs, WTs, or DGs use machines that generate electricity. Such machines include squirrel cage induction machines (SCIMs), doubly fed induction machines (DFIMs), synchronous machines (SMs), etc. Many hybrid off-grid configurations employ less power converters based on R&NR-ESs that use different electrical machines [1, 2]. All off-grid configurations proposed in [1, 2] involve DGs which serve as emergency ESs, saving fuel, extending their lifespan, and reducing carbon emissions.

For remote areas, NR-ESs are considered reliable ESs and should be present in the installation to ensure an uninterrupted power supply when RESs are not present. To replace a conventional fixed-speed diesel generator, an off-grid variable-speed diesel generator (VSDG) based on DFIM is suggested in [3]. In [4], a permanent magnet synchronous generator (PMSG) is utilized that is based on VSDGs off-grid. To compensate for the slow dynamic of the VSDG, a supercapacitor is added to the off-grid configuration in [4]. Under 11 kW load conditions, the obtained results in [4] show that fuel costs can be reduced by up to 25%. Based on the same concept, [5] proposes an application for remote areas, and the obtained results show satisfactory fuel-saving. A proposal in [6, 7] suggests that the use of R-ESs in remote communities in the Canadian Arctic could be integrated with VSDG. A detailed techno-economic feasibility study of the integration of R-ESs and NR-ESs is presented in [6], which demonstrates substantial cost savings and a reduction in CO_2 emissions. Therefore, hybridizing R-ESs with NR-ESs in off-grid systems can increase the reliability of the DG, ensuring clean energy to the connected loads and reducing costs. In off-grid systems, R-ESs and NR-ESs are integrated into three groups: AC, DC, and hybrid AC/DC off-grid configurations. As shown in Fig. 1a, the R&NR-ESs are integrated into the AC bus through their power converters. As shown in Fig. 1b, the DC off-grid configuration uses the same concept, where all R&NR-ESs are connected to the DC bus through their power converters [7]. An example of a hybrid AC/DC off-grid configuration is shown in Fig. 1c. R&NR-ESs are integrated with the DC bus through their power converters, while AC R&NR-ESs are integrated with the AC bus through their power converters. To connect the AC and DC buses, an interfacing power converter is used. Considering the comparison study presented in [8], the hybrid AC/DC off-grid configuration offers the advantages of both DC and AC. The hybrid AC/DC off-grid configuration is employed in [9] to integrate R&NR-ESs. A

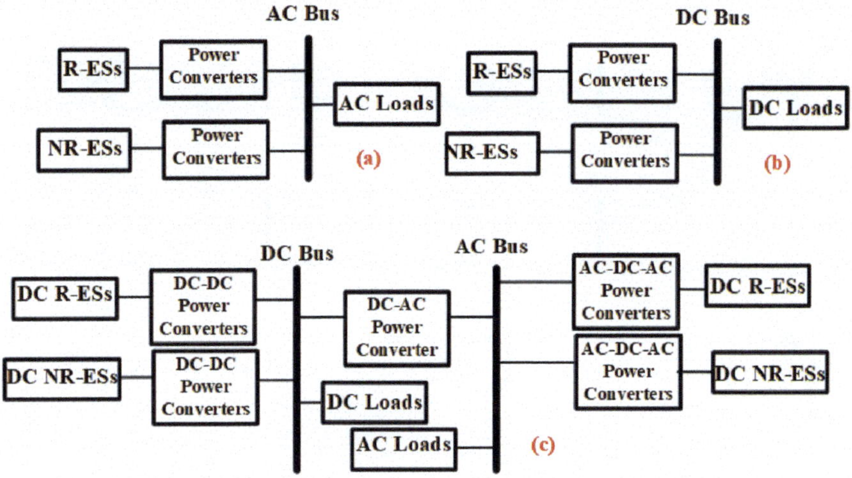

Fig. 1 **a** AC off-grid configuration, **b** DC Off-grid configuration, and **c** AC–DC hybrid off-grid configuration

common interfacing inverter connects the variable-speed wind turbine (VSWT) and the variable-speed generator (VSDG). As a result of simulations and experiments, fuel consumption can be reduced by up to 40%. Compared to PMSM, DFIMs can be used instead of PMSMs in [5], thus the rotors can be connected to the common DC bus using a less rating power converter since the stator of each DFIG is connected to the PCC without using power converters. An off-grid hybrid system that uses VSDG with DFIG of 2 MW and a 150 kWh battery system is proposed in [10]. As a result of the obtained results, it is found to be a more efficient method by saving 26.42% of fuel in comparison with existing fixed-speed generators. In [9, 10], the VSDG the stators of the DFIGs are always connected to the system. This can cause active and reactive power losses.

2 Hybrid Off-Grid Configurations Based on R&NR-ESs

Figure 2a–d shows four proposed hybrid off-grid configurations based on NR-ESs as DGs and batteries (BES) as well as R-ESs as WT and SPAs. As an electrical machine, DFIM is used for R&NR-ESs as WT and DE. As compared to existing hybrid off-grid configurations in the literature [1, 9, 10], the proposed configurations presented in Fig. 2 employ fewer power converters. The stator and rotor of the DFIG are disconnected when the VSDG is turned off. In Fig. 2a, the DFIGs for the WT and DG are connected to the DC bus through AC–DC converters, and the BES is connected to the DC bus through buck–boost converters.

A DC-AC interfacing inverter connects the R&NRE-ESs to the AC bus. The DFIG stator for the WT side is directly connected to the PCC, while the DFIG stator for the

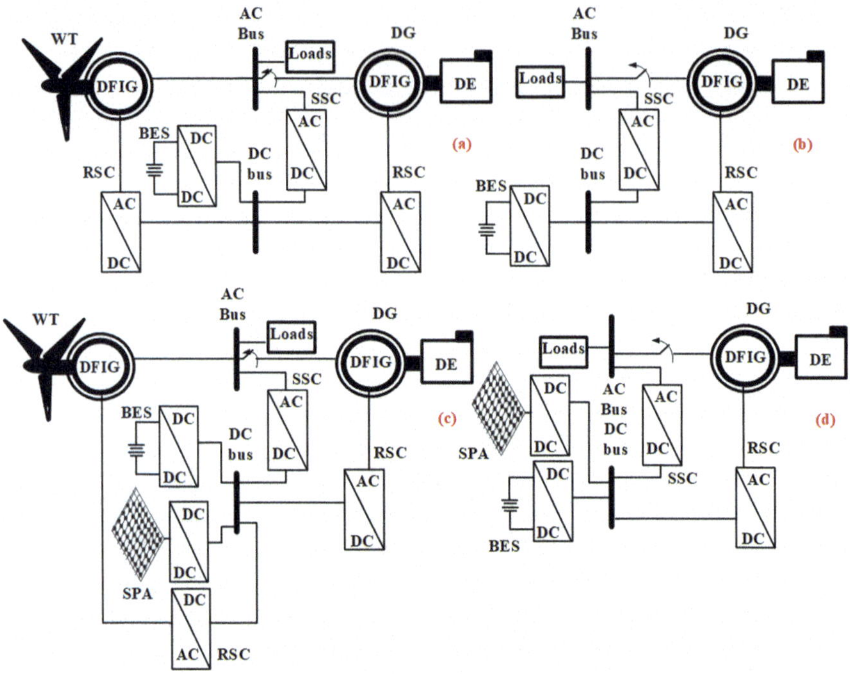

Fig. 2 Hybrid off-grid configurations; **a** WT-DG-BES, **b** DG-BES, **c** WT-DG-BES-SPA, and **d** DG-SPA-BES

DG side is connected to the PCC through a controlled switch. All DGs in the proposed hybrid off-grid configurations adhere to this concept. An off-grid configuration also contains a DC-type RES, which is coupled to the DC bus through a boost converter in Fig. 2c. Concerning the off-grid configurations shown in Figs. 2b and d, the R&NR-ESs are connected to the common DC bus employing their DC–DC power converters and to the PCC employing a DC–AC interfacing inverter. The AC load is connected to the PCC in all hybrid off-grid configurations shown in Fig. 2a–d.

3 Hierarchical Control and Operating of Hybrid Off-Grid Configurations Based on R&NR-ESs

The hierarchical control (HC) of an off-grid system, as shown in Fig. 3, is divided into three levels [11, 12]. Primary HC levels are generally responsible for cooperating with existing power converters.

As part of the hierarchical layer, control is sent to the power electronics converters at intervals of varying milliseconds to (1) regulate voltage and frequency, (2) share power between different R&NR-ESs, and (3) deal with the PQ issue at the PCC via the hierarchical layer. By sending modified power references to R&NR-ESs, the

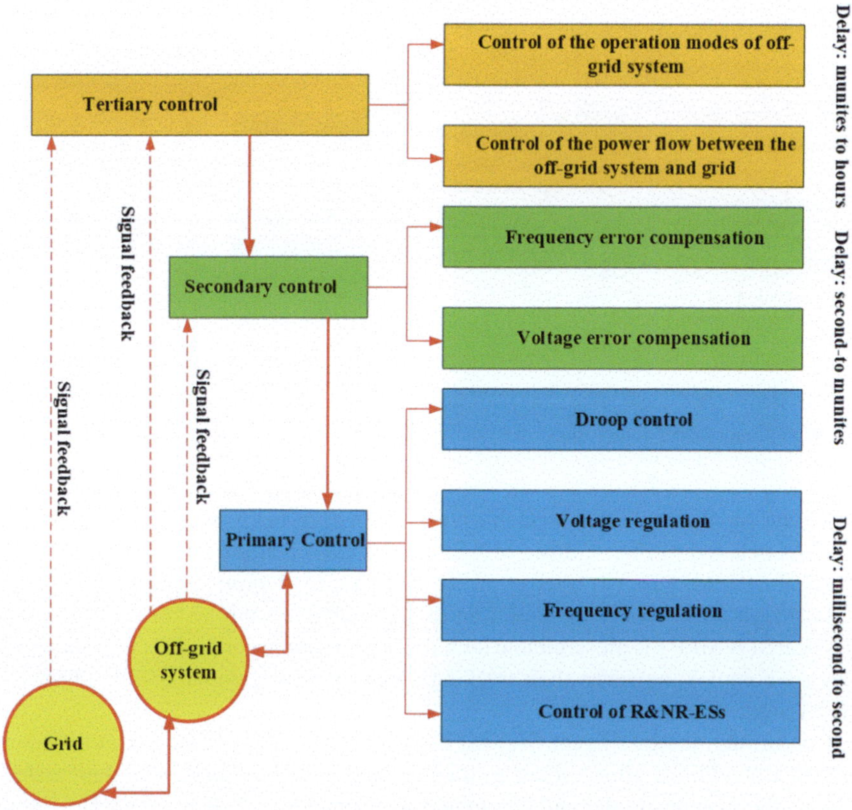

Fig. 3 HC for an off-grid system based on R&NR-ESs

second level of control corrects the voltage and frequency deviations to maintain stable operation of the off-grid. As the final control in the HC, the tertiary control is the highest and slowest. A central function of this part of HC is to define the optimal power references for each R&NR-ES, and from which the off-grid system can be connected to the grid for a balance between local energy consumption and production. This level of control as well as the information and communication are not considered in this book chapter. the proposed off-grid configurations as detailed in Fig. 2, are based on RESs as WT and SPA and N-Res as GD and BES, and the AC ESs employed DFIM as a generator. In all off-grid configurations loads are supplied through the interfacing DC–AC inverter and the stators of the DFIGs. Generally, the WT, DG, and SPA are controlled using the appropriate control algorithms to get desired performance during weather and load variation. As detailed in Fig. 3, these algorithms should be localized in the first level of the HC. Many control algorithms are suggested in the literature to operate optimally the R&NREs and always maintain the voltage as well as the system constant.

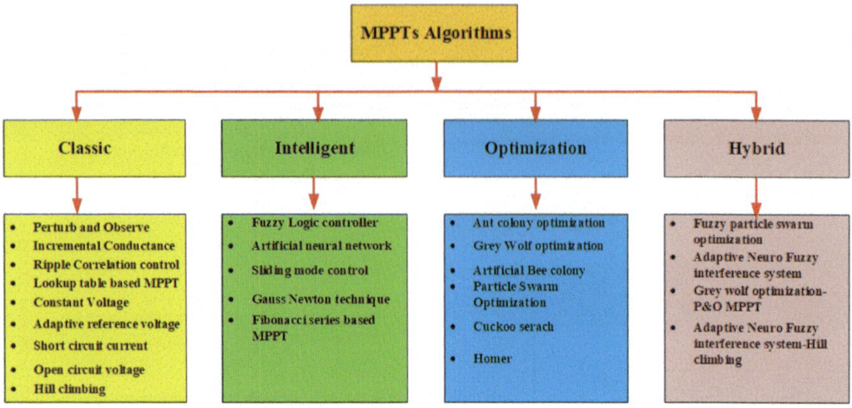

Fig. 4 Different MPPTs algorithms for SPAs

Regarding, the maximum available power from SPA, many MPPTs are employed [13]. As detailed in Fig. 4, these MPPTs can be classified into four categories: (1) classical SPA MPPTs, (2) intelligent SPA MPPTs, (3) Optimization SPA MPPTs, and (4) Hybrid SPA MPPTs. Based on the comparative study detailed in [13], the classical SPA MPPT is most reliable for uniform irradiation conditions. Furthermore, they are complex; however, their tracking speeds are low compared with advanced MPPT algorithms.

Concerning, the Intelligent SPA MPPTs algorithms, their reliability at different radiation conditions allows them to track, sense, and store enormous data faster, making the system simpler and less reliant on mathematical calculations. However, the optimization SPA MPPTs techniques are suitable for any system regardless of the PV panel parameters. Regarding the hybrid SPA MPPTs algorithms, they are highly efficient in tracking and with limited calculation. They are faster and more accurate.

As demonstrated in Fig. 5, the MPPTs algorithms for WTs are divided into four groups; (1) indirect power control (IPC)-based MPPT, (2) direct power control-based MPPT (DPC), (3) hybrid MPPT algorithms, and (4) smart MPPT algorithms. Based on the comparative study detailed in [14], there is a considerable difference in speed between TSR, OT, and PSF algorithm responses, but the latter algorithms are highly reliable and require very little implementation time. Based on the results of this study, it appears that MPPT algorithms based on adaptive and soft computing are reliable and flexible in their performance.

As detailed in Fig. 1a–d, the NR-ESs are the VSDG and BES. Various control algorithms are suggested in [15–20] to operate these two ESs. The BES is linked to the DC bus through their buck–boost converters. These devices are controlled using the appropriate control algorithms to maintain stable power in the off-grid configurations by charging and discharging the BES. To accomplish these objectives, double-loop control with a classical proportional–integral controller is suggested in [18]. In [21], the double-loop control for the buck–boost converter is reinforced by a voltage synchronization loop and virtual inertia control loop to achieve many tasks

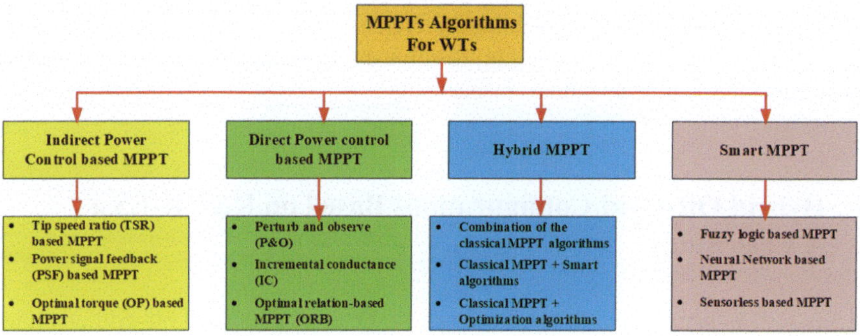

Fig. 5 Different MPPTs algorithms for WTs

simultaneously. The obtained experimental results show satisfactory performance when several buck–boost converters operate in parallel in DC microgrid. Regarding the control of the VRSDG, in [22], a lookup table is employed to determine the rotor speed references of DG based on PMSG according to the load power demand. This concept is adopted in [9] for VRSDG based on DFIG, and vector control is suggested to regulate the rotor current.

The voltage and frequency regulation at the AC bus are also localized at the first level of the HC as presented in Fig. 3. For these tasks, the interfacing inverter control by supplying a given P&Q set-point is expressed in Eq. (1), or by feeding the connected load with predefined values for voltage and frequency (i.e., 120 V/60 Hz, 220 V/50 Hz) [22].

$$\begin{cases} \omega_1 = \omega_{PCC} + k_P.P_1 \\ V_1 = V_{PCC} + k_Q.Q_1 \end{cases}, \tag{1}$$

where P_1, Q_1, ω_{PCC}, V_{PCC}, k_P, and k_Q represent the desired active and reactive powers, the PCC angular frequency and voltage, and the droop slopes, respectively.

$$\begin{cases} \omega_0 = \omega_0 - k_P.P \\ V_0 = V_0 - k_Q.Q \end{cases}, \tag{2}$$

where P, Q, ω_0, V_0, k_P, and k_Q represent the interfacing inverter active and reactive power outputs, the desired PCC angular frequency and voltage at no-load conditions, and droop slopes, respectively.

The issues such as overheating of electrical equipment, power losses, equipment failure, etc., occur due to the PQ problems in the off-grid system as detailed in [23]. R-ES penetration in an off-grid system can affect the PQ due to their intermittency. Thus, in an off-grid system, the PQ refers to the uninterruptible supply of power with fixed voltage and frequency to local loads. The permissible limit of frequency and voltage deviation as well as the permissible R-ES penetration are detailed in IEEE standards given in [24, 25]. In this context, many control algorithms such as instantaneous power (P&Q), synchronous reference frame (SRF), direct and indirect algorithms, etc., are suggested in [24] to solve the PQ issue as well as to adjust the voltage and frequency deviations, respectively.

4 Hybrid Off-Grid Configurations Based on R&NR-ESs: A Case Study

Figure 6 demonstrates the detailed scheme of the selected hybrid off-grid configuration shown in Fig. 2d. It comprises two NR-ESs; DE connected to variable speed DFIG and BES and one R-ES; SPA. The stator terminals of DFIG are linked to the PCC using a controlled switch (CS), and to the common DC bus through a rotor-side converter (RSC). The SPA and BES are linked to the common DC bus through the boost converter and buck–boost converter and to the PCC through the load-side converter (LSC). For the proposed hybrid-off-grid configuration, two ripple filters are suggested. The first RC filter is connected between the terminal stators of the DFIG and the CS and the second one is connected to the output of the LSC. A DC dump load is used to prevent overcharging of the BES. The linear and nonlinear loads are connected to PCC and can be supplied through the stator of DFIG and LSC depending on the operating mode of the hybrid off-grid configuration.

4.1 Operation Modes of Hybrid Off-Grid Configurations Based on R&NR-ESs

Figure 7 shows the operation modes of the hybrid off-grid configuration. Conditions such as the state of charge (SoC%) of BES, load, and generated powers decide the right operation mode of the hybrid off-grid configuration. If the SoC% of BES is

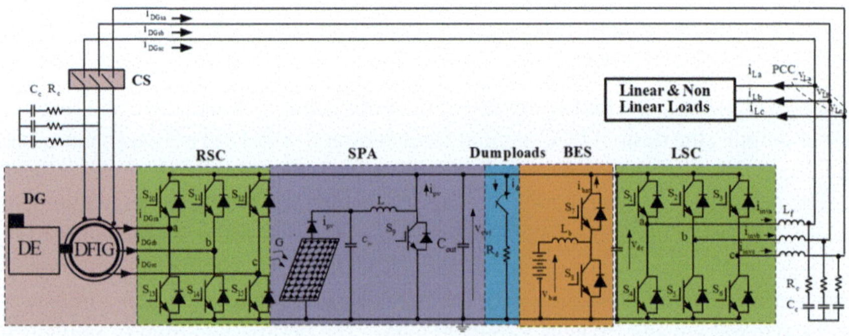

Fig. 6 Hybrid Diesel-SPA Off-grid configuration based on variable speed DFIG understudy

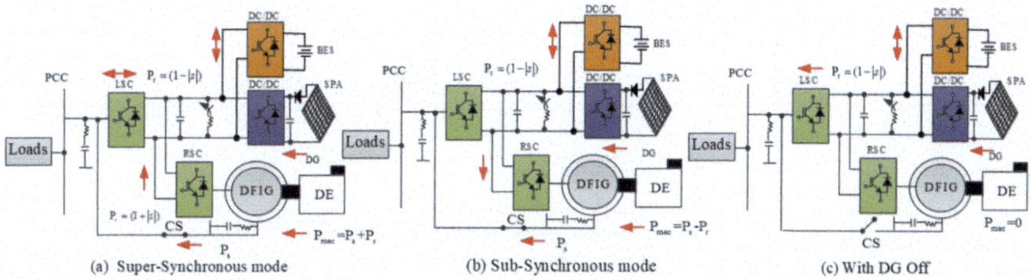

Fig. 7 Operation modes of the hybrid off-grid configuration

less than its lower limit and the generated power from the SPA is less to load power demand, DFIG operates in super-synchronous mode as presented in Fig. 7a, DG charge the BES and supply loads simultaneously. For this operation mode, the slip (s), as defined in Eq. (3), DFIG provides power to the connected loads through the stator and rotor. If the SoC% of BES is greater than 50% and the produced power from SPA is less than load power demand, DFIG operates in sub-synchronous mode as presented in Fig. 7b, the slip becomes $s > 0$, and DG provides power to the connected load through the stator and charges the BES through the LSC. For both operation modes, the required reactive power to built-up the stator voltage is supplied from the DC bus through the LSC. If the SoC% of BES is greater than 50% and the generated power from the SPA is greater than the load power needed, DG is turned off and its stator is completely isolated with help of the CS to minimize the power losses in the DFIG stator.

$$S = \frac{\omega_r}{\omega_s}, \tag{3}$$

where s, ω_r, and ω_s denote the slip, the rotor, and stator frequencies, respectively.

5 Control of the Selected Hybrid Off-Grid Configurations

In this section, the developed control algorithms for the DE, RSC, LSC, buck–boost converter, and boost converter are detailed.

5.1 Control Algorithm for DE

In Fig. 8, the coordinated control for the DE is presented. The desired speed of the DE (ω_r^*) is estimated based on the SoC% of BES, produced power from the SPA (P_{SPA}), and the load power requirement (P_L). The desired ω_r^* is compared to the measured rotor speed (ω_r) and the speed error is fed to the proportional-integral controller (PI). The output of the PI controller is supplied to the actuator to get control of the fuel

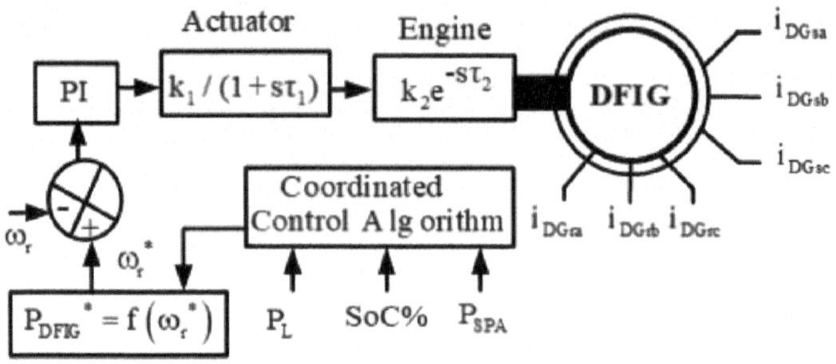

Fig. 8 Coordinated control algorithm for the DE

rack position and the output is fuel flow. The fuel flow is converted to mechanical torque using the engine block [26–28].

5.2 *Control of the Boost Converter*

The scheme of the developed control for the boost converter is presented in Fig. 9. To achieve MPPT from SPA, an improved P&O algorithm (IP&O) suggested in [27, 28], is applied in this study. This development considers a dynamic model of the DC–DC boost converter to achieve high performance under severe conditions. The current of the SPA (i_{mPV}), which is estimated by IP&O, is compared with the actual measured current (i_{PV}), and the error in the measured current is fed back to the PI controller as

$$u_i = k_p(i_{mPV} - i_{PV}) + k_i \int_0^t (i_{mPV} - i_{PV})dt, \tag{4}$$

where k_p and k_i denote PI gains of the current controller.

The control signal (d) is obtained using the output of PI controller (u_i) and dynamic model of boost converter as

$$d = 1 + \frac{u_i - V_{PV}}{V_{dc}}. \tag{5}$$

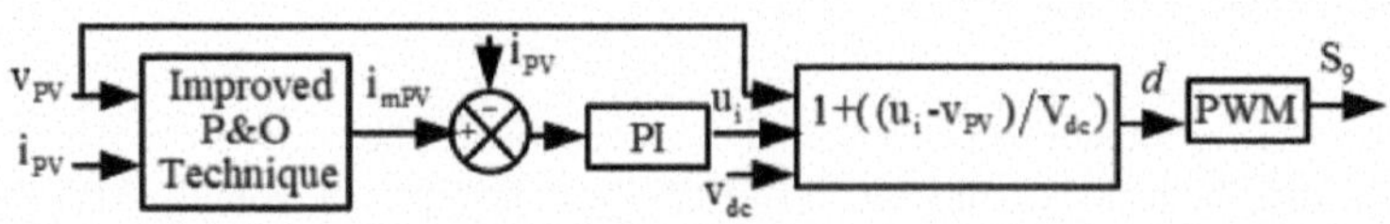

Fig. 9 Control of the boost converter

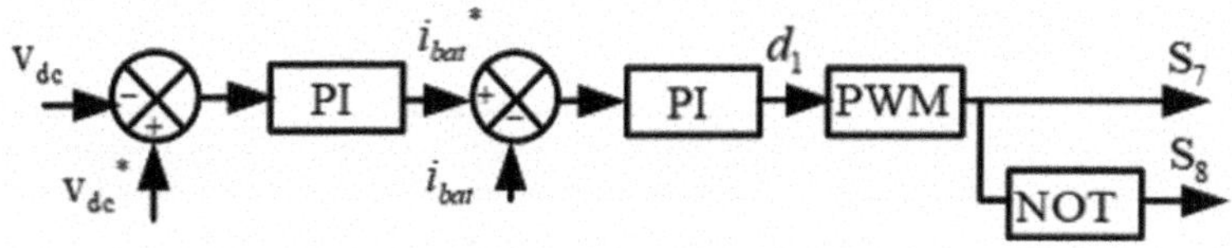

Fig. 10 Control algorithm for the buck–boost converter

To generate the switching signal for the switch S_9, the control signal is fed to the PWM controller as detailed in Fig. 9.

5.3 Control of the Buck–Boost Converter

In Fig. 10, the scheme of the developed control strategy for the boost converter is presented [28]. This algorithm has two main objectives: (1) balancing the power in the system, and (2) ensuring a continuous and stable power supply to the connected loads by compensating the SPA intermittency and transitioning between different hybrid off-grid operations modes. The desired BES current ($i_{bat}{}^*$) is calculated using (6).

$$i_{\text{bat}}^* = k_p\left(V_{dc}^* - V_{dc}\right) + k_i \int_0^t \left(V_{dc}^* - V_{dc}\right)dt, \tag{6}$$

where k_p and ki denote the PI gains of the outer control loop.

The measured BES current (i_{bat}) is compared with its desired $i_{bat}{}^*$ and the BES current error is fed to PI controller to get the control signal (d_1) to get the switching signal for S_7 and S_8 as

$$d_1 = k_p\left(i_b^* - i_b\right) + k_i \int_0^t \left(i_b^* - i_b\right)dt. \tag{7}$$

where k_p and k_i denote the proportional and integral gains of the inner control loop, respectively.

5.4 Control of the DC Dump Load

The developed control strategy for the DC dump load is demonstrated in Fig. 11. This algorithm is operating only if the SoC% is greater or equal to 100%. The DC dump load is employed to protect the BES from overcharging by dissipating the extra

Fig. 11 Control strategy for DC-dump load

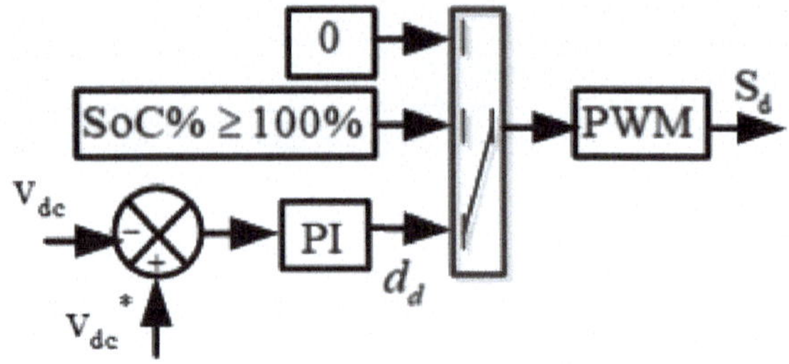

generated power from the SPA. So, the control signal (d_d) is obtained as

$$d_d = k_p\left(v_{dc}^* - v_{dc}\right) + k_i \int_0^t \left(v_{dc}^* - v_{dc}\right)dt, \tag{8}$$

where k_p and k_i denote the PI gains of the inner control loop.

A switch selector is employed to select the right signal either d_d or 0 based on the level of SoC% of BES. The obtained signal is fed to PWM to get the switching signal for S_d.

5.5 Control of the RSC

In Fig. 12, the improved vector control (IVC) proposed in [28], is presented. RSC control is designed to maintain high performance under different DE speed conditions and offer the DFIG the necessary magnetizing current.

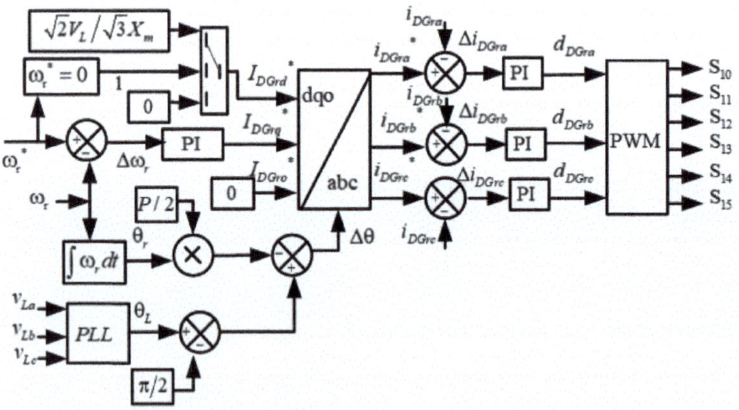

Fig. 12 Improved vector control for RSC

To calculate the quadrature axis component of RSC current ($I_{DGrq}{}^{*}$), the reference rotor speed ($\omega_r{}^{*}$) is compared with the measured value (ω_r) and the error ($\Delta\omega r$) is fed to the PI controller as

$$I_{DGrq}^{*} = k_p \Delta\omega_r + k_i \int_{0}^{t} \Delta\omega_r dt, \tag{9}$$

where k_p and k_i are the PI controller gains.

Using the direct axis component of the RSC current ($I_{DGrd}{}^{*}$) is taken as the rated magnetizing current (I_{msr}) and estimating it as [28]

$$I_{DGrd}^{*} = I_{msr} = \frac{\sqrt{2}V_L}{\sqrt{3}X_m}, \tag{10}$$

where X_m refers to the magnetizing reactance of the DFIG and V_L refers to the line voltage at the stator terminals.

A selector is added to the control scheme to optimize the power in the system by minimizing the power losses in the rotor when the VSDG is turned off. For this operation mode, the $I_{DGrd}{}^{*}$ is set equal to zero.

The references of the RSC currents ($i_{DGra}{}^{*}$, $i_{DGrb}{}^{*}$, and $i_{DGrc}{}^{*}$) are taken with help of the inverse Park's Transformation as

$$\left[i_{DGr(abc)}^{*}\right] = T_{dqo}(\Delta\theta)^{-1}\left[I_{DGr(dqo)}^{*}\right], \tag{11}$$

where $T_{dqo}(\Delta\theta)^{-1}$ represents the inverse Park's transformation which is detailed as

$$\begin{bmatrix} \cos(\Delta\theta) & \sin(\Delta\theta) & 1 \\ \cos(\Delta\theta - 2\pi/3) & \sin(\Delta\theta - 2\pi/3) & 1 \\ \cos(\Delta\theta + 2\pi/3) & \sin(\Delta\theta + 2\pi/3) & 1 \end{bmatrix}, \tag{12}$$

where $\Delta\theta$ represents the rotor transformation angle and it is estimated as

$$\Delta\theta = \left(\theta_s - \frac{\pi}{2}\right) - \left(\frac{P}{2}\right)\theta_r, \tag{13}$$

where P, θ_r, and θ_s refer to machine poles, rotor, and stator angles, respectively.

θ_s is obtained using a PLL and θ_r is obtained by integrating ω_r as

$$\theta_r = \int_{0}^{t} \omega_r dt. \tag{14}$$

The desired reference rotor currents are compared with sensed DFIG rotor currents to obtain the control signals (d_{DGra}, d_{DGrb}, and d_{DGrc}) as

$$\begin{cases} d_{DGra} = k_p \Delta i_{DGra} + k_i \int_0^t \Delta i_{DGra} dt \\ d_{DGrb} = k_p \Delta i_{DGrb} + k_i \int_0^t \Delta i_{DGrb} dt \\ d_{DGrc} = k_p \Delta i_{DGrc} + k_i \int_0^t \Delta i_{DGrc} dt \end{cases}, \tag{15}$$

where k_p and k_i are the PI controller gains.

The obtained control signals are compared with a triangular waveform to obtain a switching signal for (S_{10}–S_{15}).

5.6 *Control of the LSC*

The LSC's detailed control scheme is illustrated in Fig. 13 [28]. Stator flux oriented for LSC is intended to (1) inject power into PCC and (2) maintain constant AC voltage and system frequency. The desired stator current d-component ($i_{DGsd}{}^*$) is estimated as follows:

$$i^*_{DGsd} = k_p \Delta V_{PL} + k_i \int_0^t \Delta V_{PL} dt, \tag{16}$$

where k_p, k_i, and ΔV_{PL} are the PI gains of the controller, and error in PCC voltage, which is calculated using the desired PCC voltage ($V_{PL}{}^*$) and its measured value.

The amplitude of the PCC voltage (V_{PL}) is calculated as

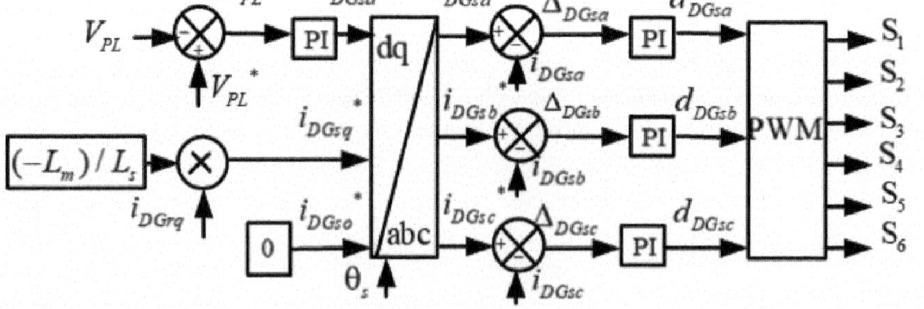

Fig. 13 Stator flux-oriented control for LSC

$$V_{PL} = \sqrt{\left(\frac{2}{3}\right)\left(v_{La}^2 + v_{Lb}^2 + v_{Lc}^2\right)}, \quad (17)$$

where v_{La}, v_{Lb}, and v_{Lc} denote the measured phase PCC voltages.

The stator current in the q-axis ($i_{DGsq}{}^*$) is estimated using (18)

$$i_{DGsq}^* = -(L_m/L_s)i_{DGrq}, \quad (18)$$

where L_m, L_s, and i_{DGrq} denote mutual and stator inductances and RSC current in the q-axis.

To regulate the frequency, LSC must operate at 60 Hz, as follows:

$$\theta_s = 2\pi f_s t, \quad (19)$$

where t and f_s denote the sampling time and the stator frequency, respectively.

With help of the inverse Park Transformation matrix as detailed in (20), one can obtain the desired stator currents of the DFIG ($i_{DGsa}{}^*$, $i_{DGsb}{}^*$, and $i_{DGsc}{}^*$) as

$$\left[i_{DGsabc}^*\right] = T_{dqo}(\theta_s)^{-1}\left[I_{DGsdqo}^*\right], \quad (20)$$

where $T_{dqo}(\theta_s)^{-1}$ represents inverse Park's transformation.

A comparison of the measured stator currents with references is made, and errors in stator currents are fed into PI current controllers to get signal controls (d_{DGsa}, d_{DGsb}, and d_{DGsc}) as follows:

$$\begin{cases} d_{DGsa} = k_p \Delta i_{DGsa} + k_i \int_0^t \Delta i_{DGsa} dt \\ d_{DGsb} = k_p \Delta i_{DGsb} + k_i \int_0^t \Delta i_{DGsb} dt \\ d_{DGsc} = k_p \Delta i_{DGsc} + k_i \int_0^t \Delta i_{DGsc} dt \end{cases} \quad (21)$$

To obtain switching signals for (S_1–S_6), control signals are compared with triangular waveforms.

6 Results and Discussion

With the help of MATLAB/Simulink, simulations are carried out to test the performance of the hybrid off-grid configuration in Fig. 6 and its control algorithms in Figs. 8, 9, 10, 11, 12 and 13. Many tasks as the regulation of AC voltage and frequency, harmonic compensation and stabilization of stator current, stable power supply to the connected load, control of the DG according to load requirements, control of

SoC% of BES, and generated power from SPA are validated in presence of severe conditions as load and weather conditions change, as well as when BES reaches its lower and higher limits.

6.1 *Dynamic Performance Under Load and Solar Irradiations Change When the SoC is Less Between 100 and 50%*

The performance of the PCC voltages (v_{DGs}), stator currents (i_{DGs}), rotor currents (i_{DGr}), frequency (*fs*), rotor speed (ω_r) and its reference ($\omega_r{}^*$), SPV current (i_{PV}) and its desired value ($i_{PV}{}^*$), load currents (i_L), LSC currents (i_{inv}), DC-link voltage (V_{dc}), and SoC% of the BES is presented in Fig. 14. As a result of these conditions, the VSDG is turned off in this mode of operation. The rotor and stator currents are equal to zero, which means that the stator is completely isolated and $I_{DGrd}{}^*$ is set equal to zero as detailed in Fig. 12. Thus, power losses in the stator and rotor of DFIG are equal to zero. As solar irradiation changes, the SPA current decreases at $t = 0.5$ s and more at $t = 1.1$ s. As can be seen, the SPA current follows its reference, confirming IP&O's good performance. A BES discharges between $t = 0$ s and $t = 0.7$ s and between $t = 1.1$ s and $t = 1.7$ s, and charges between $t = 0.7$ s and $t = 1.1$ s to balance power in the system during transitions (increase and decrease in loads and decrease in solar irradiation). During all simulations, the PCC voltage and system frequency are kept constant, and the load is supplied without interruptions. This confirms the robustness of the proposed control algorithms for LSC, RSC, boost, and buck-boost converters. It can be seen that VSDG is completely turned off and loads are supplied using only power generated from SPA and BES. This proves the fuel-saving in this operation mode.

6.2 *Dynamic Performance Under Less Solar Irradiations and with SoC Less Than 50%*

In Fig. 15, the dynamic performance when the SoC% is less than 50% and SPA provides less power, is presented. One observes at $t = 0.5$ s, the load is increased, SoC% of BES is less than 50% and SPA generates insufficient power. To ensure an uninterruptable power supply to the connected loads and charge the BES simultaneously, the coordinated algorithm estimate rotor speed ($\omega_r{}^*$) equal to 175 rad/s. This speed is estimated based on the SoC, generated power from SPA, and load power demand and with help of the optimal characteristic ($P_{DFIG}{}^* = f(\omega_r{}^*)$). One observes that DFIG rotor speed and output SPA current follow their references and DC link voltage is maintained constant during the transition, the PCC voltage and frequency are maintained constant, and load is supplied without interruption. This confirms that

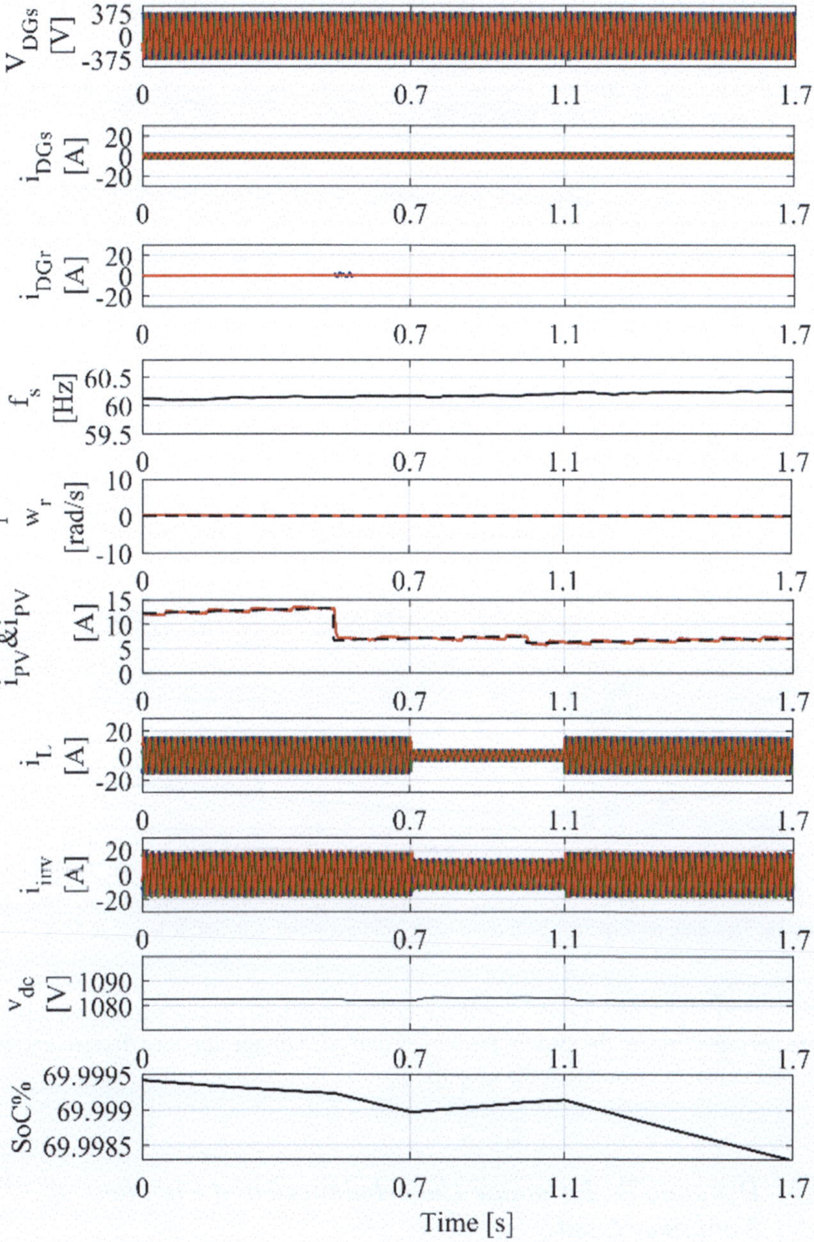

Fig. 14 Dynamic performance when the SoC% is between 100 and 50%

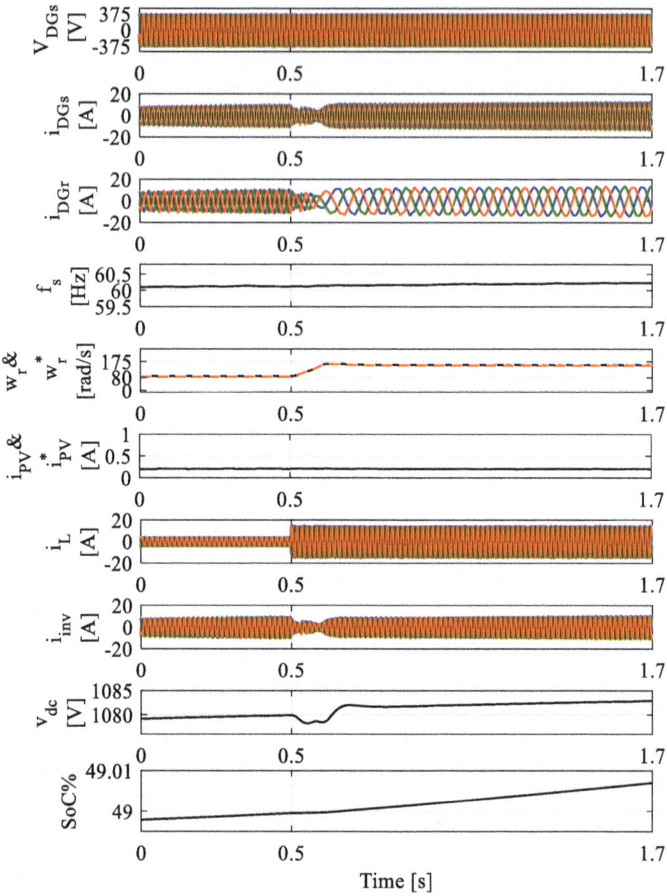

Fig. 15 Dynamic performance under less solar irradiation and with SoC% less than 50%

the control algorithms operate well under load and rotor speed variations. Also, in this operation mode, the system frequency and AC voltage are not affected, and they remain constant throughout the transition.

6.3 Dynamic Performance Underbalanced and Unbalanced Nonlinear Loads

The dynamic waveform under balanced and unbalanced loads, as shown in Fig. 16a, b, are displayed, along with their zoomed waveforms between 0.7 and 0.9 s. Based on the following conditions, simulations were performed: (1) SoC% is less than 50% and (2) SPA might not be able to provide the necessary power to the loads connected

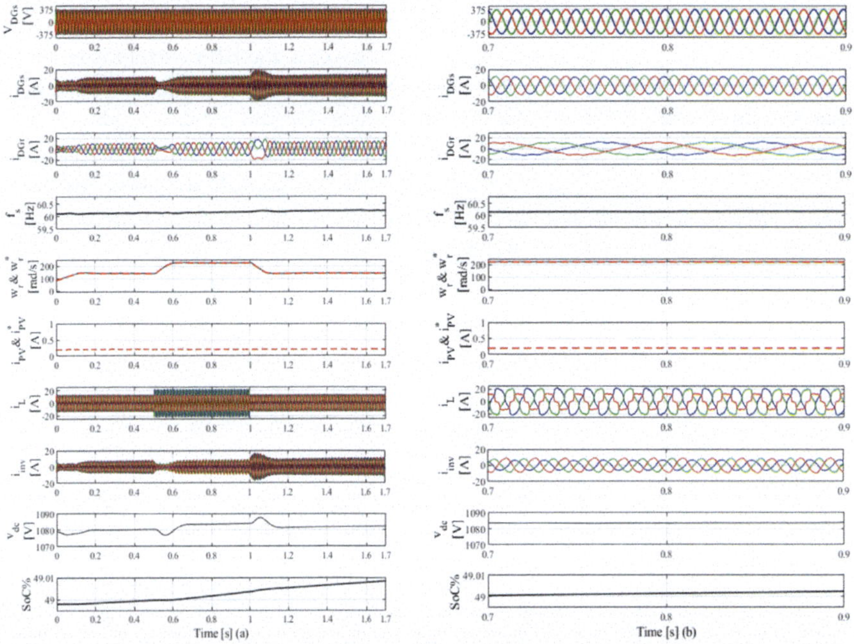

Fig. 16 **a** Dynamic performance under unbalanced linear loads, **b** zoomed waveforms of **a** between $t = 0.7$ s and $t = 0.9$ s

to the SPA. The system is subjected to a sudden increase in load between $t = 0.5$ and 1 s. Balanced and unbalanced nonlinear loads are considered in this test. To balance the load power demand and simultaneously charge the BES, the coordinated control determined that 230 rad/s should be achieved by the DFIG rotor speed. It is for this reason that DFIG switches from sub-synchronous to super-synchronous mode at $t =$ 0.5 s. As we observe during transitions, PCC voltage and system frequency remain constant. DFIG stator currents appear perfectly balanced instead of being influenced by an unbalanced nonlinear load in Fig. 16b. As a result, LSC is proved to be capable of acting as a shunt active filter.

6.4 Dynamic Performance When BES is Fully Charged

Figure 17 illustrates the performance of the PCC voltage, frequency, SoC% of BES, and the power generated by the SPA as well as the power generated by the VSDG, RSC, BES, DC dump load, and load. The SoC% in this test is 100%, and only SPAs are used to power the connected loads and charge the BES. The VSDG is turned off for this test. It can be observed that SoC% is less than 100% between $t = 0$ and 0.28 s. BES charges at $t = 0.28$ s, and where SoC% equals 100% at $t = 0.28$ s, the DC dump load is turned off. As a precaution, the DC dump load is turned on to prevent BES from overcharging. In the period of $t = 0.28–0.8$ s, the BES current is equal to zero, and the extra power is dissipated in the DC dump load while the BES current is equal to zero. From $t = 0.8$ to 1.4 s, the load power demand increases, which leads to the DC dump load being turned off and BES providing the necessary power to maintain a stable hybrid off-grid system. Observations confirm the good operation of all proposed algorithms and the reliability of the off-grid configuration based on the PCC voltage and the system frequency.

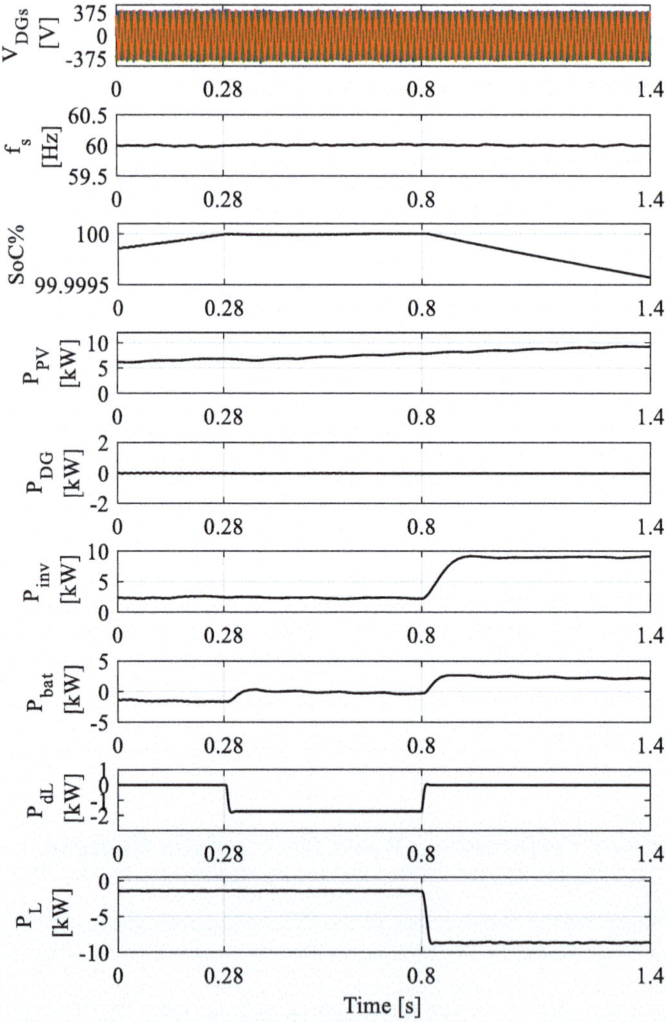

Fig. 17 Dynamic performance when batteries are fully charged

7 Conclusion

This chapter discusses hybrid off-grid configurations based on R&NR-ESs and their control strategies. A review of four new hybrid off-grid configurations based on fewer power converters is presented. All configurations employ the VSDG as the emergency ES. It operates only when the BES state of charge is less than its lower limit of charge and load power demand exceeds the power generated by REs. A hybrid off-grid configuration based on SPA and DG-driven variable speed DFIG is proposed with five control algorithms. It has been demonstrated that the use of VSDG helps reduce fuel consumption and that the integration of the controlled switch at the stator side of DFIG, as well as being added to the RSC algorithm, solves the problem of power losses in the rotor and stator when VSDG is disabled. It has been demonstrated that the LSC control algorithm performs well during transitions. The load is supplied without interruption with constant voltage and frequency. Additionally, the LSC acts as an active filter, balancing stator DFIG currents and mitigating harmonics, and protecting the BES from overcharging. As long as severe conditions exist, the SPA current follows its reference, and the DC link remains constant.

References

1. Rezkallah M, Chandra A, Singh B, Singh S (2019) Microgrid: configurations, control, and applications. IEEE Trans Smart Grid 10(2):1290–1302
2. Rezkallah M, Singh S, Chandra A, Singh B, Ibrahim H (2020) Off-grid system configurations for coordinated control of renewable energy sources. Energies 13(18):4950
3. Wang DH, Nayar CV, Wang C (2010) Modeling of a stand-alone variable speed diesel generator using doubly-fed induction generator. In: The 2nd international symposium on power electronics for distributed generation systems. IEEE, pp 1–6
4. Lee J-H, Lee S-H, Sul S-K (2009) Variable-speed engine generator with supercapacitor: isolated power generation system and fuel efficiency. IEEE Trans Ind Appl 45(6):2130–2135
5. Zhou Z, Camara MB, Dakyo B (2016) Coordinated power control of variable-speed diesel generators and lithium-battery on a hybrid electric boat. IEEE Trans Veh Technol 66(7):5775–578
6. Das I, Cañizares CA (2019) Renewable energy integration in diesel-based microgrids at the Canadian arctic. Proc IEEE 107(9):1838–1856
7. Wang C, Nehrir H, Lin F, Zhao J (2010) From hybrid energy systems to microgrids: Hybridization techniques, configuration, and control. In: IEEE PES general meeting. IEEE, pp 1–4
8. Unamuno E, Barrena JA (2015) Hybrid ac/dc microgrids—part I: review and classification of topologies. Renew Sustain Energy Rev 52:1251–1259
9. Pena R, Cárdenas R, Proboste J, Clare J, Asher G (2008) Wind–diesel generation using doubly fed induction machines. IEEE Trans Energy Convers 23(1):202–214
10. Kumar BA, Selvaraj R, Desingu K, Chelliah TR, Upadhyayula RS (2020) A coordinated control strategy for a diesel-electric tugboat system for improved fuel economy. IEEE Trans Ind Appl 56(5):5439–5451
11. Ho Q-D, Le-Ngoc T, Obaidat MS, Anpalagan A, Woungang I (2012) Smart grid communications networks: wireless technologies protocols issues and standards. In: Handbook of green information and communication systems. Elsevier, Amsterdam, The Netherlands pp 115–146

12. Yamashita DY, Vechiu I, Gaubert J-P (2020) A review of hierarchical control for building microgrids. Renew Sustain Energy Rev 118:109523
13. Bollipo RB, Mikkili S, Bonthagorla PK (2020) Hybrid, optimal, intelligent and classical PV MPPT techniques: a review. CSEE J Power Energy Syst 7(1):9–33
14. Pande J, Nasikkar P, Kotecha K, Varadarajan V (2021) A review of maximum power point tracking algorithms for wind energy conversion systems. J Mar Sci Eng 9(11):1187
15. Mercier P, Cherkaoui R, Oudalov A (2009) Optimizing a battery energy storage system for frequency control application in an isolated power system. IEEE Trans Power Syst 24(3):1469–1477
16. Wali SB, Hannan MA, Reza MS, Ker PJ, Begum RA, Abd Rahman MS, Mansor M (2021) Battery storage systems integrated renewable energy sources: a biblio metric analysis towards future directions. J Energy Storage 35:102296
17. Rezkallah M, Singh S, Singh B, Chandra A, Ibrahim H, Ghandour M (2020) Implementation of two-level coordinated control for seamless transfer in standalone microgrid. IEEE Trans Ind Appl 57(1):1057–1068
18. Rezkallah M, Ibrahim H, Bekbouti Y, Chandra A (2021) Performance analysis of low-cost configuration based on single-phase inverter for microgrid application. In: Proceedings of the IEEE industry applications society annual meeting (IAS). IEEE, pp 1–5
19. Mobarra M, Rezkallah M, Ilinca A (2022) Variable speed diesel generators: performance and characteristic comparison. Energies 15(2):592
20. Zhou Z, Camara MB, Dakyo B (2016) Coordinated power control of variable-speed diesel generators and lithium-battery on a hybrid electric boat. IEEE Trans Veh Technol 66(7):5775–5784
21. Cheng Z, Li Z, Li S, Gao J, Si J, Das HS, Dong W (2020) A novel cascaded control to improve stability and inertia of parallel buck-boost converters in DC microgrid. Int J Electr Power Energy Syst 119:105950
22. Lopes JAP, Moreira CL, Madureira AG (2006) Defining control strategies for microgrids islanded operation. IEEE Trans Power Syst 21(2):916–924
23. Chawda GS, Shaik AG, Mahela OP, Padmanaban S, Holm-Nielsen JB (2020) Comprehensive review of distributed FACTS control algorithms for power quality enhancement in utility grid with renewable energy penetration. IEEE Access 8:107614–107634
24. Singh B, Chandra A, Al-Haddad K (2014) Power quality: problems and mitigation techniques. Wiley
25. Green HJ, Wind T (2000) The IEEE grid interconnection standard: how will it affect wind power? National Renewable Energy Laboratory, Golden, CO, USA, technical report NREL/CP-500-28409
26. Rezkallah M, Chandra A (2011) Wind diesel battery hybrid system with power quality improvement for remote communities. In: Proceedings of the IEEE industry applications society annual meeting. IEEE, pp 1–6
27. Kollimalla SK, Mishra MK (2014) Variable perturbation size adaptive P&O MPPT algorithm for sudden changes in irradiance. IEEE Trans Sustain Energy 5(3):718–728
28. Rezkallah M, Dubuisson F, Singh S, Singh B, Chandra A, Ibrahim H, Ghandour M (May 2022) Coordinated control strategy for hybrid off-grid system based on variable speed diesel generator. IEEE Trans Ind Appl. Early Access

Chapter 9
Control for Grid Synchronization of Single-Phase Voltage Source Converter

Shailendra Kumar Sharma and Pemendra Kumar Pardhi

Abstract There is a progressive enhancement of sustainable energy sources with the electricity grid, the power converters play an important role to deliver power from sustainable energy sources to the utility grid. With increased penetration of uncertain sustainable energy sources and proliferation of large power electronics-based load on the electrical network, the power quality issues are introduced with the grid supply. The power quality issues such as impulsive, oscillatory, voltage sag, voltage swell, interruption, DC offset, harmonics, interharmonics, notching, etc. The extraction of grid synchronization signals from the grid voltage under these power quality disturbances is a big challenge. This chapter provides various control strategies for extracting the synchronization signals. These control strategies are based on orthogonal signal (OS) generator such as delay-based OS, second-order generalized integrator-based (SOGI), multilayer SOGI (mLSOGI), modified SOGI (mSOGI), enhanced SOGI (ESOGI). Third-order generalized integrator (TOGI), mixed second-order–third-order generalized integrator (MSTOGI), enhanced TOGI (ETOGI), fourth-order generalized integrator (F_0OGI), fifth-order generalized integrator (F_iOGI), multilayer F_iOGI (MLF$_i$OGI), and modified multilayer fifth-order generalized integrator (mMLF$_i$OGI). It is also shown that these generalized integrators can be used in proportional resonant (PR) controllers.

1 Introduction

In the single-phase grid-interfaced voltage source converter (VSC) system, the VSC is synchronized with the utility grid or other VSC. Its control also performs the function of power factor improvement (i.e., reactive power control), power quality improvement, ensuring smooth start-up of VSC, and smooth connection of VSC with

S. K. Sharma (✉) · P. K. Pardhi
Department of Electrical Engineering, Shri G. S. Institute of Technology and Science Indore, Indore 452003, Madhya Pradesh, India
e-mail: sshrama@sgsits.ac.in

A. K. Giri et al. (eds.), *Power Quality: Infrastructures and Control*, Studies in Infrastructure and Control,
https://doi.org/10.1007/978-981-19-7956-9_9

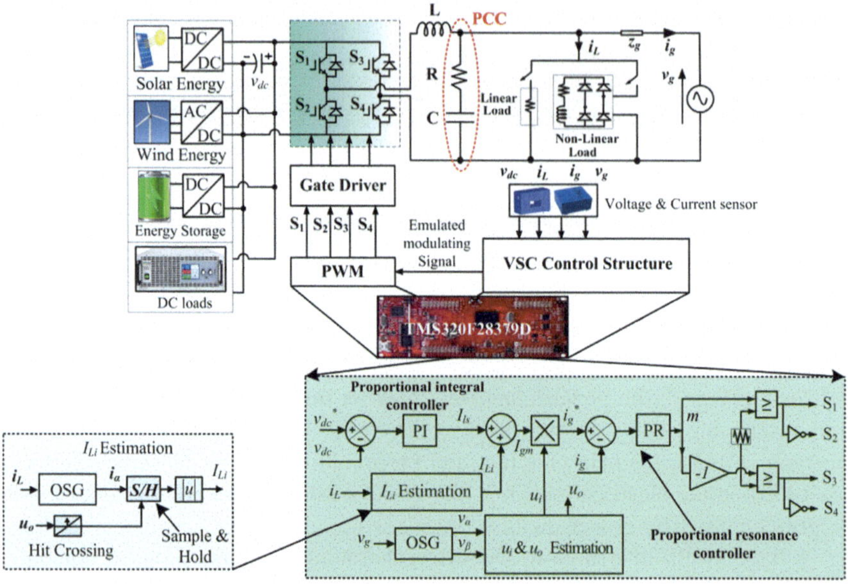

Fig. 1 Configuration of single-phase grid-interfaced voltage source converter with synchronization control

the utility grid. To ensure the above control function of VSC, an orthogonal signal generator (OSG) of sensed grid supply voltage plays an important role with implementing phase locked loop (PLL) and frequency locked loop (FLL) [1]. Figure 1 shows the single-phase VSC interfaced with the utility grid as a practical implementation. The VSC control is implemented on a microcontroller using actual sensed electrical signals. As shown in Fig. 1, the grid voltage is sensed and fed to the OSG block. Using its output signals, the unit template of "in phase" (u_i) and its orthogonal of phase (u_o) are obtained. The u_i and u_o are used for synchronization of VSC with the utility grid, active reactive power control, power quality improvement, and smooth start-up of VSC. To estimate u_i and u_o, there are various OSG schemes reported in literature [2–13]. From basic structure to advanced OSG schemes are described in the next section of this chapter.

2 Basic Structure of Orthogonal Signal Generator

The delay-based OSG scheme is the simplest method. To create it, $T/4$ time period delay is provided into a sensed grid voltage signal (here T is the fundamental time period of supply voltage) [1]. In the realistic system, sensed signals are processed into the digital controller and it worked in discrete time domain for reducing their memory size. In the continuous time domain, sensed grid supply voltage is given in

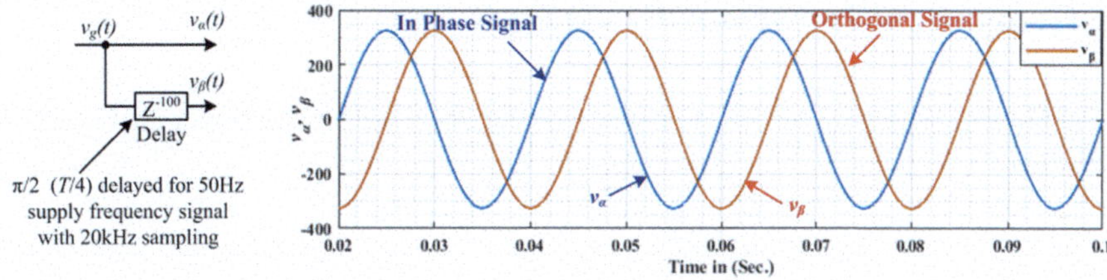

Fig. 2 In and orthogonal component of sensed grid supply voltage

Eq. 1 and its orthogonal signal (OS) is expressed in Eq. 2. The OS is obtained by providing 5 ms (1/4 of 20 ms) time delay for 50 Hz supply frequency.

$$v_g(t) = v_\alpha(t) = V_t \mathrm{Sin}(\omega t) \tag{1}$$

$$v'_g(t) = v_\beta(t) = V_t \mathrm{Sin}\left(\omega t - \frac{\pi}{2}\right) = V_t \mathrm{Cos}(\omega t) \tag{2}$$

In discrete time domain, continuous time signal is discretized with sampling time. The $T/4$, i.e., 5 ms time delay for 50 Hz supply frequency signal with 20 kHz sampling frequency, $Z^{-n} = Z^{-100}$ delay cycle is given to obtain Eq. 2 in discrete time domain. The obtaining process for n is given in Eq. 3. The "in phase" and its OS are shown in Fig. 2.

$$\left(\frac{T}{4} = 5\,\mathrm{ms}\right) = \left(nT_s = n\frac{1}{20\,\mathrm{kHz}}\right) \tag{3}$$

The magnitude (V_t) of sensed grid supply voltage is obtained from Eqs. 1 and 2 and expressed as

$$V_t = \sqrt{v_\alpha^2(t) + v_\beta^2(t)} \tag{4}$$

The unit template, i.e., u_i and u_o are obtained as expressed in Eq. 5, these are used in VSC control as discussed in the introduction section of this chapter.

$$u_i = \frac{v_\alpha(t)}{V_t} \;\&\; u_o = \frac{v_\beta(t)}{V_t} \tag{5}$$

On the one hand, it is the simplest method; however, it is unable to extract accurate sinusoidal synchronization signals (u_i and u_o) under disturbances in grid supply voltage such as dc offset, harmonics, notching, and frequency variation. Therefore to extract accurate synchronization signal, filtering approaches are used by researchers as discussed in the next section of the chapter.

3 Generalized Integrator-Based OSG

In the filtering approach of OSG to extract synchronization signals under grid supply disturbances, a generalized integrator-based techniques are preferred. This section gives overview about implementation of the various generalized integrators with their advantages and limitations in detail.

3.1 Second-Order Generalized Integrator (SOGI)

Second-order generalized integrator (SOGI) is most commonly used filter to extract synchronization signal. Figure 3a shows the structure of SOGI and Fig. 3b shows the extracted synchronization signals u_i and u_o.

To obtain proper gain parameters for SOGI, its characteristic expression is obtained.

From Fig. 3a, the key expression are given as

$$E_1 = v_g - v_\alpha \tag{6}$$

$$E_2 = kE_1 - v_\beta \tag{7}$$

$$v_\alpha = \frac{\omega}{s} E_2 \tag{8}$$

$$v_\beta = \frac{\omega}{s} v_\alpha \tag{9}$$

By solving Eqs. 6–9, transfer functions for alpha, beta, and error components are obtained as

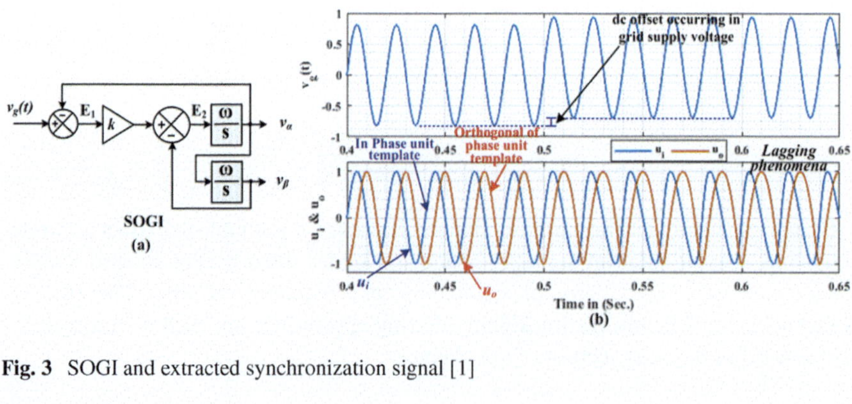

Fig. 3 SOGI and extracted synchronization signal [1]

$$\frac{v_\alpha}{v_g} = \frac{k\omega s}{s^2 + k\omega s + \omega^2} \tag{10}$$

$$\frac{v_\beta}{v_g} = \frac{k\omega^2}{s^2 + k\omega s + \omega^2} \tag{11}$$

$$\frac{E_1}{v_g} = \frac{s^2 + \omega^2}{s^2 + k\omega s + \omega^2} \tag{12}$$

where ω is the natural frequency and k is the error gain. Error gain can be obtained after comparing the standard second-order characteristic equation which is given in Eq. 13 as

$$s^2 + 2\xi\omega s + \omega^2 = 0 \tag{13}$$

Mostly practical systems are under damped and damping ratio (ξ) for such system lies between 0 and 1.

For Eqs. 10, 11, and 12, Bode plot is given in Fig. 4. It is obtained at $\omega = 314.15$ rad/s and error gain $k = 1.414$.

The “in phase” and “orthogonal” synchronization signal is obtained from SOGI which is given in Fig. 3b. From Fig. 3b, it is clear that under the dc offset conditions occurring on the sensed v_g, extracted u_i and u_o suffer with lagging phenomena. When the sensed v_g signal suffers from lower order (3rd, 5th and both 3rd + 5th) harmonics, extracted u_i signal from SOGI also carry harmonics contents. Due to harmonic contents in the u_i, modulating signal also carry harmonics content and hence introduce power quality issues.

Figures 5 and 6 show the harmonics analysis of v_g and u_i. Figure 5 shows the total harmonic distortions (THDs) of v_g and u_i under 3rd harmonics contents in v_g. The THDs of v_g is 8.70%, however at same time THDs of u_i is 2.11% that is lesser than THDs in v_g but harmonics are presented in u_i and same condition is noticed at presence of both 3rd and 5th harmonics in v_g as shown in Fig. 6.

3.2 *Multilayer Second-Order Generalized Integrator (mLSOGI)*

As discussed in previous section, SOGI is unable to extract pure sinusoidal synchronization signal under presence of harmonics in v_g. To overcome the problem associated with SOGI regarding harmonic elimination from synchronization signal, a multilayer has been created as shown in Fig. 7. Here the 2nd layer eliminates the 3rd harmonics component and the 3rd layer eliminates the 5th harmonics component. Further lower order harmonics such as 7th, 9th, and so on can be eliminated from sensed signals with adding the layer for 7th, 9th, and so on.

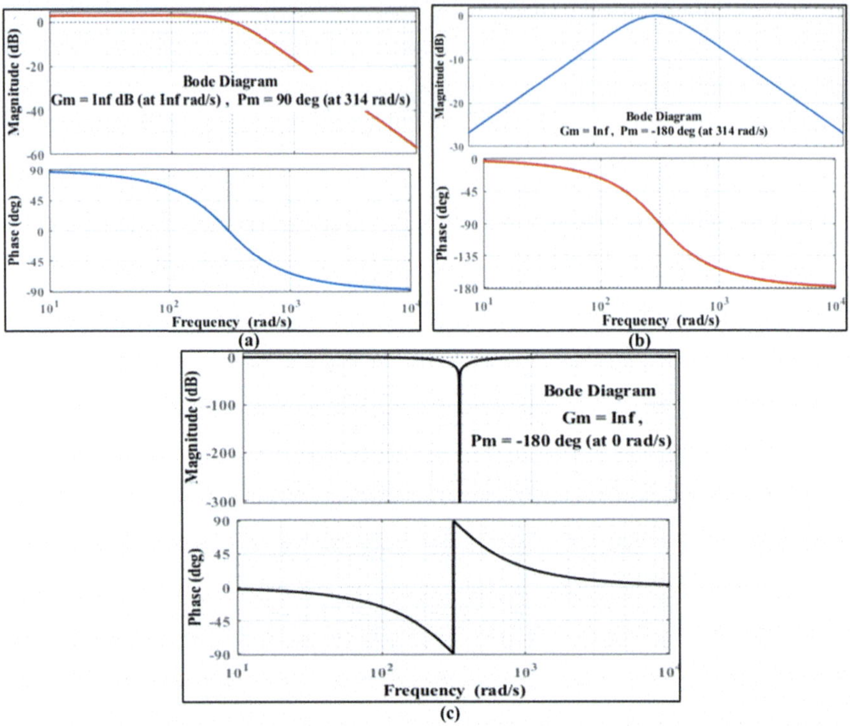

Fig. 4 Bode plot for SOGI, **a** in-phase voltage component, **b** orthogonal of phase voltage component, and **c** error component

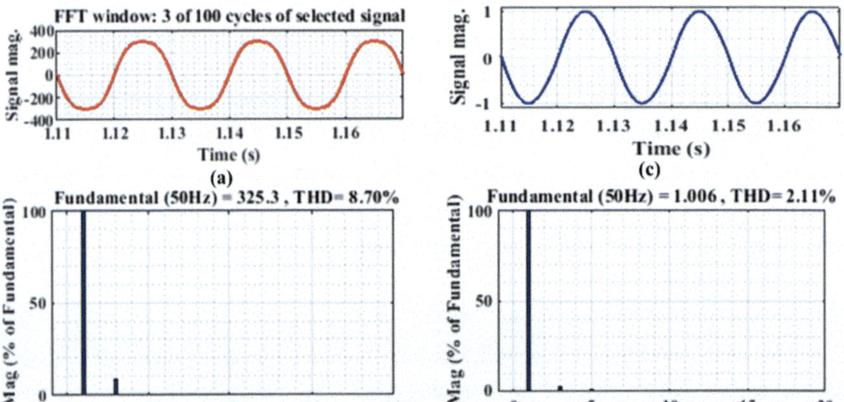

Fig. 5 Harmonic spectra analysis for SOGI under 3rd harmonics, **a** v_g contains 3rd harmonics, **b** harmonic spectra of v_g, **c** extracted u_i from SOGI, and **d** harmonic spectra of u_i

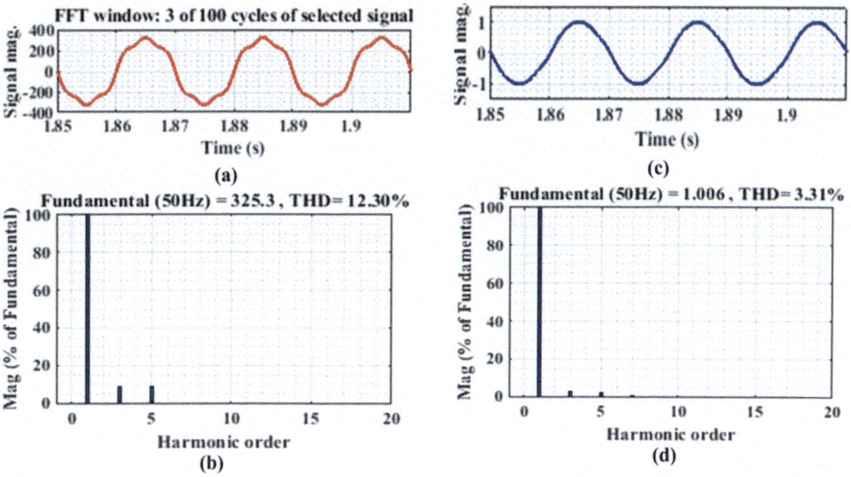

Fig. 6 Harmonic spectra analysis for SOGI under 3rd and 5th harmonics, **a** v_g contains 3rd and 5th harmonics, **b** harmonic spectra of v_g, **c** extracted u_i from SOGI, and **d** harmonic spectra of u_i

Fig. 7 Structure of mLSOGI

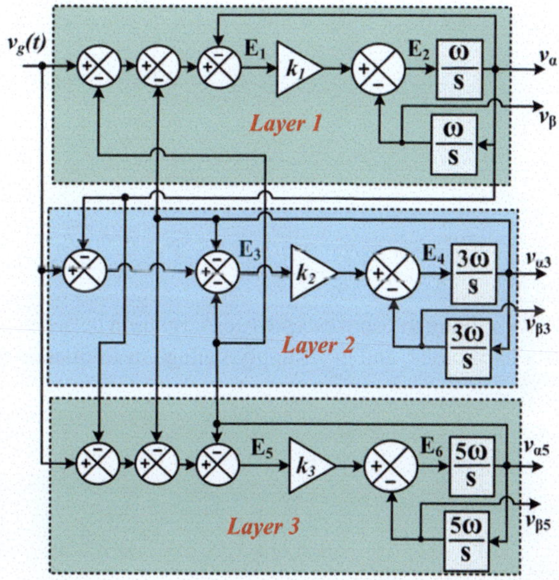

Figure 7 shows the structure of multilayer second-order generalized integrator (mLSOGI), the governing expression for this integrator is given as

$$E_1 = v_g - v_\alpha - v_{\alpha 3} - v_{\alpha 5} \tag{14}$$

$$E_2 = k_1 E_1 - \frac{\omega}{s} v_\alpha \tag{15}$$

$$v_\beta = \frac{\omega}{s} v_\alpha \tag{16}$$

$$v_\alpha = \frac{\omega}{s} E_2 \tag{17}$$

$$E_3 = v_g - v_\alpha - v_{\alpha 3} - v_{\alpha 5} \tag{18}$$

$$E_4 = k_2 E_3 - \frac{3\omega}{s} v_{\alpha 3} \tag{19}$$

$$v_{\alpha 3} = \frac{3\omega}{s} E_4 \tag{20}$$

$$v_{\beta 3} = \frac{3\omega}{s} v_{\alpha 3} \tag{21}$$

$$E_5 = v_g - v_\alpha - v_{\alpha 3} - v_{\alpha 5} \tag{22}$$

$$E_6 = k_3 E_5 - \frac{5\omega}{s} v_{\alpha 5} \tag{23}$$

$$v_{\alpha 5} = \frac{5\omega}{s} E_6 \tag{24}$$

$$v_{\beta 5} = \frac{5\omega}{s} v_{\alpha 5} \tag{25}$$

By solving the above expressions, relation between fundamental, third harmonics, fifth harmonics, and grid supply voltage in frequency (s) domain is given as

$$\frac{v_\alpha}{v_g} = \frac{s^2}{\Phi_1(s)} \left(\frac{k_1 \omega}{s} - \frac{k_1 \omega}{s} \times \frac{v_{\alpha 3}}{v_g} - \frac{k_1 \omega}{s} \times \frac{v_{\alpha 5}}{v_g} \right) \tag{26}$$

$$\frac{v_{\alpha 3}}{v_g} = \frac{s^2}{\Phi_3(s)} \left\{ k_2 \frac{3\omega}{s} - k_2 \frac{3\omega}{s} \times \frac{v_\alpha}{v_g} - k_2 \frac{3\omega}{s} \times \frac{v_{\alpha 5}}{v_g} \right\} \tag{27}$$

$$\frac{v_{\alpha 5}}{v_g} = \frac{s^2}{\Phi_5(s)} \left(\frac{5k_3\omega}{s} - \frac{5k_3\omega}{s} \frac{v_\alpha}{v_g} - \frac{5k_3\omega}{s} \frac{v_{\alpha 3}}{v_g} \right) \tag{28}$$

where $\Phi_1(s)$, $\Phi_3(s)$, and $\Phi_5(s)$ are second-order expression for layers 1st, 2nd, and 3rd, respectively, and given as

$$\Phi_1(s) = s^2 + k_1 s \omega + \omega^2 \tag{29}$$

$$\Phi_3(s) = s^2 + 3k_2 s\omega + 3^2\omega^2 \tag{30}$$

$$\Phi_5(s) = s^2 + 5k_3 s\omega + 5^2\omega^2 \tag{31}$$

After solving the expressions (26–28), transfer function for alpha and beta component is obtained as

$$\frac{v_\alpha}{v_g} = \frac{k_1 s\omega\delta_1(s)\Phi_5(s) - 3k_1k_2s^2\omega^2\Phi_5^2(s) - 5k_1k_3s^2\omega^2\delta_1(s) - 75k_1k_2k_3^2s^4\omega^4}{\delta_1(s)\Phi_1(s)\Phi_5(s) - 3k_1k_2s^2\omega^2\Phi_5^2(s) - 5k_1k_3s^2\omega^2\delta_1(s) - 75k_1k_2k_3^2s^4\omega^4} \tag{32}$$

$$\frac{v_\beta}{v_g} = \frac{k_1\omega^2\delta_1(s)\Phi_5(s) - 3k_1k_2s\omega^3\Phi_5^2(s) - 5k_1k_3s\omega^3\delta_1(s) - 75k_1k_2k_3^2s^3\omega^5}{\delta_1(s)\Phi_1(s)\Phi_5(s) - 3k_1k_2s^2\omega^2\Phi_5^2(s) - 5k_1k_3s^2\omega^2\delta_1(s) - 75k_1k_2k_3^2s^4\omega^4} \tag{33}$$

where $\delta_1(s) = \Phi_3(s)\Phi_5(s) - 15k_2k_3s^2\omega^2$.

For Eqs. 32 and 33, Bode plot is given in Fig. 8. It is obtained at $\omega = 314.15$ rad/s and error gains $k_1 = k_2 = k_3 = 1.414$. From the Bode plot analysis, the phase voltage component magnitude is higher than 0 dB at near fundamental frequency as shown in Fig. 8a. It indicates that, it is able to reject dc offset with small magnitude. However, this integrator is unable to eliminate the dc offset component from the extracted orthogonal phase component. Therefore, under the dc offset in sensed grid voltage lagging phenomena is occurring in synchronization signals as shown in Fig. 9.

Figures 10 and 11 show the harmonics analysis of v_g and u_i. Figure 10 shows the THDs of v_g and u_i under presence of 3rd harmonics in v_g. The THD of v_g is 8.70%; however, at the same time THDs of u_i is 0.00% due to the 2nd layer of the mLSOGI. The harmonics contents analysis of u_i under both 3rd and 5th harmonics present in

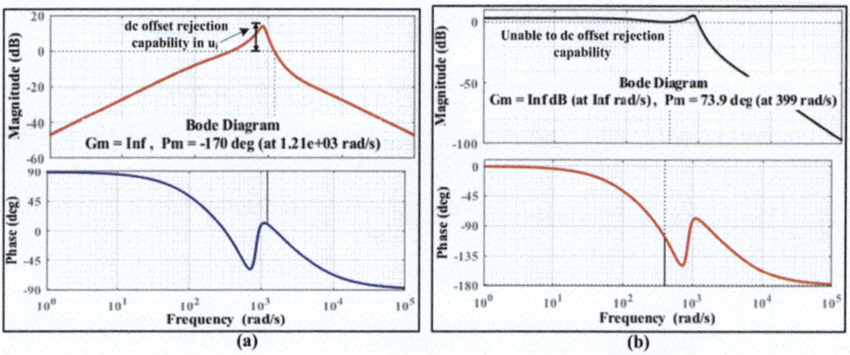

Fig. 8 Bode plot for mLSOGI, **a** in-phase voltage component and **b** orthogonal of phase voltage component

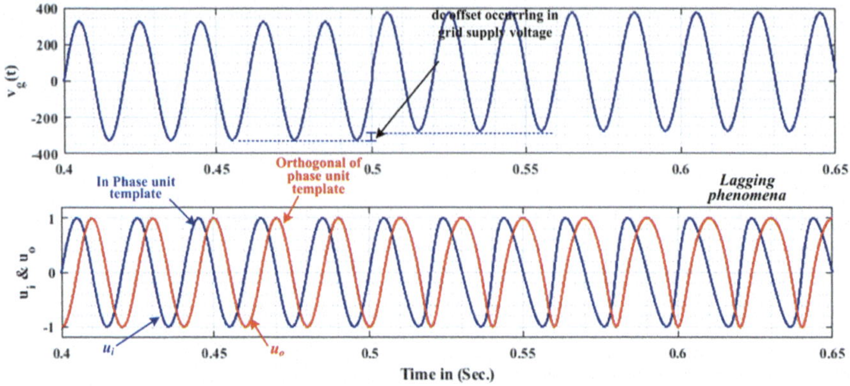

Fig. 9 Extracted synchronization signal from mLSOGI under dc offset in v_g

sensed grid voltage is shown in Fig. 11. Under these conditions, THD of u_i is 0.02%, i.e., sinusoidal in nature due to both 2nd and 3rd layer of mLSOGI.

The mLSOGI is able to extract pure sinusoidal signal even under lower order harmonics (3rd and 5th) presented in the sensed grid voltage, but from Fig. 9 it is clear that the mLSOGI is unable to extract pure sinusoidal synchronization signal under presence of dc offset condition in v_g. The mLSOGI provides lagging phenomena in extracted synchronization signals.

To overcome the problems of lagging phenomena associated with SOGI and mLSOGI; under dc offset conditions, these generalized integrators are further improved by adding dc offset rejection loop. The detailed structure of generalized integrator with dc offset rejection loop is discussed in the next section of the chapter.

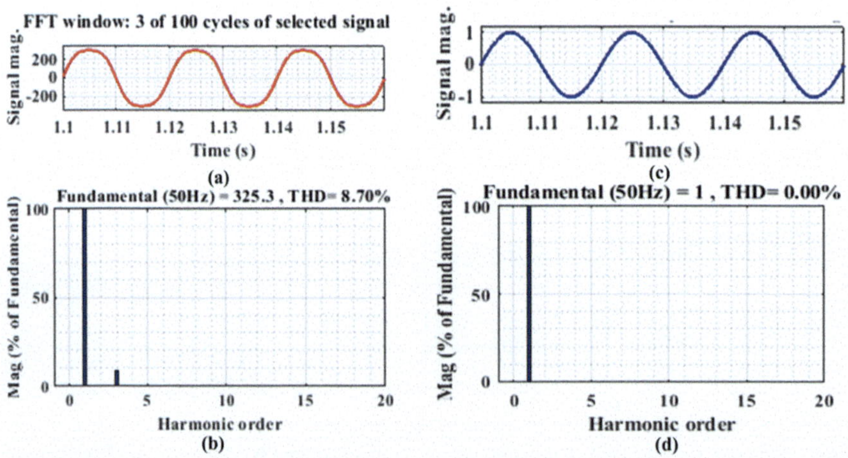

Fig. 10 Harmonic spectra analysis for mLSOGI under 3rd harmonics, **a** v_g contains with 3rd harmonics, **b** harmonic spectra of v_g, **c** extracted u_i from mLSOGI, and **d** harmonic spectra of u_i

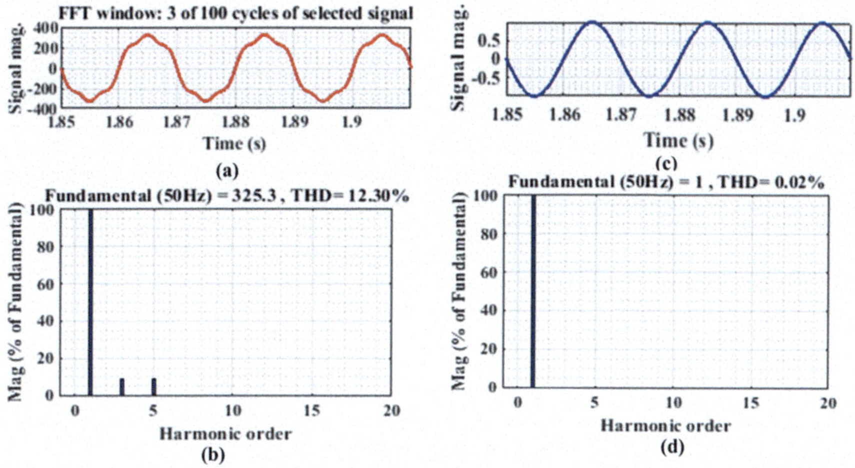

Fig. 11 Harmonic spectra analysis for mLSOGI under with 3rd and 5th harmonics, **a** v_g contains with 3rd and 5th harmonics, **b** harmonic spectra of v_g, **c** extracted u_i from mLSOGI, and **d** harmonic spectra of u_i

3.3 *Modified Second-Order Generalized Integrator (mSOGI)*

As discussed, the SOGI and mLSOGI provide lagging phenomena in extraction of synchronization signal; under sensed v_g suffers with dc offset condition. To overcome the lagging phenomena in u_i and u_o, a dc offset elimination loop is added in SOGI as depicted in Fig. 12.

The main expressions for mSOGI are given as

$$E_1 = v_g - E_d \tag{34}$$

$$E_2 = kE_1 - v_\beta \tag{35}$$

$$v_\alpha = \frac{\omega}{s} E_2 \tag{36}$$

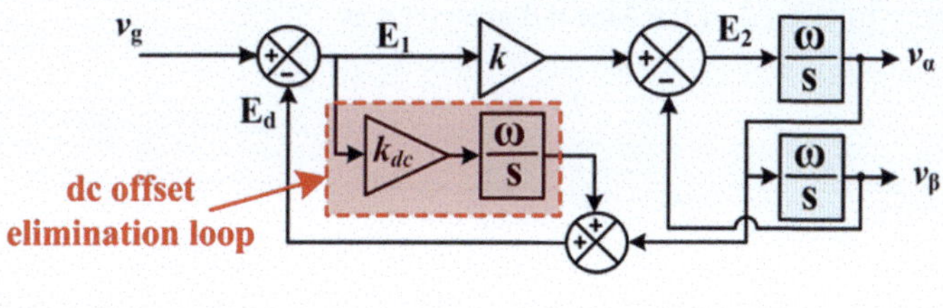

Fig. 12 Structure of mSOGI [3]

$$v_\beta = \frac{\omega}{s} v_\alpha \tag{37}$$

$$E_d = \frac{\omega}{s} k_{dc} E_1 + v_\alpha \tag{38}$$

Solving expressions (34–38), transfer functions for alpha, beta, and error components are obtained as

$$\frac{v_\alpha}{v_g} = \frac{k\omega s^2}{\left\{s^3 + (k_{dc} + k)s^2\omega + s\omega^2 + k_{dc}\omega^3\right\}} \tag{40}$$

$$\frac{v_\beta}{v_g} = \frac{k\omega^2 s}{\left\{s^3 + (k_{dc} + k)s^2\omega + s\omega^2 + k_{dc}\omega^3\right\}} \tag{41}$$

$$\frac{E_1}{v_g} = \left[\frac{s^4 + k_{dc}s^3\omega + s^2\omega^2 + k_{dc}s\omega^3}{s^4 + (2k_{dc} + k)s^3\omega + \left(k_{dc}^2 + kk_{dc} + 1\right)s^2\omega^2 + 2k_{dc}s\omega^3 + k_{dc}^2\omega^4}\right] \tag{42}$$

For Eqs. 40 and 42, Bode plot is given in Fig. 13. It is obtained at $\omega = 314.15$ rad/s, error gain $k = 1.414$, and dc offset error gain $k_{dc} = 0.4$. From the Bode plot analysis, it is clear that alpha component is able to reject dc offset with small magnitude and also the orthogonal component provides a higher dc offset rejection capability. Due to the dc offset rejection loop, it is able to extract pure sinusoidal synchronization signal without lagging phenomena even under dc offset in vg as shown in Fig. 14.

On the one end, mSOGI mitigates the lagging phenomena in extracted synchronization signals but it is unable to provide pure sinusoidal synchronization signals, i.e., u_i and u_o signals as shown in Fig. 15 under harmonics present in v_g.

3.4 Modified Second-Order Generalized Integrator (mSOGI)

The mSOGI is reported in [3] as discussed in previous section of the chapter. The dc offset rejection loop is introduced with an in-phase voltage component for feedback signal and used three integrators, to achieve dc offset rejection capability. In [4], authors have developed a new modified second-order generalized integrator (MSOGI) as shown in Fig. 16. There is a dc offset rejection loop introduced with an orthogonal voltage component and only two integrators are required.

The expressions used in the MSOGI are given as

$$E_1 = v_g - v_\alpha \tag{43}$$

$$E_2 = kE_1 - v_\beta \tag{44}$$

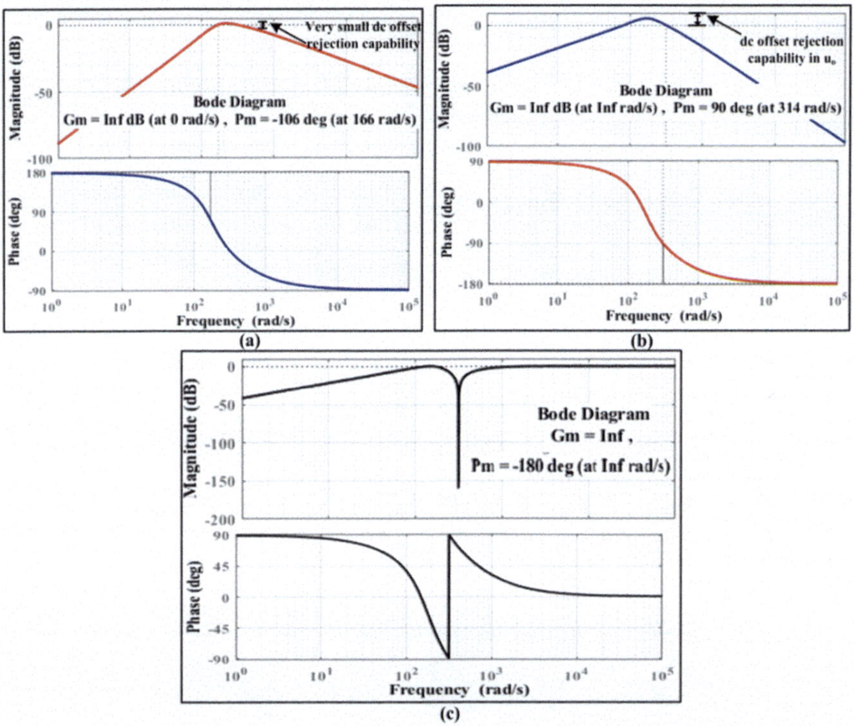

Fig. 13 Bode plot for mSOGI, **a** in-phase voltage component, **b** orthogonal of phase voltage component, and **c** error component

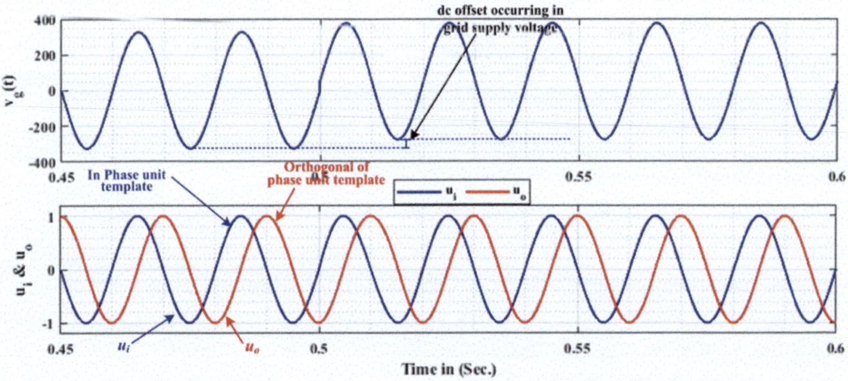

Fig. 14 Extracted synchronization signal from mSOGI under dc offset in v_g

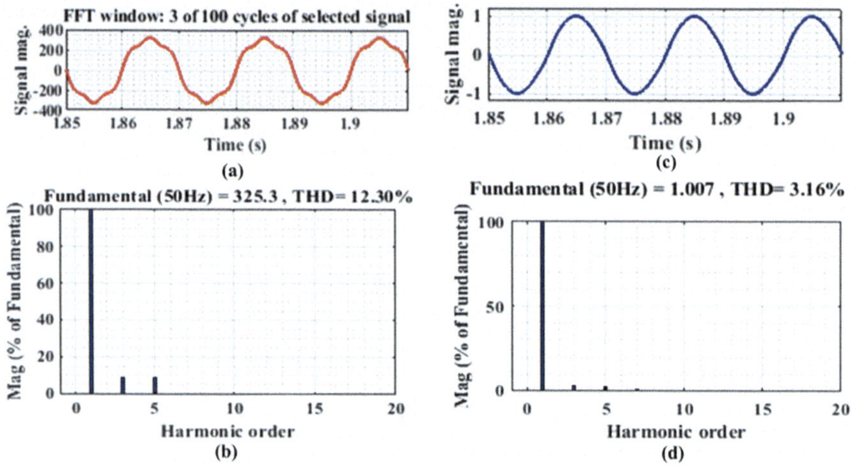

Fig. 15 Harmonic spectra analysis for mSOGI under with 3rd and 5th harmonics, **a** v_g contains with 3rd and 5th harmonics, **b** harmonic spectra of v_g, **c** extracted u_i from mSOGI, and **d** harmonic spectra of u_i

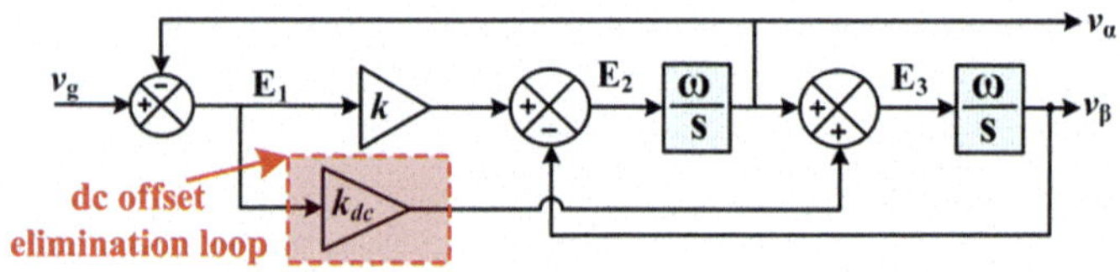

Fig. 16 Structure of MSOGI [4]

$$v_\alpha = \frac{\omega}{s} E_2 \tag{45}$$

$$E_3 = k_{dc} E_1 + v_\alpha \tag{46}$$

$$v_\beta = \frac{\omega}{s} E_3 \tag{47}$$

$$E_3 = \frac{s}{\omega} v_\beta \tag{48}$$

By solving expressions (42–48), transfer functions for alpha, beta, and error components are obtained as

$$\frac{v_\alpha}{v_g} = \frac{ks\omega - \omega^2 k_{dc}}{s^2 + ks\omega + (1 - k_{dc})\omega^2} \tag{49}$$

$$\frac{v_\beta}{v_g} = \frac{k_{dc}s\omega + k\omega^2}{s^2 + ks\omega + (1 - k_{dc})\omega^2} \tag{50}$$

$$\frac{E_1}{v_g} = \frac{s^2 + \omega^2}{s^2 + ks\omega + (1 - k_{dc})\omega^2} \tag{51}$$

For Eqs. 49 and 51, Bode plot is plotted as shown in Fig. 17. It is obtained at $\omega =$ 314.15 rad/s, error gain $k = 2$, and dc offset error gain $k_{dc} = -1$. From the Bode plot analysis, it is clear that, alpha component is able to reject dc offset with small magnitude and the orthogonal component provides a higher dc offset rejection capability. Due to the dc offset rejection loop, it is able to extract sinusoidal synchronization signal without lagging phenomena even under dc offset in v_g as shown in Fig. 18.

The MSOGI is able to mitigate the lagging phenomena in extracted synchronization signals but it is unable to provide pure sinusoidal synchronization signals, i.e., u_i and u_o signals as shown in Fig. 19 under harmonics present in v_g. There are only two integrators required but its harmonic mitigation is poor than mSOGI reported in Ref. [3].

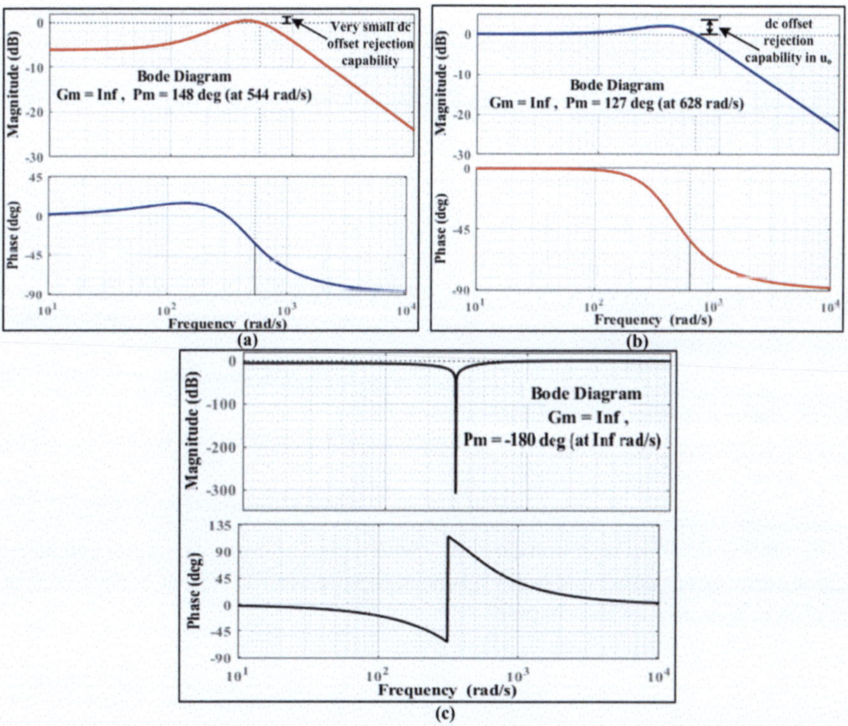

Fig. 17 Bode plot for MSOGI, **a** in-phase voltage component, **b** orthogonal of phase voltage component, and **c** error component

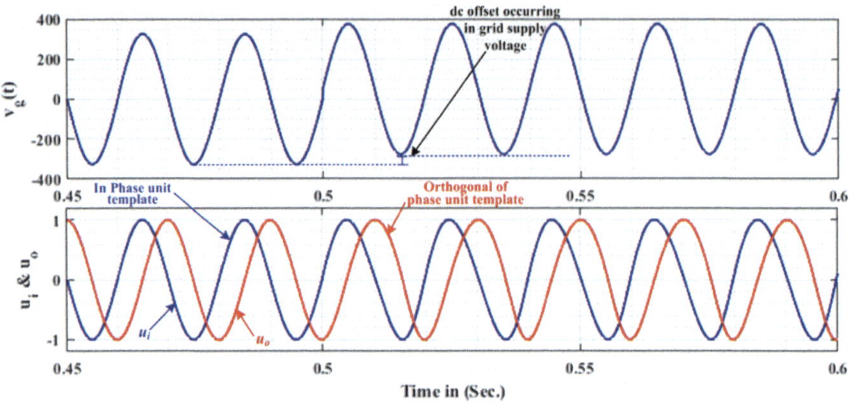

Fig. 18 Extracted synchronization signal from MSOGI under dc offset in v_g

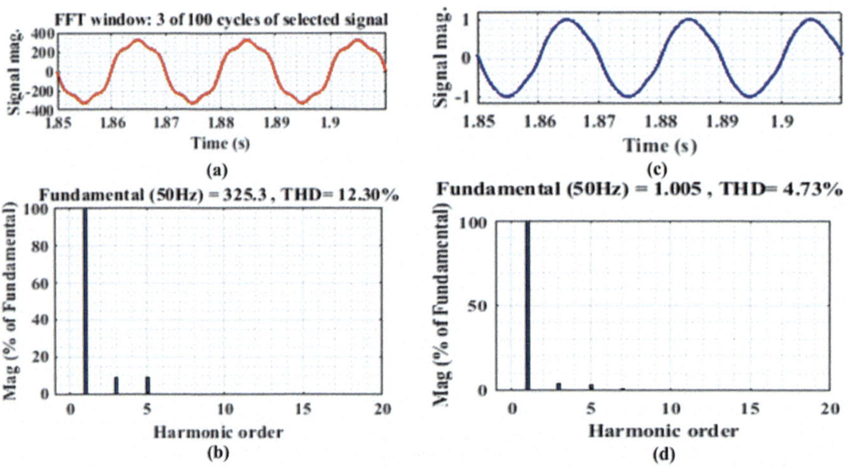

Fig. 19 Harmonic spectra analysis for MSOGI with 3rd and 5th harmonics, **a** v_g contains with 3rd and 5th harmonics, **b** harmonic spectra of v_g, **c** extracted u_i from MSOGI, and **d** harmonic spectra of u_i

To improve dc offset rejection capability with higher filtering, capability enhanced second-order generalized integrator has been reported in Ref. [5]. It is described in detail in next section of the chapter.

3.5 Enhanced Second-Order Generalized Integrator (ESOGI)

Enhanced second-order generalized integrator (ESOGI) is improved in both versions of mSOGI. Its structure is given in Fig. 20.

The expressions used in the ESOGI are given as

$$E_1 = v_g - E_d \tag{52}$$

$$E_2 = kE_1 - v_\beta \tag{53}$$

$$v_\alpha = \frac{\omega}{s}E_2 \tag{54}$$

$$E_3 = v_\alpha + k_{dc}E_1 \tag{55}$$

$$v_\beta = \frac{\omega}{s}E_3 \tag{56}$$

$$E_d = \frac{\omega}{s}k_{dc}E_1 + v_\alpha \tag{57}$$

By solving expressions (52–57), transfer functions for alpha, beta, and error components are obtained as

$$\frac{v_\alpha}{v_g} = \frac{\left(ks^2\omega - k_{dc}s\omega^2\right)}{s^3 + (k_{dc} + k)s^2\omega + (1 - k_{dc})s\omega^2 + k_{dc}\omega^3} \tag{58}$$

$$\frac{v_\beta}{v_g} = \frac{k_{dc}s^2\omega + ks\omega^2}{\left\{s^3 + (k_{dc} + k)s^2\omega + (1 - k_{dc})s\omega^2 + k_{dc}\omega^3\right\}} \tag{59}$$

$$\frac{E_1}{v_g} = \frac{s^3 + s\omega^2}{s^3 + (k_{dc} + k)s^2\omega + (1 - k_{dc})s\omega^2 + k_{dc}\omega^3} \tag{60}$$

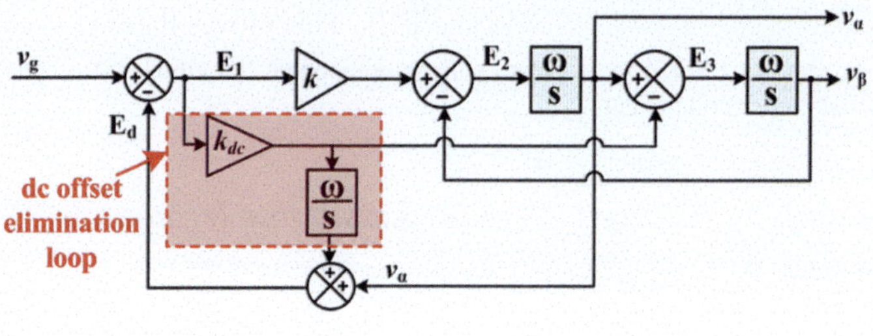

Fig. 20 Structure of ESOGI [5]

For Eqs. 58, 59, and 60, Bode plot is obtained as shown in Fig. 21 with ω = 314.15 rad/s frequency, error gain k = 1.414, and dc offset error gain k_{dc} = 0.5. From the Bode plot analysis it is clear that, alpha and beta components are able to reject dc offset conditions. Due to the dc offset rejection loop it is able to extract pure sinusoidal synchronization signal without lagging phenomena even under dc offset in v_g as shown in Fig. 22.

The ESOGI is able to mitigate the lagging phenomena in extracted synchronization signals and also provide improved sinusoidal synchronization signals, i.e., u_i and u_o signals as shown in Fig. 23 under presence of harmonics in v_g.

Further to improve the filtering capability, order of generalized integrator must increase as follows.

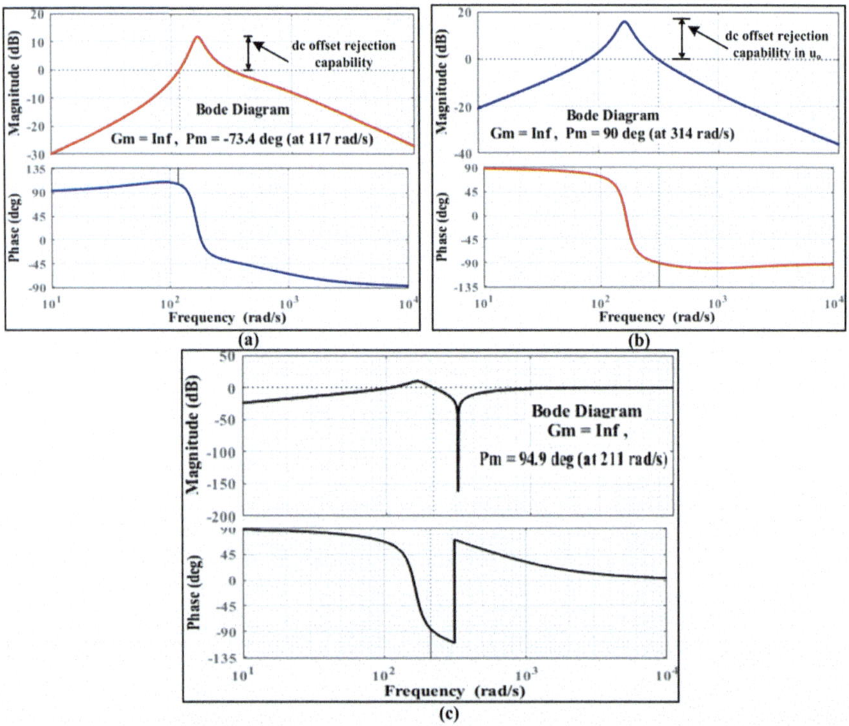

Fig. 21 Bode plot for ESOGI, **a** in-phase voltage component, **b** orthogonal of phase voltage component, and **c** error component

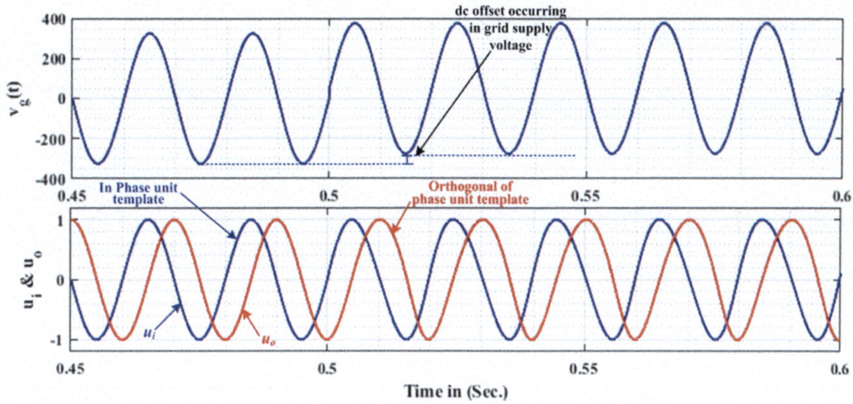

Fig. 22 Extracted synchronization signal from ESOGI under dc offset in v_g

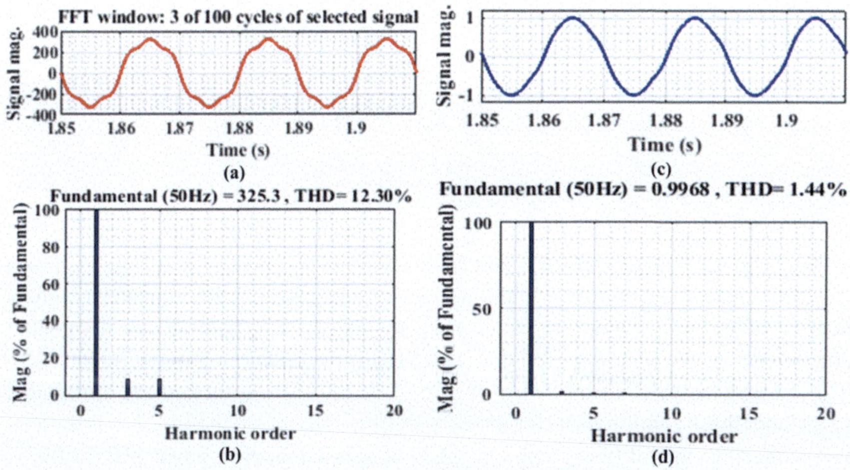

Fig. 23 Harmonic spectra analysis for ESOGI with 3rd and 5th harmonics, **a** v_g contains with 3rd and 5th harmonics, **b** harmonic spectra of v_g, **c** extracted u_i from ESOGI, and **d** harmonic spectra of u_i

3.6 Third-Order Sinusoidal Integrator (TOSSI)

Figure 24 shows the structure of third-order sinusoidal integrator (TOSSI). The TOSSI ensures accurate tracking of the sinusoidal reference without any steady-state error, and it is adjusted with a certain frequency. To mitigate the harmonics component from extracted synchronization signal, selective frequency layer may be added same as mLSOGI. Thus the sinusoidal integrator can be used to improve the total harmonic distortion (THD) performance.

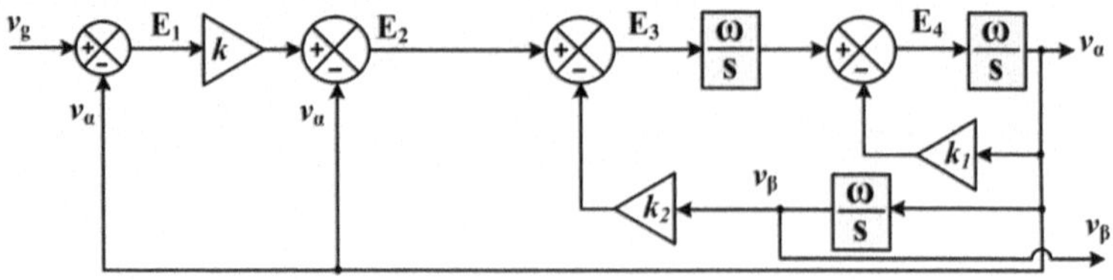

Fig. 24 Structure of TOSSI [6]

The transfer function expressions for in-phase, orthogonal phase voltage component and error component of TOSSI are given as

$$\frac{v_\alpha}{v_g} = \frac{ks\omega^2}{s^3 + k_1 s^2\omega + (1+k)s\omega^2 + k_2\omega^3} \tag{61}$$

$$\frac{v_\beta}{v_g} = \frac{k\omega^3}{s^3 + k_1 s^2\omega + (1+k)s\omega^2 + k_2\omega^3} \tag{62}$$

$$\frac{E_1}{v_g} = \frac{s^3 + k_1 s^2\omega + s\omega^2 + k_2\omega^3}{s^3 + k_1 s^2\omega + (1+k)s\omega^2 + k_2\omega^3} \tag{63}$$

$$\frac{E_2}{v_g} = \frac{ks^3 + kk_1 s^2\omega + kk_2\omega^3}{s^3 + k_1 s^2\omega + (1+k)s\omega^2 + k_2\omega^3} \tag{64}$$

The Bode plot for Eqs. 61, 62, and 63 is obtained and depicted in Fig. 25. From Fig. 25a, it is noticed that it is unable to extract sinusoidal synchronization signals under dc offset in v_g. Under dc offset in v_g, the extracted synchronization signal is depicted in Fig. 26. It is observed that TOSSI introduced lagging phenomena.

To validate the filtering capability of TOSSI, 3rd and 5th harmonics are introduced in v_g. Under these conditions its harmonic spectrum is depicted in Fig. 27. It is noticed that its filtering capability is good, i.e., THD of u_i is 1.1%, even when THDs of v_g is 12.3%.

To improve the dc offset rejection capabilities, other generalize integrators reported.

3.7 *Mixed Second- and Third-Order Generalized Integrator*

Figure 28 shows the structure of mixed second- and third-order generalized integrator (MSTOGI). The MSTOGI ensures accurate tracking of the reference synchronization signal without lagging phenomena under dc offset in sensed v_g.

The transfer function expressions for in-phase, orthogonal of phase voltage component, and error component of MSTOGI are given as

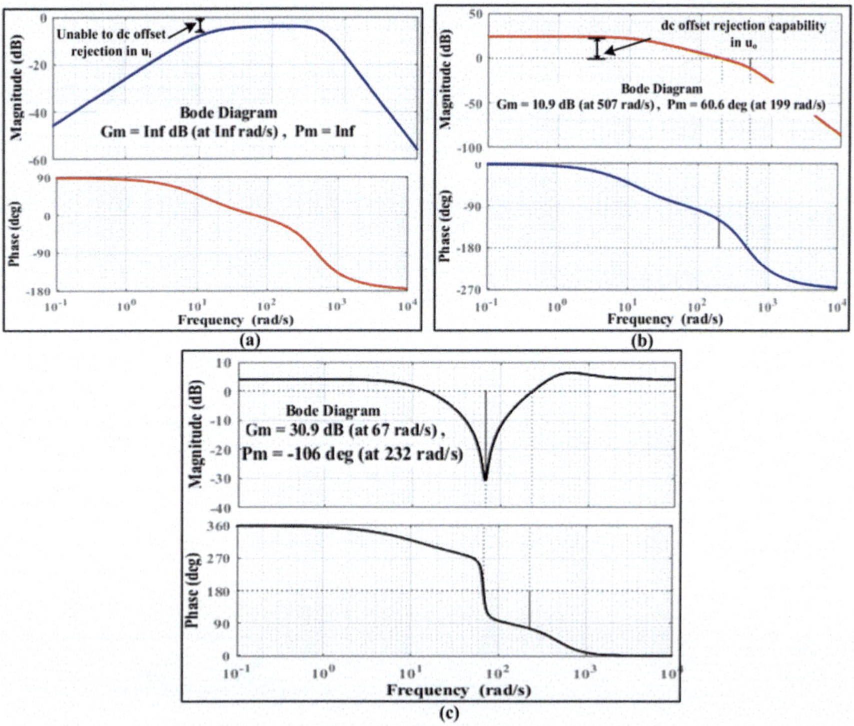

Fig. 25 Bode plot for TOSSI, **a** in-phase voltage component, **b** orthogonal of phase voltage component, and **c** error component

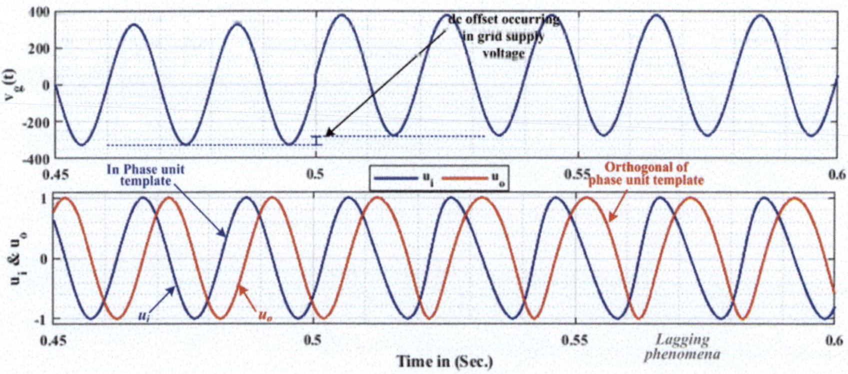

Fig. 26 Extracted synchronization signal from TOSSI under dc offset in v_g

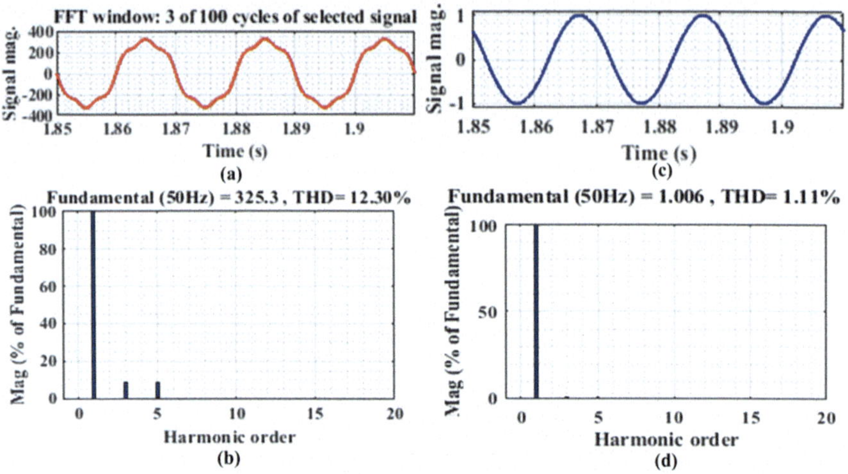

Fig. 27 Harmonic spectra analysis for TOSSI with 3rd and 5th harmonics, **a** v_g contains with 3rd and 5th harmonics, **b** harmonic spectra of v_g, **c** extracted u_i from TOSSI, and **d** harmonic spectra of u_i

Fig. 28 Structure of MSTOGI [7]

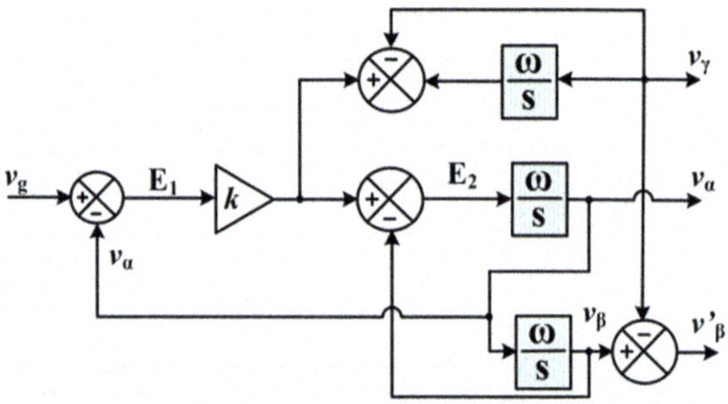

$$\frac{v_\alpha}{v_g} = \frac{ks\omega}{s^2 + ks\omega + \omega^2} \tag{65}$$

$$\frac{v'_\beta}{v_g} = \frac{-ks^2\omega + ks\omega^2}{s^3 + (1+k)s^2\omega + (1+k)s\omega^2 + \omega^3} \tag{66}$$

$$\frac{E_1}{v_g} = \frac{s^2 + \omega^2}{s^2 + ks\omega + \omega^2} \tag{67}$$

The Bode plot for Eqs. 65 and 66 is depicted in Fig. 29. From Fig. 29, it is noticed that it is able to extract sinusoidal synchronization signals under dc offset in vg. Under dc offset in v_g, the extracted synchronization signal is depicted in Fig. 30. Also to validate its filtering capability harmonic spectrum analysis is depicted in

Fig. 31. From Figs. 30 and 31, it is concluded that MSTOGI is able to reject dc offset, however its filtering capability is poorer than TOSSI.

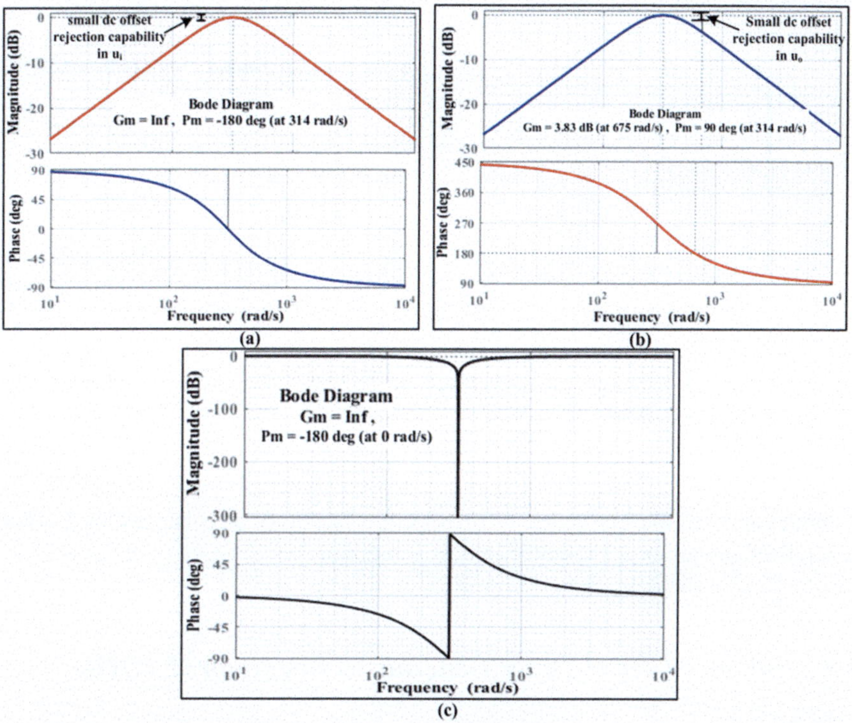

Fig. 29 Bode plot for MSTOGI, **a** in-phase voltage component, **b** orthogonal of phase voltage component, and **c** error component

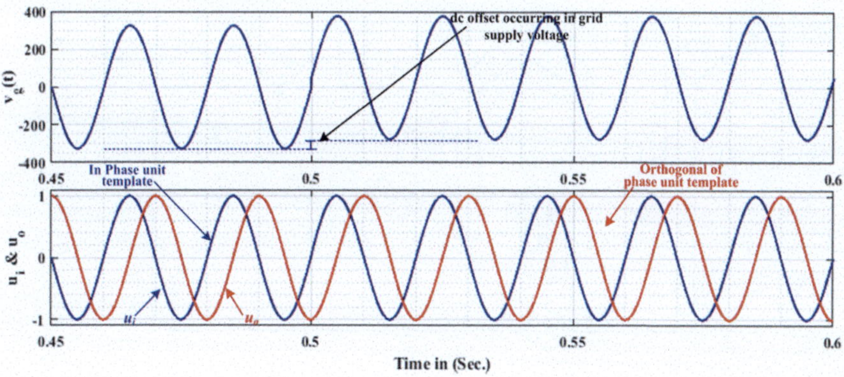

Fig. 30 Extracted synchronization signal from MSTOGI under dc offset in v_g

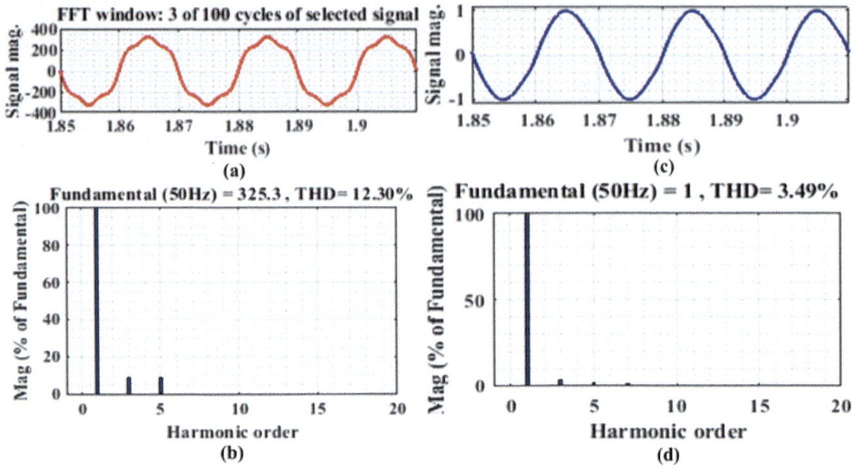

Fig. 31 Harmonic spectra analysis for MSTOGI with 3rd and 5th harmonics, **a** v_g contains with 3rd and 5th harmonics, **b** harmonic spectra of v_g, **c** extracted u_i from MSTOGI, and **d** harmonic spectra of u_i

With improved filtering and dc offset rejection capability, an enhanced third-order generalized integrator is presented in next subsection.

3.8 Enhanced Third-Order Generalized Integrator (ETOGI)

Figure 32 shows the structure of enhanced third-order generalized integrator (ETOGI). The ETOGI ensures accurate tracking of the reference synchronization signal without lagging phenomena under dc offset in sensed v_g.

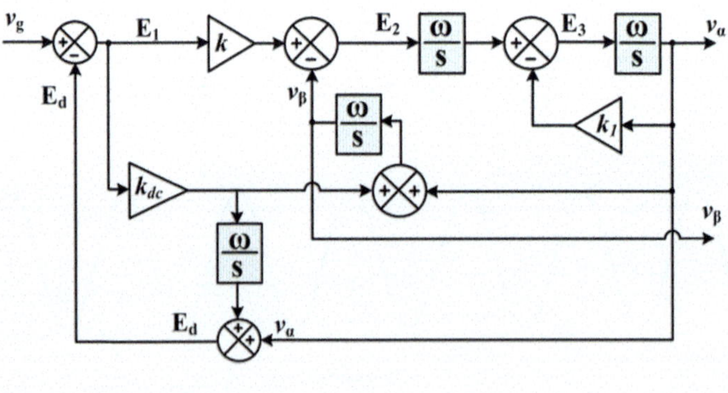

Fig. 32 Structure of ETOGI [8]

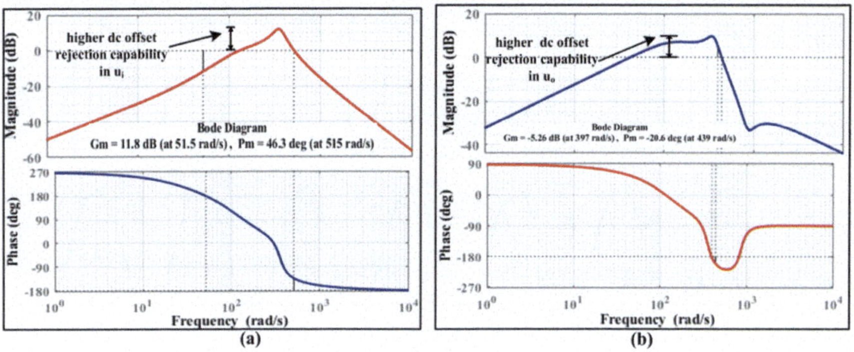

Fig. 33 Bode plot for ETOGI, **a** in-phase voltage component and **b** orthogonal of phase voltage component

The transfer function expressions for in-phase, orthogonal phase voltage component, and error component of ETOGI are given as

$$\frac{v_\alpha}{v_g} = \left[\frac{(ks^2 - k_{dc}s\omega)\omega^2}{s^4 + (k_{dc} + k_1)s^3\omega + (k_{dc}k_1 + k)s^2\omega^2 + (1 - k_{dc})s\omega^3 + k_{dc}\omega^4}\right] \tag{68}$$

$$\frac{v_\beta}{v_g} = \left(\frac{k_{dc}s^3\omega + k_{dc}k_1s^2\omega^2 + ks\omega^3}{s^4 + (k_{dc} + k_1)s^3\omega + (k_{dc}k_1 + k)s^2\omega^2 + (1 - k_{dc})s\omega^3 + k_{dc}\omega^4}\right) \tag{69}$$

$$\frac{E_1}{v_g} = \frac{s^4 + k_1s^3\omega + s\omega^3}{s^4 + (k_{dc} + k_1)s^3\omega + (k_{dc}k_1 + k)s^2\omega^2 + (1 - k_{dc})s\omega^3 + k_{dc}\omega^4} \tag{70}$$

The Bode plot for Eqs. 68 and 69 is depicted in Fig. 33. From Fig. 33, it is noticed that it is able to extract sinusoidal synchronization signals under higher dc offset in v_g. Under dc offset in v_g, the extracted synchronization signal is depicted in Fig. 34. Also to validate its filtering capability harmonic spectra analysis is depicted in Fig. 35. From Figs. 33, 34 and 35, it is concluded that ETOGI has higher dc offset rejection and filtering capability than MSTOGI. Moreover, its filtering capability is poorer than TOSSI.

Further to improve the filtering capability, fourth-order generalize integrator is presented in the next subsection.

3.9 Fourth-Order Generalized Integrator (F_oOGI)

Figure 36 shows the structure of fourth-order generalized integrator (F_oOGI). The F_oOGI extracts reference synchronization signal without lagging phenomena under

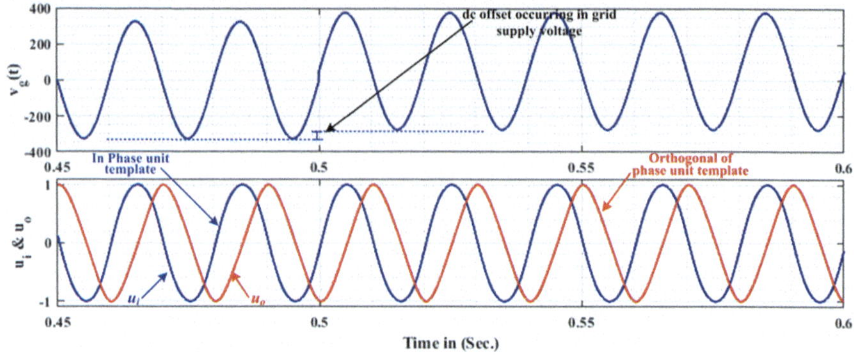

Fig. 34 Extracted synchronization signal from ETOGI under dc offset in v_g

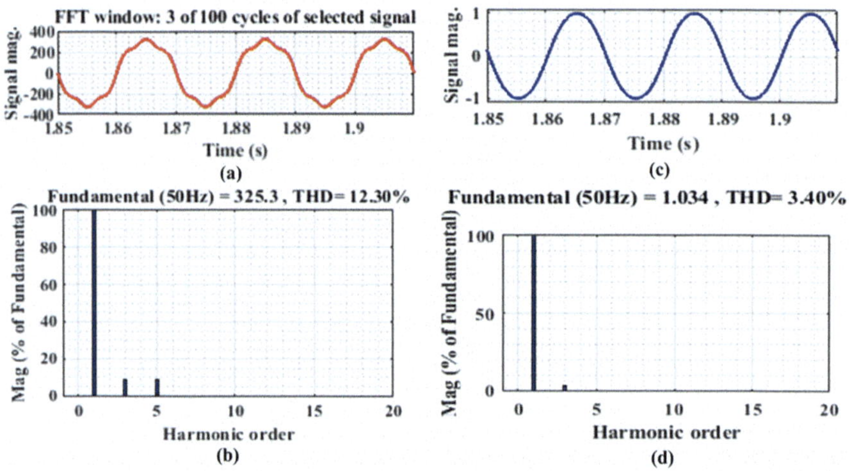

Fig. 35 Harmonic spectra analysis for ETOGI with 3rd and 5th harmonics, **a** v_g contains with 3rd and 5th harmonics, **b** harmonic spectra of v_g, **c** extracted u_i from ETOGI, and **d** harmonic spectra of u_i

dc offset in sensed v_g and shows higher filtering capability than other previously discussed generalized integrator except mLSOGI and TOSSI.

The transfer function expressions for in-phase, orthogonal phase voltage component, and error component of F_oOGI are given as

$$\frac{v_\alpha}{v_g} = \frac{k_1 k_2 s^2 \omega^2}{s^4 + k_1 \omega s^3 + (k_1 k_2 + 2)s^2 \omega^2 + k_1 s \omega^3 + \omega^4} \tag{71}$$

$$\frac{v_\beta}{v_g} = \frac{k_1 k_2 s \omega^3}{s^4 + k_1 \omega s^3 + (k_1 k_2 + 2)s^2 \omega^2 + k_1 s \omega^3 + \omega^4} v_\alpha \tag{72}$$

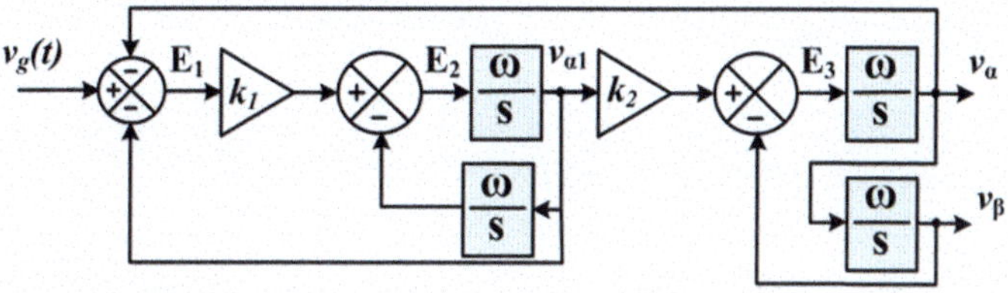

Fig. 36 Structure of F_0OGI [9]

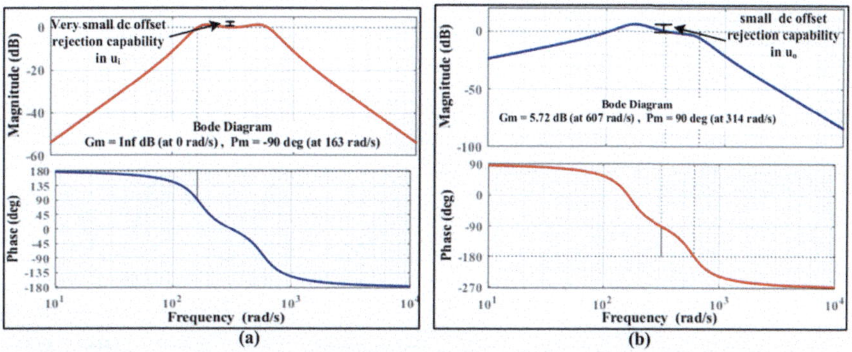

Fig. 37 Bode plot for FoOGI, **a** in-phase voltage component and **b** orthogonal of phase voltage component

$$\frac{E_1}{v_g} = -k_1\left(\frac{s^5\omega + (k_1 + k_1k_2)s^4\omega^2 + \left(2 + k_1^2k_2\right)s^3\omega^3 + (k_1 + k_1k_2)s^2\omega^4 + s\omega^5}{\begin{array}{c} s^6 + 2k_1s^5\omega + \left(k_1^2 + k_1k_2 + 3\right)s^4\omega^2 + \left(k_1 + k_1^2k_2 + 3k_1\right)s^3\omega^3 \\ +\left(k_1k_2 + 3 + k_1^2\right)s^2\omega^4 + 2k_1s\omega^5 + \omega^6 \end{array}}\right) \tag{73}$$

The Bode plot for Eqs. 71 and 72 is depicted in Fig. 37. From Fig. 37, it is noticed that it is capable of extracting synchronization signals under higher dc offset in v_g. But its dc offset rejection capability is worse than ETOGI. Under dc offset in v_g, the extracted synchronization signal is depicted in Fig. 38. Its filtering capability is validated at 3rd and 5th harmonic components present in v_g. Its harmonic spectra analysis is depicted in Fig. 39.

3.9.1 Fifth-Order Generalized Integrator (F_iOGI)

Figure 40 shows the structure of fifth-order generalized integrator (F_iOGI). The F_iOGI ensures higher dc offset rejection (except ETOGI) and filtering capability than other previously discussed generalized integrator except mLSOGI, ESOGI, TOSSI, and F_0OGI.

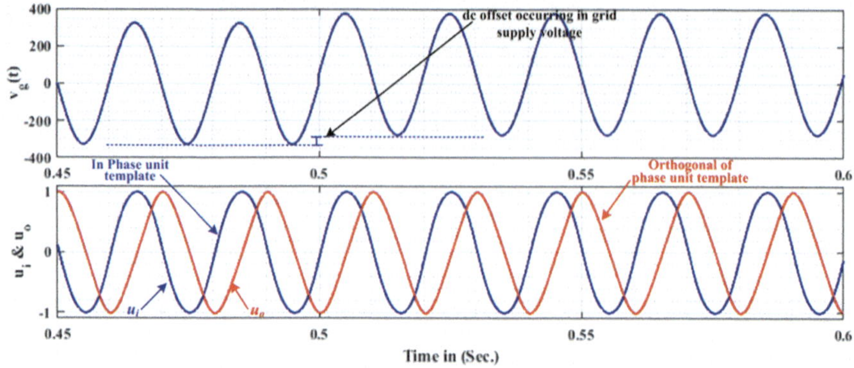

Fig. 38 Extracted synchronization signal from FoOGI under dc offset in v_g

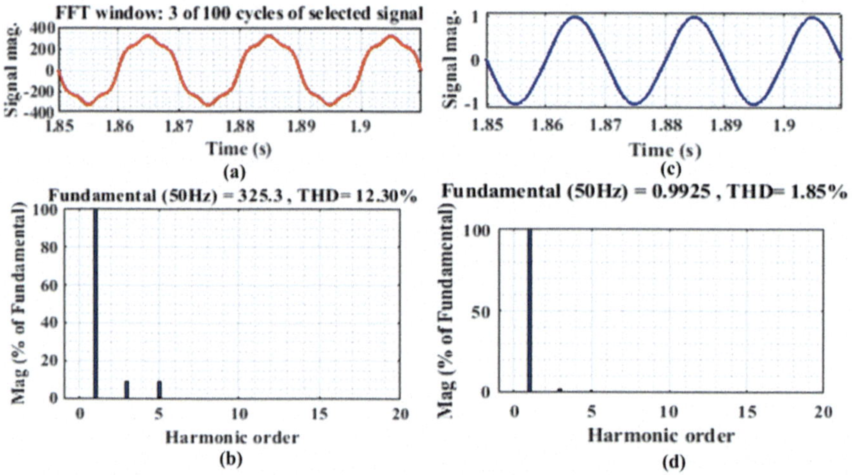

Fig. 39 Harmonic spectra analysis for F_oOGI with 3rd and 5th harmonics, **a** v_g contains with 3rd and 5th harmonics, **b** harmonic spectra of v_g, **c** extracted u_i from F_oOGI, and **d** harmonic spectra of u_i

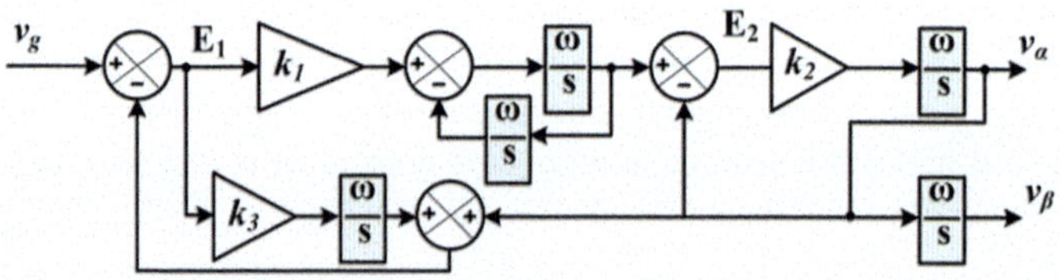

Fig. 40 Structure of F_iOGI [10]

The transfer function expressions for in-phase, orthogonal of phase voltage component, and error component of F_iOGI are given as

$$\frac{v_\alpha}{v_g} = \frac{k_1 k_2 \omega^2 s^3}{s^5 + k_3 \omega s^4 + (1 + k_2 + k_1 k_2)\omega^2 s^3 + (k_2 + k_2 k_3)\omega^3 s^2 + k_2 \omega^4 s + k_1 k_2 \omega^5} \quad (74)$$

$$\frac{v_\beta}{v_g} = \frac{k_1 k_2 \omega^3 s^2}{s^5 + k_3 \omega s^4 + (1 + k_2 + k_1 k_2)\omega^2 s^3 + (k_2 + k_2 k_3)\omega^3 s^2 + k_2 \omega^4 s + k_1 k_2 \omega^5} \quad (75)$$

$$\frac{E_1}{v_g} = \left(\frac{s^6 + k_3 \omega s^5 + (1 + k_2)\omega^2 s^4 + (k_2 + k_2 k_3)\omega^3 s^3 + k_2 \omega^4 s^2 + k_1 k_2 \omega^5 s}{\begin{array}{l} s^6 + 2k_3 \omega s^5 + (1 + k_2 + k_1 k_2 + k_3^2)\omega^2 s^4 + (1 + 2k_2 + k_1 k_2 + k_2 k_3)\omega^3 s^3 \\ + (k_2 + k_2 k_3 + k_2 k_3^2)\omega^4 s^2 + (k_2 k_3 + k_1 k_2)\omega^5 s + k_1 k_2 k_3 \omega^6 \end{array}} \right) \quad (76)$$

The Bode plot for Eqs. 74 and 75 is depicted in Fig. 41. From Fig. 41, it is observed that it is capable of extracting synchronization signals under higher dc offset present in v_g. Under these conditions, the extracted synchronization signal is depicted in Fig. 42. Its filtering capability is validated at 3rd and 5th harmonic components present in v_g. Its harmonic spectra analysis is depicted in Fig. 43.

Its dc offset rejection capability is better, however its filtering capability is poor. To enhance its filtering ability further, its modified version with multilayer is reported in next subsection of the chapter.

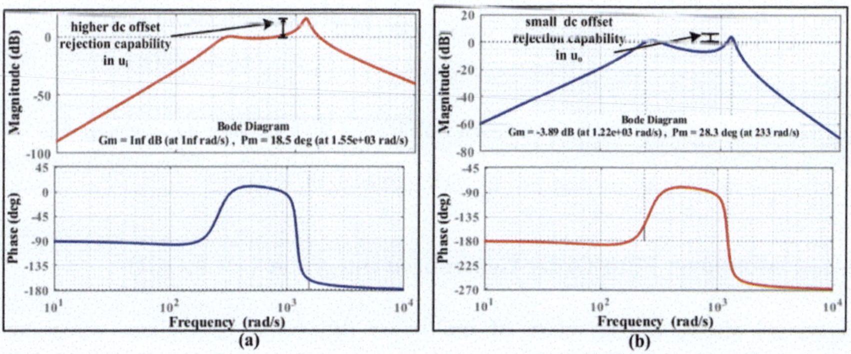

Fig. 41 Bode plot for F_iOGI, **a** in-phase voltage component and **b** orthogonal of phase voltage component

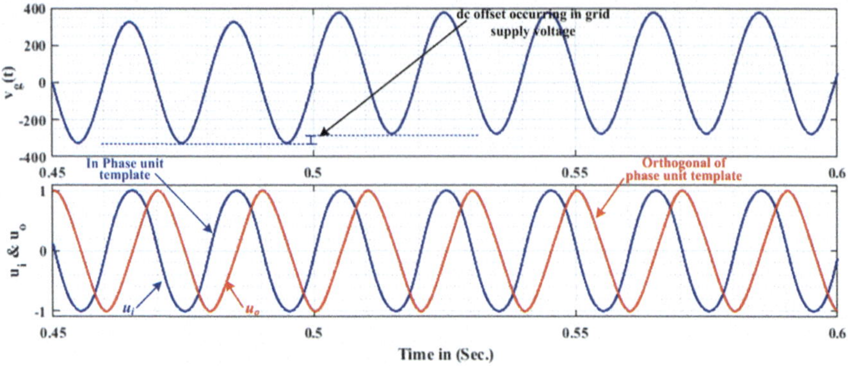

Fig. 42 Extracted synchronization signal from F_iOGI under dc offset in v_g

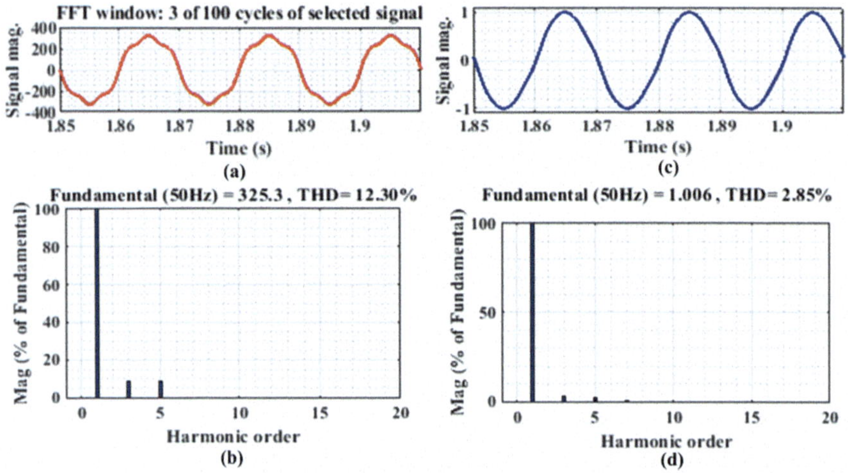

Fig. 43 Harmonic spectra analysis for F_iOGI with 3rd and 5th harmonics, **a** v_g contains with 3rd and 5th harmonics, **b** harmonic spectra of v_g, **c** extracted u_i from F_iOGI, and **d** harmonic spectra of u_i

3.9.2 Multilayer Fifth-Order Generalized Integrator (MLF_iOGI)

Figure 44 shows the structure of multilayer fifth-order generalized integrator (MLF_iOGI). The MLF_iOGI ensures higher dc offset rejection and harmonic component filtering capability than other previously discussed generalized integrator.

The expressions used for MLF_iOGI are given as

$$\frac{v_\alpha}{v_g} = \left\{\frac{k_1 k_2 s^3 \omega^2}{\Psi_\alpha}\right\}\left\{1 - \frac{v_{\alpha 3}}{v_g} - \frac{v_{\alpha 5}}{v_g}\right\} \tag{77}$$

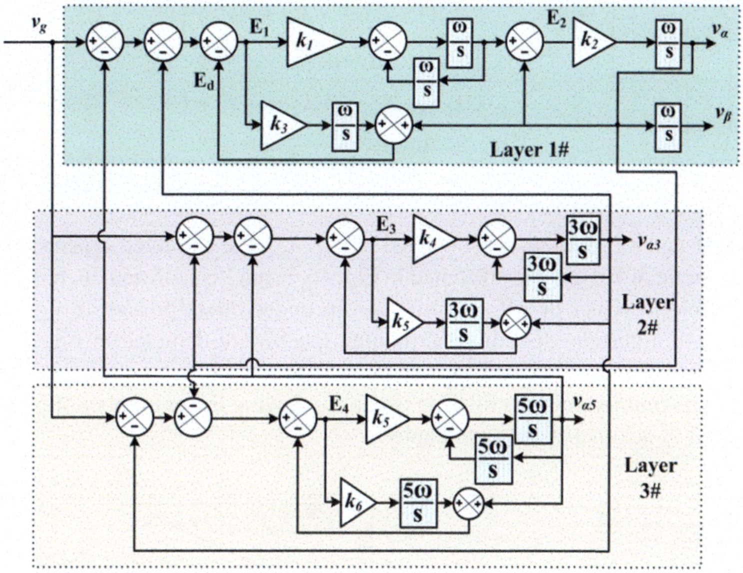

Fig. 44 Structure of MLFiOGI [11]

$$\frac{v_\beta}{v_g} = \left\{\frac{k_1 k_2 s^3 \omega^3}{\Psi_\alpha}\right\}\left\{1 - \frac{v_{\alpha 3}}{v_g} - \frac{v_{\alpha 5}}{v_g}\right\} \tag{78}$$

$$\frac{v_{\alpha 3}}{v_g} = \left\{\frac{k_4 s^2 (3\omega)}{\Psi_{\alpha 3}}\right\}\left\{1 - \frac{v_\alpha}{v_g} - \frac{v_{\alpha 5}}{v_g}\right\} \tag{79}$$

$$\frac{v_{\alpha 5}}{v_g} = \left\{\frac{k_6 s^2 (5\omega)}{\Psi_{\alpha 5}}\right\}\left\{1 - \frac{v_\alpha}{v_g} - \frac{v_{\alpha 3}}{v_g}\right\} \tag{80}$$

where $\psi_\alpha = s^5 + (k_2 + k_3)\omega s^4 + (2 + k_1 k_2 + k_2 k_3)\omega^2 s^3 + (k_2 + 2k_3)\omega^3 s^2 + (1 + k_2 k_3)\omega^4 s + k_3 \omega^5$ $\psi_{\alpha 3} = s^3 + (k_4 + k_5)(3\omega)s^2 + (3\omega)^2 s + k_5 (3\omega)^3$ and $\psi_{\alpha 5} = s^3 + (k_6 + k_7)(5\omega)s^2 + (5\omega)^2 s + k_7 (5\omega)^3$

From above equations, transfer functions of alpha and beta components are given as

$$\frac{v_\alpha}{v_g} = \left(\frac{k_1 k_2 \omega^2 s^3 \begin{bmatrix} \Psi_{\alpha 3}\Psi_{\alpha 5} - k_4 k_6 (15\omega^2) s^4 - k_4 (3\omega) s^2 * \left\{\Psi_{\alpha 5} - k_6 (5\omega) s^2\right\} \\ -k_6 (5\omega) s^2 \left\{\Psi_{\alpha 3} - k_4 (3\omega) s^2\right\} \end{bmatrix}}{\begin{bmatrix} \Psi_\alpha \left\{\Psi_{\alpha 3}\Psi_{\alpha 5} - 15 k_4 k_6 \omega^2 s^4\right\} - 3 k_1 k_2 k_4 \omega^3 s^5 * \left\{\Psi_{\alpha 5} - 5 k_6 \omega s^2\right\} \\ -5 k_1 k_2 k_6 \omega^3 s^5 \left\{\Psi_{\alpha 3} - 3 k_4 \omega s^2\right\} \end{bmatrix}}\right) \tag{81}$$

$$\frac{v_\beta}{v_g}=\left(\frac{k_1k_2\omega^3s^2\begin{bmatrix}\Psi_{\alpha3}\Psi_{\alpha5}-k_4k_6\left(15\omega^2\right)s^4-k_4(3\omega)s^2*\left\{\Psi_{\alpha5}-k_6(5\omega)s^2\right\}\\-k_6(5\omega)s^2\left\{\Psi_{\alpha3}-k_4(3\omega)s^2\right\}\end{bmatrix}}{\begin{bmatrix}\Psi_\alpha\left\{\Psi_{\alpha3}\Psi_{\alpha5}-15k_4k_6\omega^2s^4\right\}-3k_1k_2k_4\omega^3s^5*\left\{\Psi_{\alpha5}-5k_6\omega s^2\right\}\\-5k_1k_2k_6\omega^3s^5\left\{\Psi_{\alpha3}-3k_4\omega s^2\right\}\end{bmatrix}}\right) \tag{82}$$

For Eqs. 81 and 82, Bode plot is depicted in Fig. 45. The extracted synchronization signal under dc offset in vg is presented in Fig. 46. From Figs. 45 and 46, it is noticed that it is able to reject dc offset from u_i even under offset present in v_g. Further to validate its filtering capability, harmonics spectra are depicted in Fig. 47 at v_g contained with 3rd and 5th harmonic components.

Further its improvement in filtering capability having lesser number of integrator is described in next subsection of chapter.

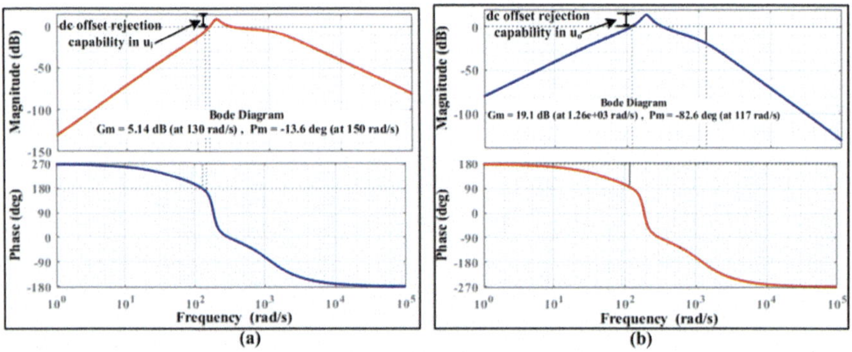

Fig. 45 Bode plot for MLF$_i$OGI, **a** in-phase voltage component and **b** orthogonal of phase voltage component

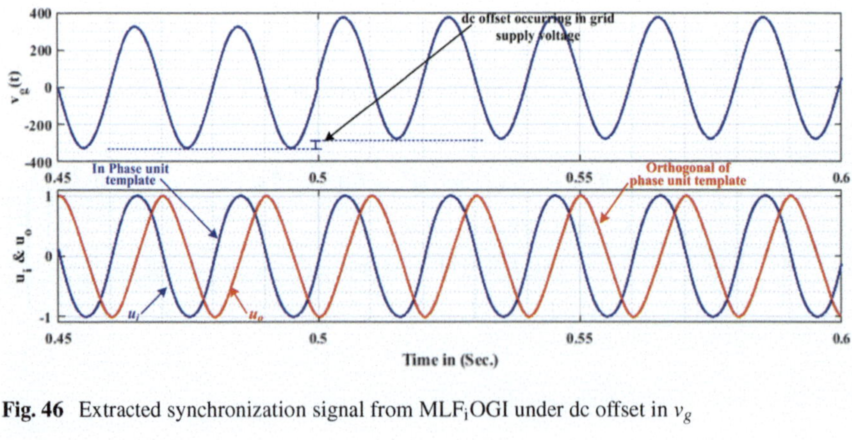

Fig. 46 Extracted synchronization signal from MLF$_i$OGI under dc offset in v_g

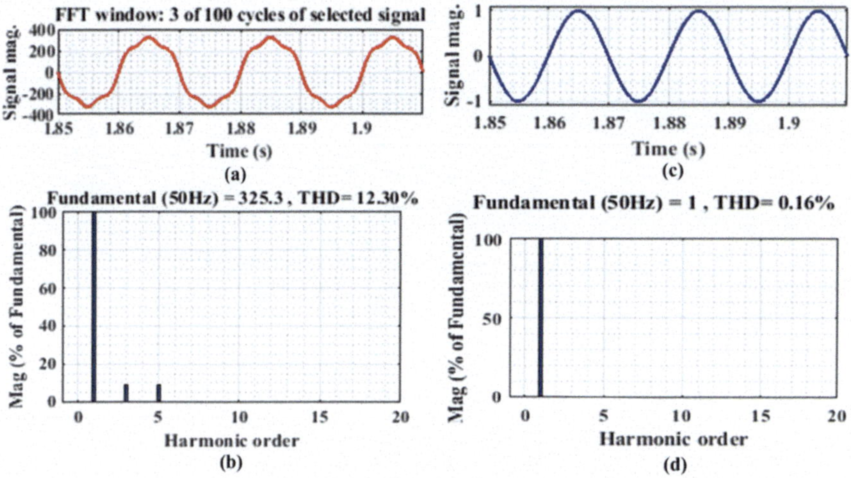

Fig. 47 Harmonic spectra analysis for MLF$_i$OGI with 3rd and 5th harmonics, **a** v_g contains with 3rd and 5th harmonics, **b** harmonic spectra of v_g, **c** extracted u_i from MLF$_i$OGI, and **d** harmonic spectra of u_i

3.9.3 Modified Multilayer Fifth-Order Generalized Integrator

Figure 48 shows the structure of modified multilayer fifth-order generalized integrator (mMLF$_i$OGI).

The main expressions used for mMLF$_i$OGI are given as

$$E_1 = v_g - E_d - v_{\alpha 3} - v_{\alpha 5} \tag{83}$$

$$E_2 = \frac{k_1 s\omega}{s^2 + \omega^2} E_1 - v_\alpha \tag{84}$$

$$v_\alpha = \frac{k_2 s\omega}{s^2 + \omega^2} E_2 \tag{85}$$

$$E_d = v_\alpha + E_1 \frac{k_3 \omega}{s} \tag{86}$$

$$\frac{v_\alpha}{v_g} = \left\{ \frac{\mathrm{k_1 k_2 s^3 \omega^2}}{\Psi_\alpha} \right\} \left\{ 1 - (\frac{v_{\alpha 3}}{v_g} - \frac{v_{\alpha 5}}{v_g} \right\} \tag{87}$$

$$\frac{v_\beta}{v_g} = \left\{ \frac{k_1 k_2 s^3 \omega^3}{\Psi_\alpha} \right\} \left\{ 1 - \frac{v_{\alpha 3}}{v_g} - \frac{v_{\alpha 5}}{v_g} \right\} \tag{88}$$

$$E_3 = v_g - v_\alpha - v_{\alpha 5} - v_{\alpha 3} \tag{89}$$

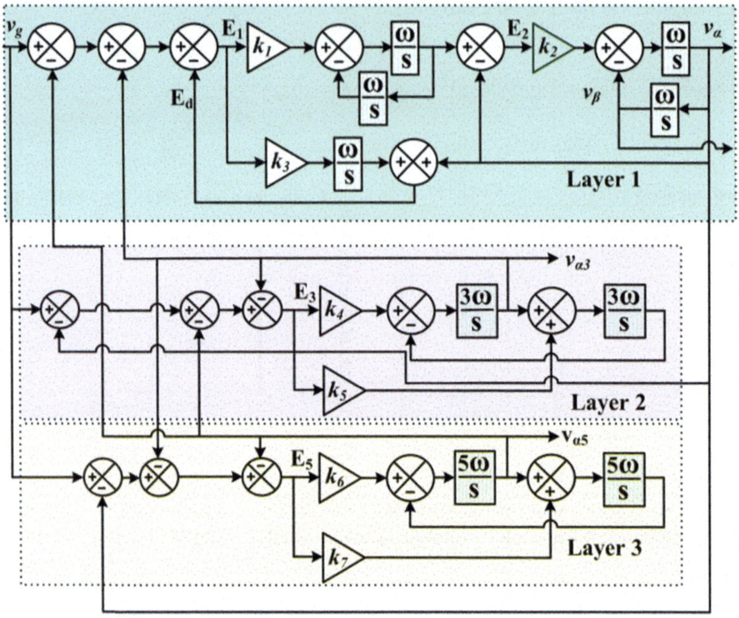

Fig. 48 Structure of mMLF$_i$OGI [12]

$$v_{\alpha 3} = \left[\frac{k_4(3\omega)s - k_5(3\omega)^2}{s^2 + (3\omega)^2}\right] E_3 \tag{90}$$

$$\frac{v_{\alpha 3}}{v_g} = \left\{\frac{k_4 s(3\omega) - k_5(3\omega)^2}{\Psi_{\alpha 3}}\right\}\left\{1 - \frac{v_\alpha}{v_g} - \frac{v_{\alpha 5}}{v_g}\right\} \tag{91}$$

$$E_5 = v_g - v_\alpha - v_{\alpha 3} - v_{\alpha 5} \tag{92}$$

$$v_{\alpha 5} = \frac{k_6(5\omega)s - k_7(5\omega)^2}{s^2 + (5\omega)^2} E_5 \tag{93}$$

$$\frac{v_{\alpha 5}}{v_g} = \left\{\frac{k_6 s(5\omega) - k_7(5\omega)^2}{\Psi_{\alpha 5}}\right\}\left\{1 - \frac{v_\alpha}{v_g} - \frac{v_{\alpha 3}}{v_g}\right\} \tag{94}$$

$$\frac{v_\alpha}{v_g} = \frac{k_1 k_2 \omega^2 s^3 \left[\Psi_{\alpha 3}\Psi_{\alpha 5} + \left\{-k_4(3\omega)s + k_5(3\omega)^2\right\}\Psi_{\alpha 5} + \left\{-k_6(5\omega)s + k_7(5\omega)^2\right\}\Psi_{\alpha 3}\right]}{\Psi_\alpha \Psi_{\alpha 3}\Psi_{\alpha 5}} \tag{95}$$

$$\frac{v_\beta}{v_g} = \frac{k_1 k_2 \omega^3 s^2 \left[\Psi_{\alpha 3}\Psi_{\alpha 5} + \left\{-k_4(3\omega)s + k_5(3\omega)^2\right\}\Psi_{\alpha 5} + \left\{-k_6(5\omega)s + k_7(5\omega)^2\right\}\Psi_{\alpha 3}\right]}{\Psi_\alpha \Psi_{\alpha 3}\Psi_{\alpha 5}} \tag{96}$$

where $\psi_\alpha = s^5 + (k_2 + k_3)\omega s^4 + (2 + k_1 k_2 + k_2 k_3)\omega^2 s^3 + (k_2 + 2k_3)\omega^3 s^2 + (1 + k_2 k_3)\omega^4 s + k_3 \omega^5$

$\psi_{\alpha 3} = s^2 + k_4(3\omega)s + (1 - k_5)(3\omega)^2$ and $\psi_{\alpha 5} = s^2 + k_6(5\omega)s + (1 - k_7)(5\omega)^2$.

The Bode plot for Eqs. 95 and 96 is depicted in Fig. 49. However, Fig. 50 shows the extracted synchronization signal under dc offsets in v_g. Further to validate its filtering capability, its harmonics spectra are presented in Fig. 51 at v_g which contains 3rd and 5th harmonics.

Based on above discussed OSG schemes, a comparative analysis is presented in Table 1.

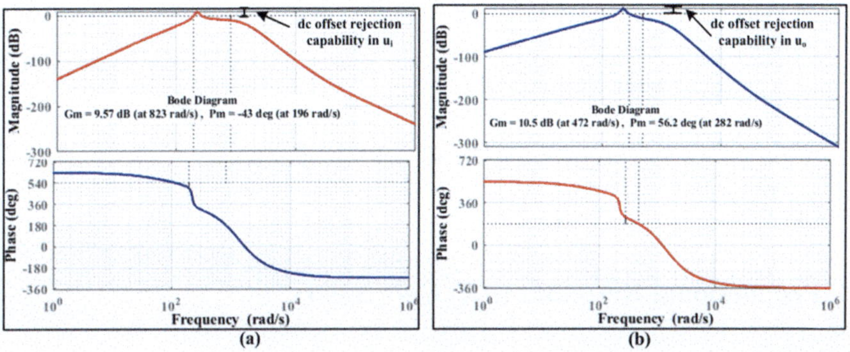

Fig. 49 Bode plot for mMLF$_i$OGI, **a** in-phase voltage component and **b** orthogonal of phase voltage component

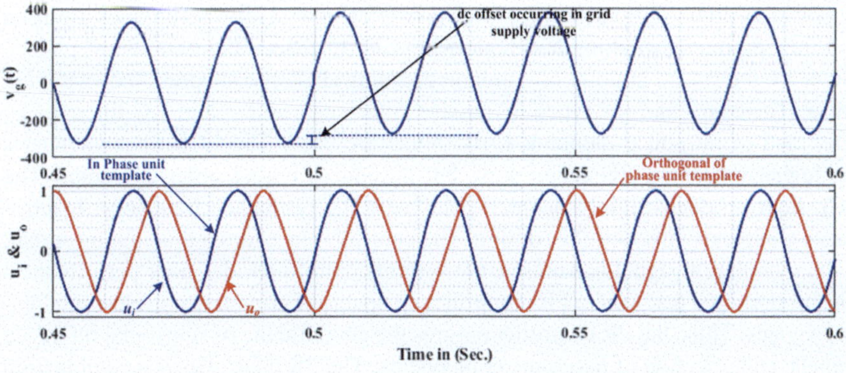

Fig. 50 Extracted synchronization signal from mMLF$_i$OGI under dc offset in v_g

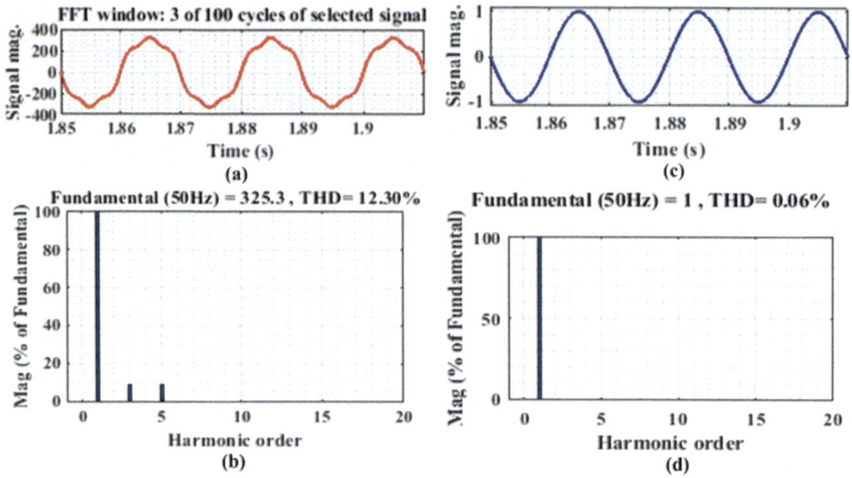

Fig. 51 Harmonic spectra analysis for mMLF$_i$OGI with 3rd and 5th harmonics, **a** v_g contains with 3rd and 5th harmonics, **b** harmonic spectrum of v_g,, **c** extracted u_i from mMLF$_i$OGI, and **d** harmonic spectrum of u_i

4 Proportional Resonant (PR) Controllers

With the integration of renewable energy sources and power electronic converters with the utility grid, the utility grid becomes more complex and it introduces difficulty in its control. In the distributed power generation infrastructure, the interfacing of different power generation systems with the utility grid must comply standards of power system. To control the power converter, modulation techniques are used and they produce switching harmonics. In order to reduce the harmonics, passive filters are interfaced at point of common coupling (PCC) to suppress the switching noise. There are various types of passive filter such as L-filter, LC-filter, LCL-filter, and to introduce damping L-RC filters are used. The LCL-filter attenuates the higher switching frequency noise but introduces higher order dynamics. The LCL-filter produces lightly damped poles with the resonance frequency. It introduces complexity in controller design. The controller provides a modulating signal; if the modulation signal has noises, the system control provides poor performance. The proportional resonance (PR) controller is capable of providing noiseless modulation signals. Thus, the PR controller ensures the good tracing of sinusoidal reference signals, and rejection of harmonic components. The PR controller is a combination of proportional and resonance controller as shown in Fig. 52.

Table 1 Comparative analysis of OSG schemes

S. no	OSG ↓	Properties →	Frequency variance capability	DC offset rejection capability	Filtering capability	THDs of unit template at v_g contains 3rd and 5th harmonics	Number of integrators
1.	Delay		No	No	No	12.30%	0
2.	SOGI		Yes	No	Yes (small)	3.31%	2
3.	mLSOGI		Yes	No	Yes (very high)	0.02%	6
4.	mSOGI		Yes	Yes(small)	Yes	3.16%	3
5.	mSOGI		Yes	Yes (high)	Yes (small)	4.73%	2
6.	ESOGI		Yes	Yes (high)	Yes (high)	1.44%	3
7.	TOSSI		Yes	No	Yes (high)	1.11	3
8.	MSTOGI		Yes	Yes (small)	Yes (small)	3.49%	3
9.	ETOGI		Yes	Yes (very high)	Yes (small)	3.40%	4
10.	$\mathrm{F_0}$OGI		Yes	Yes (small)	Yes (high)	1.85%	4
11.	$\mathrm{F_i}$OGI		Yes	Yes (high)	Yes (small)	2.85%	5
12.	$\mathrm{MLF_i}$OGI		Yes	Yes (high)	Yes (very high)	0.16%	11
13.	$\mathrm{mMLF_i}$OGI		Yes	Yes (very high)	Yes (very high)	0.06%	9

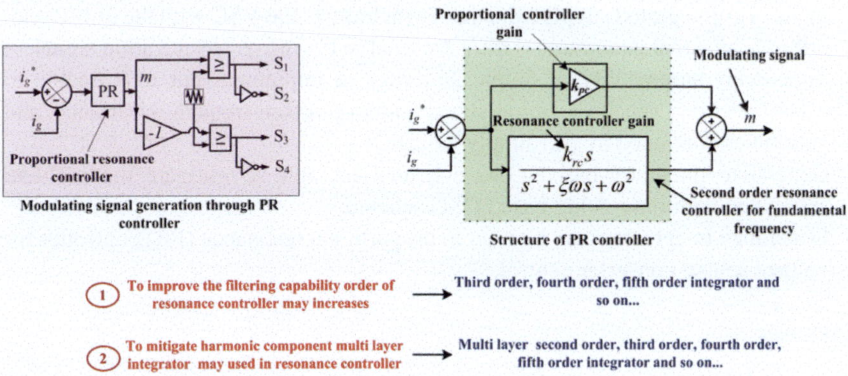

Fig. 52 Structure of PR controller

5 VSC Control

This section shows the performance analysis of VSC. Whereas VSC is used in single-phase grid-tied photovoltaic (PV) supply systems. Its control structure is shown in the introduction section of the chapter. The F_iOGI synchronization techniques are used to obtain the VSC performance. The structure of F_iOGI is described in Sect. 3.10. The rating of PV power is considered 600VA and its performances are obtained at 110 V, 50 Hz grid supply conditions with non-linear nature of load at the PCC. Figure 53 shows the power quality analysis of electrical signals at the end of the utility grid. Under this condition utility grid voltage is sinusoidal having total harmonics distortion of 2.1% as shown in Fig. 53g. The non-linear load demands power of 105 W and its current is non-sinusoidal having THDs of 27%; moreover, at the same time, current injected to the grid is pure sinusoidal having THDs of 3.8% as shown in Fig. 53f. It is obtained using a synchronization control structure.

Further modulation signal condition is observed under disturbance in grid supply as shown in Fig. 54. There harmonics are noticed in grid supply voltage, however extracted synchronization signal is sinusoidal in nature.

6 Conclusion

This book chapter presents various orthogonal signal generators with their features and limitation. The main focus of the chapter is structure of orthogonal signal generator and its applications. Further generalized integrator used in proportional resonance controller is discussed to mitigate noises in extracted modulating signal.

Key Points

- In the grid-integrated supply system, to synchronize the VSC with the utility grid, orthogonal signal generators (OSG) are used to obtain synchronization signal.
- Time-delay-based OSG is simple and easy to implement but it is unable to extract accurate synchronization signal under dc offset, frequency variance, and harmonics contained in grid supply voltage.
- Generalize integrator-based OSG schemes are able to mitigate the problem associated with time-delay-based OSG scheme.
- Generalize integrators may be used in proportional resonance (PR) controller for voltage source converter control.

Exercises

Question 1. Obtain OS in discrete time domain

a. For 60 Hz grid supply frequency with 10 kHz sampling frequency.
b. For 400 Hz frequency signal with 40 kHz sampling frequency.

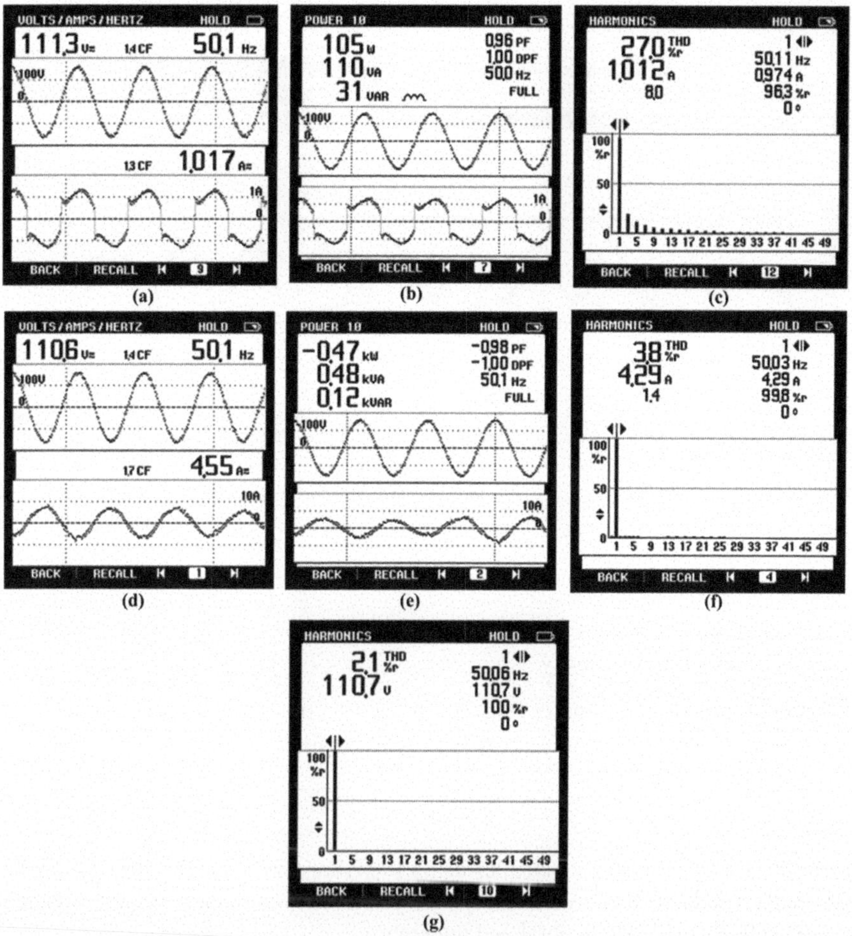

Fig. 53 VSC performance with non-linear load **a** v_g, and i_L, **b** P_L, **c** THDs of i_L, **d** v_g and i_g, **e** P_g, **f** THDs of i_g, and**g** THDs of v_g

Question 2. Obtain the Bode plot analysis for SOGI, mSOGI, and ESOGI with considering

a. 60 Hz natural frequency and 0.8 damping ratio.
b. 400 Hz natural frequency and 0.707 damping ratio.

And comment on dc offset and harmonics rejection capability of SOGI, mSOGI, and ESOGI.

Question 3. Obtain the discrete time domain model of TOSSI, MSTOGI, and ETOGI with considering

a. 10 kHz sampling frequency.

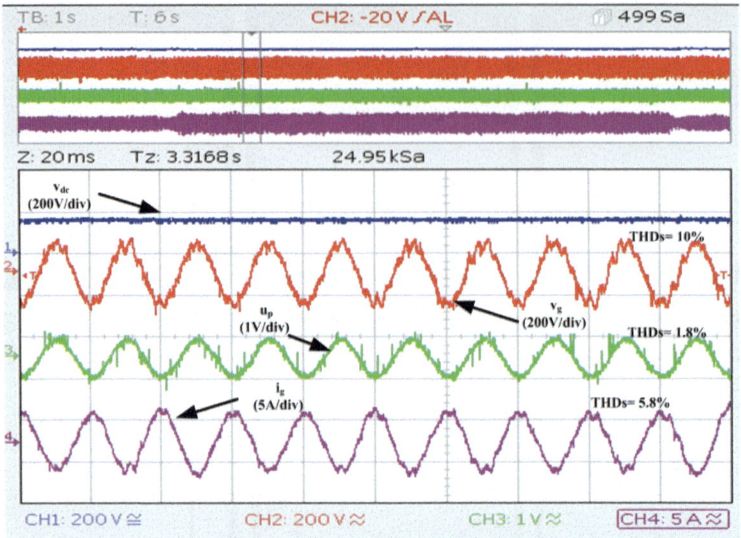

Fig. 54 Performance of F_iOGI under presence of harmonics in v_g

b. 50 kHz sampling frequency.

Question 4. Obtain the transfer function in discrete time domain for in-phase and orthogonal phase component of grid supply voltage for,

a. F_oOGI.
b. F_iOGI.

Question 5. Develop a PR controller using F_iOGI and MLF_iOGI for single-phase grid-integrated solar photovoltaic supply system and comment on grid-injected current THDs under both generalized integrators.

References

1. Ciobotaru M, Teodorescu R, Blaabjerg F (2006) A new single-phase PLL structure based on second order generalized integrator. IEEE Power Electron Spec Conf, 1–6
2. Han Y, Luo M, Zhao X, Guerrero JM, Xu L (2016) Comparative performance evaluation of orthogonal-signal-generators-based single-phase PLL algorithms—a survey. IEEE Trans Power Electron 31(5):3932–3944
3. Xie M, Wen H, Zhu C, Yang Y (2017) DC offset rejection improvement in single-phase SOGI-PLL algorithms: methods review and experimental evaluation. IEEE Access 5:12810–12819
4. Hackl CM, Landerer M (2020) Modified second-order generalized integrators with modified frequency locked loop for fast harmonics estimation of distorted single-phase signals. IEEE Trans Power Electron 35(3):3298–3309
5. Nazir FU, Kumar N, Pal BC, Singh B, Panigrahi BK (2020) Enhanced SOGI controller for weak grid integrated solar PV system. IEEE Trans Energy Conv 35(3):1208–1217

6. Kumar S, Patel LN, Singh B, Vyas AL (2020) Self-adjustable step-based control algorithm for grid-interactive multifunctional single-phase PV-battery system under abnormal grid conditions. IEEE Trans Ind Appl 56(3):2978–2987
7. Zhang C, Zhao X, Wang X, Chai X, Zhang Z, Guo X (2018) A grid synchronization PLL method based on mixed second- and third-order generalized integrator for DC offset elimination and frequency adaptability. IEEE J Emerg Stele. Topics Power Electron 6(3):1517–1526
8. Pardhi PK, Sharma SK (2021) High Gain Non Isolated DC converter employed in single-phase grid-tied solar photovoltaic supply system. IEEE Trans Ind Appl 57(5):5170–5182
9. Saxena V, Kumar N, Singh B, Panigrahi BK (2021) An MPC based algorithm for a multipurpose grid integrated solar PV system with enhanced power quality and PCC voltage assist. IEEE Trans Energy Conv 36(2):1469–1478
10. Pardhi PK, Sharma SK, Chandra A (2020) Solar photovoltaic roof-top system design and control employing high gain DC converter. IEEE Ind Appl Soc Ann Meet, 1–6
11. Pardhi PK, Sharma SK, Chandra A (2019) Control of single phase solar photovoltaic supply system. IEEE Ind Appl Soc Ann Meet, 1–14
12. Pardhi PK, Sharma SK, Chandra A (2020) Control of single-phase solar photovoltaic supply system. IEEE Trans Ind Appl 56(6):7132–7144
13. Ebrahim MA, Aziz BA, Nashed MNF, Osman FA (2021) Optimal design of proportional-resonant controller and its harmonic compensators for grid-integrated renewable energy sources based three-phase voltage source inverters. IET Gene Transmits Distr 15(8):1371–1386

Chapter 10
Reduced Switch Multilevel Inverter with Solar Photovoltaic System for Enhanced Power Quality

Veramalla Rajagopal and Salkuti Surender Reddy

Abstract In this chapter, a 127-level multilevel inverter (MLI) which is designed for a solar photovoltaic energy unit is proposed and the configurations of 15, 31, and 63 MLIs are also compared with the same topology. The proposed technique while compared with other popular MLIs, it is clear that this model uses very less semiconductor switches to get the most number of steps in the output AC voltage. This topology uses a square wave switching in the place of pulse width modulation, which helps to reduce the switching losses. Since the switches used are minimum, the cost of the system is considerably reduced. The harmonics and the total harmonic distortion are found to be less in the output waveform. This 127-level MLI is simulated in MATLAB/SIMULINK R2009a and the results are compared with 15-level, 31-level, and 63-level MLIs.

Keywords Power quality · Reduced switch multilevel inverter · Cost saving · Solar photovoltaic system

1 Introduction

As the world is getting advanced, the electric power demand is also increasing. The conventional energy sources, which are used at present, create many environmental issues. The reduction in the availability of fossil fuels considerably affects the electricity generation. Thus, the situation demands the proper utilization of those energy sources, which are non-conventional like hydro, wind, biomass, and solar, which are eco-friendly.

V. Rajagopal (✉)
Department of Electrical and Electronics Engineering, Kakatiya Institute of Technology and Science, Warangal 506015, India
e-mail: vrg.eee@kitsw.ac.in

S. Surender Reddy
Department of Railroad and Electrical Engineering, Woosong University, Daejeon 34606, Korea
e-mail: surender@wsu.ac.kr

A. K. Giri et al. (eds.), *Power Quality: Infrastructures and Control*,
Studies in Infrastructure and Control,
https://doi.org/10.1007/978-981-19-7956-9_10

The earth receives nearly 173 Terawatts of solar power, almost thousands of times more than the whole world's demand. If it is possible to convert all the energy that falls on the earth's surface to electric energy, then the whole world can comfortably rely on the solar energy conversion to meet the energy demand. The solar power has the advantages of being clean and pollution-free. As the technology is progressed, the solar panels and the associated equipment have become affordable [1]. The photovoltaic (PV) cells produce a low value of DC voltage. In order to interface the same with the grid, the voltage is to be increased to a higher value and to be converted to AC. For increasing the DC-generated voltage, many series-connected PV cells are used [2]. Conversion of high DC voltage to AC or stepping up of AC requires highly rated conversion equipment and transformer (step-up) which in turn enhance the overall size, weight and losses, and system cost [3]. Avoiding the transformer is one way to reduce these problems.

A multilevel inverter (MLI) is a combination of power semiconductor devices which helps to modify a staircase waveform into a sine waveform. As the traditional two-level inverters have the disadvantages of being less efficient due to the high switching losses and thus exhibit poor power quality [4, 5], multilevel inverters are good options for giving quality power output in a solar PV system. Different types of inverters like MLI-flying capacitor, MLI-cascaded (H-bridge), and MLI-neutral clamped are reported in [6–9]. The difficulty in the design and the increased count of power switches are the disadvantages of MLIs. Advanced versions of MLIs go for a limited number of switches [10, 11]. Less number of switches causes fewer losses and improves the efficiency of the MLIs [12]. A configuration, which uses a smaller number of switches to produce multiple output voltage steps for different DC input voltages, is proposed in [13, 14]. Compared to a single-phase MLI technique for minimum switch count, three-phase MLI technique is more difficult [15–17]. A more approachable technique for three-phase MLI with a lesser number of semiconductor devices [18]. The bi-directional power IGBT switches cause difficulty in the design of MLIs with minimum switches [19]. A three-phase hybrid control-based MLI operates with the least switching losses [20]. Control with soft switching cells for PV applications is suggested in [21], using which the inconvenience by hard switching is avoided. Ando et al. [22] suggest to have a transportable solar power generator with inverter. An MLI which is designed especially for the purpose of integration of a renewable energy source with the grid is suggested in [23]. The recalculation and modulation of the pulse width and height help in reducing the number of switching states and the losses. Gopal et al. [24] have reported an MLI 9-level H-bridge with two methods for the MPPT. [25–28] have different structures of MLIs with a reduced number of switches. A quazi-Z-source inverter is suitable for solar PV applications [29]. Kakar et al. [30] reported an MLI 1-Φ having less stress on the power semiconductor devices. Sathik et al. [31] explain a topology novel MLI which uses a diode half-bridge circuit. Rawa et al. [32] suggest an MLI where the switching control is designed with a reduced number of DC sources. A parking lot for the electric vehicles powered by solar with improved efficiency along with reduced cost [33].

A 17-level inverter uses a smaller number of components, but with reduced voltage stress [34]. Mukundan et al. [35] explain an H-bridge MLI which is designed for the interconnection of a solar energy unit with distributed system.

The proposed system in this paper is a 127-level MLI designed for a solar PV system. This configuration uses only ten Insulated Gate Bipolar Transistor (IGBT) switches. This number is very less as compared to other types of MLIs like flying capacitor type, diode clamped, and cascaded type. The 127–level MLI design is inexpensive as the number of IGBT switches, and thus the gate circuits, are the minimum. Here, the solar PV units with DC to DC converters act as six distinct DC sources to produce a 127-level output. This topology decreases harmonics and improves the power quality 15-, 31-, and 63-level MLI configurations are also explained using the same topology. This proposed MLI is also implemented in isolated energy systems. The efficiency of the proposed MLI is validated using MATLAB/Simulink 2009a.

2 Solar Power-Generating Unit with Battery

The solar module consists of 84 cells connected in series. Each cell has a voltage open circuit of 60 V (Voc), current short circuit of 7.8A (Isc), Rsh = 360 Ω, and Rse = 0.18 Ω. Series connected modules (six in number) are for adding up the voltages and six parallel connected modules are for increasing the current capacity.

The battery unit has six batteries numbered from V1 to V6, each having an internal resistance of 0.1 Ω. Each battery with the series resistance is in series and connected parallel with a 50,000F capacitor and a 10kΩ resistor.

3 15-Level Multilevel Inverter System

The 15-level MLI uses only 7 switches for constructing the output waveform.

3.1 Circuit Diagram

The circuit diagram for 15-level MLI is shown in Fig. 1.

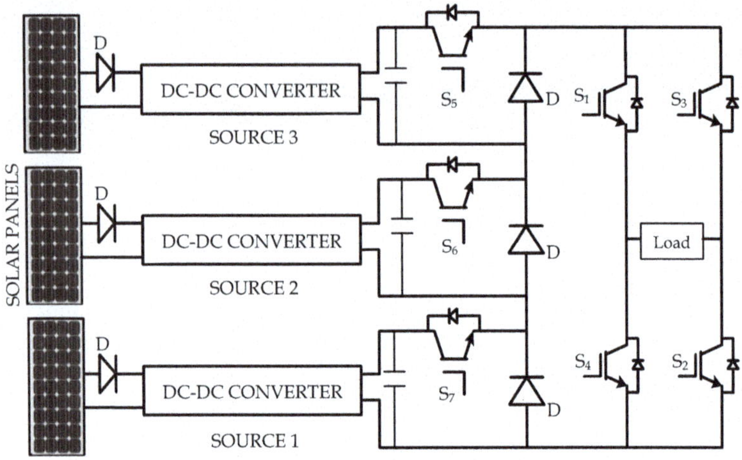

Fig. 1 Circuit diagram of 15-level MLI

3.2 *Principle of Operation*

In this 15-level MLI, three solar panels along with DC–DC converters provide input. DC–DC converters control the output voltage magnitude according to the requirement. This system uses seven IGBT switching devices which generate a 15-level output in a half cycle. The diodes are connected to allow the current to flow only in one direction. The switching devices are triggered in such a manner to acquire the output voltage. The switches S1, S2, S3, and S4 form the inverter circuit for the conversion of DC to AC voltage and SW5, SW6, and SW7are for adjusting the output voltage magnitude of the solar panel. S1 and S2 conduct to give the half cycle (positive) and S3 and S4 conduct to give the half cycle (negative) of the output voltage. With this configuration, the harmonics in the output waveform are found to be very less and the output voltage waveform looks nearly sinusoidal.

It is considered that $V_1 = 10X$ V, $V_2 = 20X$ V, and $V_3 = 40X$ V, where X is the multiplication factor and is given by 325.22/70.

3.3 *Sequence of Switching in a 15-Level Multilevel Inverter*

Table 1 displays the switching sequences of the switches used in this circuit: 1–8 sequences in the half cycle (positive) and 1–8 sequences in the half cycle (negative).

Table 1 Pattern of Switching of SW1–SW7 in a 15-level MLI for positive and negative half cycles

Positive half cycle								
STEP	*S1*	*S2*	*S3*	*S4*	*S5*	*S6*	*S7*	Output (v_O)
01	0	0	0	0	0	0	0	0
02	1	1	0	0	0	0	1	$v1$
03	1	1	0	0	0	1	0	$v2$
04	1	1	0	0	0	1	1	$v2 + v1$
05	1	1	0	0	1	0	0	$v3$
06	1	1	0	0	1	0	1	$v3 + v1$
07	1	1	0	0	1	1	0	$v3 + v2$
08	1	1	0	0	1	1	1	$v3 + v2 + v1$
Negative half cycle								
01	0	0	0	0	0	0	0	0
02	0	0	1	1	0	0	1	$-(v1)$
03	0	0	1	1	0	1	0	$-(v2)$
04	0	0	1	1	0	1	1	$-(v2 + v1)$
05	0	0	1	1	1	0	0	$-(v3)$
06	0	0	1	1	1	0	1	$-(v3 + v1)$
07	0	0	1	1	1	1	0	$-(v3 + v2)$
08	0	0	1	1	1	1	1	$-(v3 + v2 + v1)$

4 31-Level Multilevel Inverter System

The 31-level MLI uses only 8 switches for constructing the output waveform.

4.1 *Circuit Diagram*

The circuit diagram for 31-level MLI is shown in Fig. 2.

4.2 *Operation and Principle*

In this 31-level MLI, 4 solar panels along with DC–DC converters provide input. DC–DC converters control the output voltage magnitude according to the requirement. This system uses 8 IGBT switching devices which generate a 31-level output in a half cycle. The diodes are connected to allow the current to flow only in one direction. The

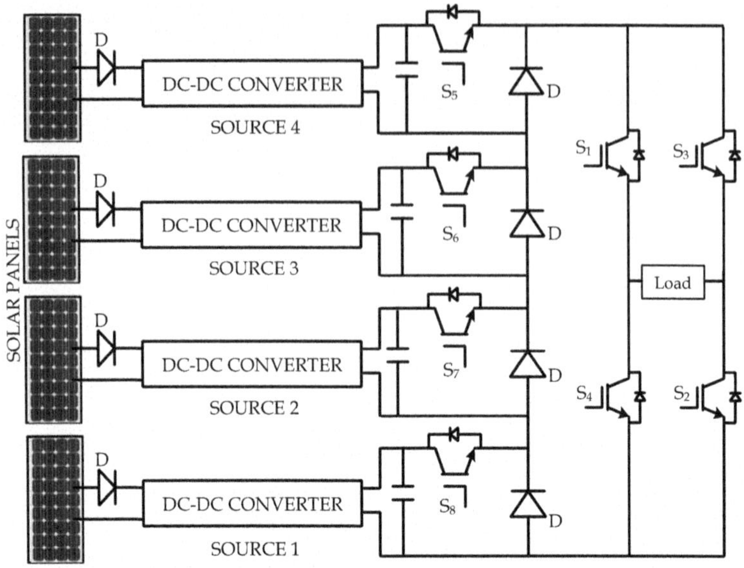

Fig. 2 Circuit diagram of level-31 MLI

switching devices are triggered accordingly to acquire the output voltage. The IGBT switches S1, S2, S3, and S4 form the inverter circuit for the conversion of DC to AC voltage and S5, S6, S7, and S8 are for adjusting the output voltage magnitude of the solar panel. S1 and S2 conduct to give the half cycle (positive) and S3 and S4 conduct to give the half cycle (negative) of the output voltage. With this configuration, the harmonics in the output waveform is found to be very less and the output voltage waveform looks nearly sinusoidal.

It is considered that $V_1 = 10X$ V, $V_2 = 20X$ V, $V_3 = 40X$V, and $V_4 = 80X$, where X is the multiplication factor and is given by 325.22/150.

4.3 Sequence of Switching in a 31-Level Multilevel Inverter

Table 2 displays the switching sequences of the switches used in this circuit: 1–16 steps in half cycle (positive) and 1–16 steps in half cycle (negative).

Table 2 Switching Patterns of SW1–SW8 in 31-level MLI for positive and negative half cycles

STEP	*S1*	*S2*	*S3*	*S4*	*S5*	*S6*	*S7*	*S8*	Output (v_0)
Positive half cycle									
01	0	0	0	0	0	0	0	0	0
02	1	1	0	0	0	0	0	1	v1
03	1	1	0	0	0	0	1	0	v2
04	1	1	0	0	0	0	1	1	(v2 + v1)
05	1	1	0	0	0	1	0	0	(v3)
06	1	1	0	0	0	1	0	1	(v3 + v1)
07	1	1	0	0	0	1	1	0	(v3 + v2)
08	1	1	0	0	0	1	1	1	(v3 + v2 + v1)
09	1	1	0	0	1	0	0	0	(v4)
10	1	1	0	0	1	0	0	1	(v4 + v1)
11	1	1	0	0	1	0	1	0	(v2 + v4)
12	1	1	0	0	1	0	1	1	(v4 + v2 + v1)
13	1	1	0	0	1	1	0	0	(v4 + v3)
14	1	1	0	0	1	1	0	1	(v4 + v3 + v1)
15	1	1	0	0	1	1	1	0	(v4 + v3 + v2)
16	1	1	0	0	1	1	1	1	(v4 + v3 + v2 + v1)
Negative half cycle									
01	0	0	0	0	0	0	0	0	0
02	1	1	0	0	0	0	0	1	-v1
03	1	1	0	0	0	0	1	0	-v2
04	1	1	0	0	0	0	1	1	−(v1 + v2)
05	1	1	0	0	0	1	0	0	−(v3)
06	1	1	0	0	0	1	0	1	−(v1 + v3)
07	1	1	0	0	0	1	1	0	−(v2 + v3)
08	1	1	0	0	0	1	1	1	−(v1 + v2 + v3)
09	1	1	0	0	1	0	0	0	−(v4)
10	1	1	0	0	1	0	0	1	−(v1 + v4)
11	1	1	0	0	1	0	1	0	−(v2 + v4)
12	1	1	0	0	1	0	1	1	−(v1 + v2 + v4)
13	1	1	0	0	1	1	0	0	−(v3 + v4)
14	1	1	0	0	1	1	0	1	−(v1 + v3 + v4)
15	1	1	0	0	1	1	1	0	−(v2 + v3 + v4)
16	1	1	0	0	1	1	1	1	−(v1 + v2 + v3 + v4)

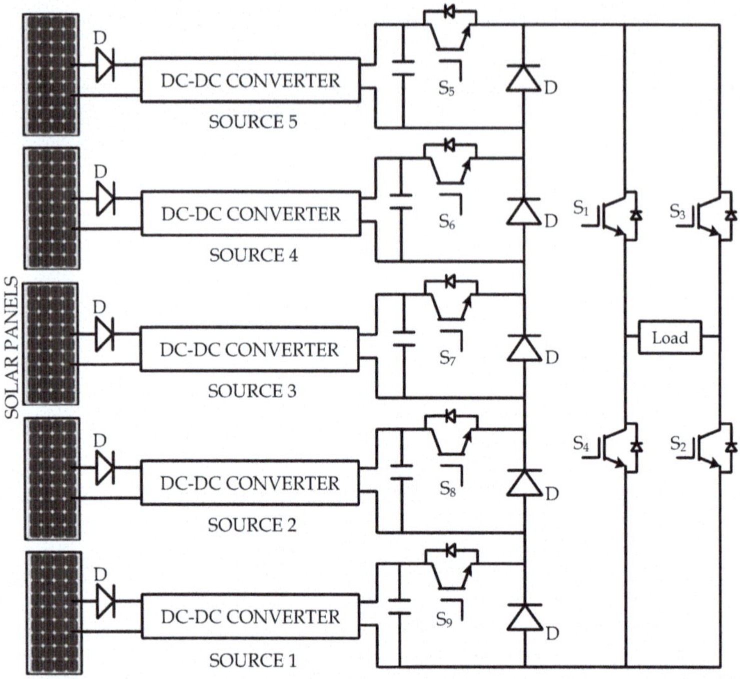

Fig. 3 Circuit diagram of 63-level multilevel inverter

5 63-Level Multilevel Inverter System

The 63-level MLI uses only 9 switches for constructing the output waveform.

5.1 Circuit Diagram

The circuit diagram for 63-level MLI is shown in Fig. 3.

5.2 Principle of Operation

In this 63-level MLI, 5 solar panels along with DC–DC converters provide input. DC–DC converters control the output voltage magnitude according to the requirement. This system uses nine IGBT switching devices which generate a 63-level output in a half cycle. The diodes are connected to allow the current to flow only in one direction. The switching devices are triggered accordingly to get the output voltage. The IGBT switches S1, S2, S3, and S4 form the inverter circuit for the conversion of DC to AC voltage and S5, S6, S7, S8, and S9 are for adjusting the output voltage magnitude of the solar panel. S1 and S2 conduct to give the half cycle (positive) and S3 and S4 conduct to give the half cycle (negative) of the output voltage. With this configuration, the harmonics in the output waveform is found to be very less and the output voltage waveform looks nearly sinusoidal.

It is considered that $V_1 = 10X$ V, $V_2 = 20X$ V, $V_3 = 40X$ V, and $V_5 = 160X$ V, where X is the multiplication factor and is given by 325.22/310.

5.3 *Sequence of Switching in a 63-Level Multilevel Inverter*

Table 3 displays the switching sequences of the switches used in this circuit:1–32 steps in the half cycle (positive) and 1–32 steps in the half cycle (negative).

6 127-Level Multilevel Inverter System

The 127-level MLI uses only 10 switches for constructing the output waveform.

6.1 *Circuit Diagram*

A 127-level MLI is shown in Fig. 4. A total of 10 switches are utilized to get 127 levels in the output voltage. As per the topology, solar modules in association with the DC to DC converters, are used as the input. The converter duty cycle is changed to get the required value of output voltage.

6.2 *Principle and Operation*

In this 127-level MLI, six solar modules and DC to DC converters provide input. DC–DC converters control the output voltage magnitude according to the requirement. This system uses 10 IGBT switching devices which generate a 127-level output in a half cycle. The diodes are connected to allow the current to flow only in one direction. The switching devices are triggered accordingly to get the output voltage. The IGBT switches S1, S2, S3, and S4 form the inverter circuit for the conversion of DC to AC voltage and S5, S6, S7, S8, S9, and S10 are for adjusting the output voltage

Table 3 Pattern of Switching of SW1–SW9 in 63-level MLI for positive and negative half cycles

Step	*S1*	*S2*	*S3*	*S4*	*S5*	*S6*	*S7*	*S8*	*S9*	Output (v_0)
Positive half cycle										
01	0	0	0	0	0	0	0	0	0	0
02	1	0	0	0	0	1	1	0	0	*v1*
03	0	1	0	0	0	1	1	0	0	*v2*
04	1	1	0	0	0	1	1	0	0	(*v1* + *v2*)
05	0	0	1	0	0	1	1	0	0	(*v3*)
06	1	0	1	0	0	1	1	0	0	(*v1* + *v3*)
07	0	1	1	0	0	1	1	0	0	(*v2* + *v3*)
08	1	1	1	0	0	1	1	0	0	(*v1* + *v2* + *v3*)
09	0	0	0	1	0	1	1	0	0	(*v4*)
10	1	0	0	1	0	1	1	0	0	(*v1* + *v4*)
11	0	1	0	1	0	1	1	0	0	(*v2* + *v4*)
12	1	1	0	1	0	1	1	0	0	(*v1* + *v2* + *v4*)
13	0	0	1	1	0	1	1	0	0	(*v3* + *v4*)
14	1	0	1	1	0	1	1	0	0	(*v1* + *v3* + *v4*)
15	0	1	1	1	0	1	1	0	0	(*v2* + *v3* + *v4*)
16	1	1	1	1	0	1	1	0	0	(*v1* + *v2* + *v3* + *v4*)
17	0	0	0	0	1	1	1	0	0	(*v5*)
18	1	0	0	0	1	1	1	0	0	(*v1* + *v5*)
19	0	1	0	0	1	1	1	0	0	(*v2* + *v5*)
20	1	1	0	0	1	1	1	0	0	(*v1* + *v2* + *v5*)
21	0	0	1	0	1	1	1	0	0	(*v3* + *v5*)
22	1	0	1	0	1	1	1	0	0	(*v1* + *v3* + *v5*)
23	0	1	1	0	1	1	1	0	0	(*v2* + *v3* + *v5*)
24	1	1	1	0	1	1	1	0	0	(*v1* + *v2* + *v3* + *v5*)
25	0	0	0	1	1	1	1	0	0	(*v4* + *v5*)
26	1	0	0	1	1	1	1	0	0	(*v1* + *v4* + *v5*)
27	0	1	0	1	1	1	1	0	0	(*v2* + *v4* + *v5*)
28	1	1	0	1	1	1	1	0	0	(*v1* + *v2* + *v4* + *v5*)
29	0	0	1	1	1	1	1	0	0	(*v3* + *v4* + *v5*)
30	1	0	1	1	1	1	1	0	0	(*v1* + *v3* + *v4* + *v5*)
31	0	1	1	1	1	1	1	0	0	(*v2* + *v3* + *v4* + *v5*)
32	1	1	1	1	1	1	1	0	0	(*v1* + *v2* + *v3* + *v4* + *v5*)
Negative half cycle										
01	0	0	0	0	0	0	0	0	0	0
02	1	0	0	0	0	1	1	0	0	−*v1*

(continued)

Table 3 (continued)

Positive half cycle

Step	*S1*	*S2*	*S3*	*S4*	*S5*	*S6*	*S7*	*S8*	*S9*	Output (v_0)
03	0	1	0	0	0	1	1	0	0	$-v2$
04	1	1	0	0	0	1	1	0	0	$-(v1 + v2)$
05	0	0	1	0	0	1	1	0	0	$-(v3)$
06	1	0	1	0	0	1	1	0	0	$-(v1 + v3)$
07	0	1	1	0	0	1	1	0	0	$-(v2 + v3)$
08	1	1	1	0	0	1	1	0	0	$-(v1 + v2 + v3)$
09	0	0	0	1	0	1	1	0	0	$-(v4)$
10	1	0	0	1	0	1	1	0	0	$-(v1 + v4)$
11	0	1	0	1	0	1	1	0	0	$-(v2 + v4)$
12	1	1	0	1	0	1	1	0	0	$-(v1 + v2 + v4)$
13	0	0	1	1	0	1	1	0	0	$-(v3 + v4)$
14	1	0	1	1	0	1	1	0	0	$-(v1 + v3 + v4)$
15	0	1	1	1	0	1	1	0	0	$-(v2 + v3 + v4)$
16	1	1	1	1	0	1	1	0	0	$-(v1 + v2 + v3 + v4)$
17	0	0	0	0	1	1	1	0	0	$-(v5)$
18	1	0	0	0	1	1	1	0	0	$-(v1 + v5)$
19	0	1	0	0	1	1	1	0	0	$-(v2 + v5)$
20	1	1	0	0	1	1	1	0	0	$-(v1 + v2 + v5)$
21	0	0	1	0	1	1	1	0	0	$-(v3 + v5)$
22	1	0	1	0	1	1	1	0	0	$-(v1 + v3 + v5)$
23	0	1	1	0	1	1	1	0	0	$-(v2 + v3 + v5)$
24	1	1	1	0	1	1	1	0	0	$-(v1 + v2 + v3 + v5)$
25	0	0	0	1	1	1	1	0	0	$-(v4 + v5)$
26	1	0	0	1	1	1	1	0	0	$-(v1 + v4 + v5)$
27	0	1	0	1	1	1	1	0	0	$-(v2 + v4 + v5)$
28	1	1	0	1	1	1	1	0	0	$-(v1 + v2 + v4 + v5)$
29	0	0	1	1	1	1	1	0	0	$-(v3 + v4 + v5)$
30	1	0	1	1	1	1	1	0	0	$-(v1 + v3 + v4 + v5)$
31	0	1	1	1	1	1	1	0	0	$-(v2 + v3 + v4 + v5)$
32	1	1	1	1	1	1	1	0	0	$-(v1 + v2 + v3 + v4 + v5)$

magnitude of the solar panel. S1 and S2 conduct to give the half cycle (positive) and S3 and S4 conduct to give the half cycle (negative) of the output voltage. With this configuration, the harmonics in the output waveform is found to be very less and the output voltage waveform looks nearly sinusoidal.

It is considered that $V_1 = 10X$ V, $V_2 = 20X$ V, $V_3 = 40X$ V, $V_4 = 80X$ V, $V_5 = 160X$ V, and $V_6 = 320\ X$ V, where multiplication factor $= X$ and is given by 325.22/630.

6.3 Sequence of Switching in the 127-Level Multilevel Inverter

Table 4 displays the switching sequences of the switches used in this circuit: 1–63 steps in the half cycle (positive) and 1–63 steps in the half cycle (negative).

6.4 Results and Discussion of 127-Level Multilevel Inverter

The output waveform of 127-level MLI is simulated using MATLAB/Simulink 2009a. The sequences of switching of SW1–SW10 and the output voltages are given in Table 4. The switching pulse generation of 10 IGBT switching devices is shown in Fig. 4. The input to these switches is from the solar panel through DC–DC converters.

Figure 5 displays the waveform of the AC output voltage as obtained from the 127-level MLI for a resistive load connected at the load terminals.

Figure 6 displays the fundamental component of AC output voltage and the THD levels. The fundamental value is 324.9 V and THD of 2.33%. The harmonic distortion exhibited is well below the limits advised by the IEEE-519 standard.

6.5 Circuit Diagram and Results of 15-, 31-, and 63-Level MLIs

Figures 7, 8 and 9 show the output voltages as per the switching pulses of 15-, 31-, and 63-level multilevel inverters using 7, 8, and 9 IGBT switches, respectively.

The fundamental component of the output voltage of 127-level MLI has a magnitude of 324.9 V. The component of the output voltage of 15-level MLI is 323 V; for 31-level, it is 323.9 V; and for 63-level, the magnitude is 324.6 V.

Figure 10 shows the THD analysis of output waveform of 15-level MLI. Figures 11 and 12 show the THD analysis of 31-level and level-63 MLIs, respectively. The THD of output voltage of 15-level MLE is found to be 5.72%; for 31-level MLI, it is 3.84%; and for 63-level MLI, THD is 3.02%.

The MLI with 15-, 31-, 63-, and 127-level MLIs are compared with respect to the number of voltage sources, number of IGBT Switches, fundamental voltage, number of diodes, and percentage of total harmonic distortion are given in Table 5.

Table 4 Switching patterns of SW1–SW10 in 127-level MLI for positive and negative half cycles

Positive half cycle

Step	*S1*	*S2*	*S3*	*S4*	*S5*	*S6*	*S7*	*S8*	*S9*	*S10*	Output (v_0)
0	0	0	0	0	0	0	0	0	0	0	0
1	1	1	0	0	0	0	0	0	0	1	*v1*
2	1	1	0	0	0	0	0	0	1	0	*v2*
3	1	1	0	0	0	0	0	0	1	1	(*v1* + *v2*)
4	1	1	0	0	0	0	0	1	0	0	(*v3*)
5	1	1	0	0	0	0	0	1	0	1	(*v1* + *v3*)
6	1	1	0	0	0	0	0	1	1	0	(*v2* + *v3*)
7	1	1	0	0	0	0	0	1	1	1	(*v1* + *v2* + *v3*)
8	1	1	0	0	0	0	1	0	0	0	(*v4*)
9	1	1	0	0	0	0	1	0	0	1	(*v1* + *v4*)
10	1	1	0	0	0	0	1	0	1	0	(*v2* + *v4*)
11	1	1	0	0	0	0	1	0	1	1	(*v1* + *v2* + *v4*)
12	1	1	0	0	0	0	1	1	0	0	(*v3* + *v4*)
13	1	1	0	0	0	0	1	1	0	1	(*v1* + *v3* + *v4*)
14	1	1	0	0	0	0	1	1	1	0	(*v2* + *v3* + *v4*)
15	1	1	0	0	0	0	1	1	1	1	(*v1* + *v2* + *v3* + *v4*)
16	1	1	0	0	0	1	0	0	0	0	(*v5*)
17	1	1	0	0	0	1	0	0	0	1	(*v1* + *v5*)
18	1	1	0	0	0	1	0	0	1	0	(*v2* + *v5*)
19	1	1	0	0	0	1	0	0	1	1	(*v1* + *v2* + *v5*)
20	1	1	0	0	0	1	0	1	0	0	(*v3* + *v5*)
21	1	1	0	0	0	1	0	1	0	1	(*v1* + *v3* + *v5*)
22	1	1	0	0	0	1	0	1	1	0	(*v2* + *v3* + *v5*)
23	1	1	0	0	0	1	0	1	1	1	(*v1* + *v2* + *v3* + *v5*)
24	1	1	0	0	0	1	1	0	0	0	(*v4* + *v5*)
25	1	1	0	0	0	1	1	0	0	1	(*v1* + *v4* + *v5*)
26	1	1	0	0	0	1	1	0	1	0	(*v2* + *v4* + *v5*)
27	1	1	0	0	0	1	1	0	1	1	(*v1* + *v2* + *v4* + *v5*)
28	1	1	0	0	0	1	1	1	0	0	(*v3* + *v4* + *v5*)
29	1	1	0	0	0	1	1	1	0	1	(*v1* + *v3* + *v4* + *v5*)
30	1	1	0	0	0	1	1	1	1	0	(*v2* + *v3* + *v4* + *v5*)
31	1	1	0	0	0	1	1	1	1	1	(*v1* + *v2* + *v3* + *v4* + *v5*)
32	1	1	0	0	1	0	0	0	0	0	(*v6*)
33	1	1	0	0	1	0	0	0	0	1	(*v1* + *v6*)
34	1	1	0	0	1	0	0	0	1	0	(*v2* + *v6*)

(continued)

Table 4 (continued)

Positive half cycle											
35	1	1	0	0	1	0	0	0	1	1	(*v1* + *v2* + *v6*)
36	1	1	0	0	1	0	0	1	0	0	(*v3* + *v6*)
37	1	1	0	0	1	0	0	1	0	1	(*v1* + *v3* + *v6*)
38	1	1	0	0	1	0	0	1	1	0	(*v2* + *v3* + *v6*)
39	1	1	0	0	1	0	0	1	1	1	(*v1* + *v2* + *v3* + *v6*)
40	1	1	0	0	1	0	1	0	0	0	(*v4* + *v6*)
41	1	1	0	0	1	0	1	0	0	1	(*v1* + *v4* + *v6*)
42	1	1	0	0	1	0	1	0	1	0	(*v2* + *v4* + *v6*)
43	1	1	0	0	1	0	1	0	1	1	(*v1* + *v2* + *v4* + *v6*)
44	1	1	0	0	1	0	1	1	0	0	(*v3* + *v4* + *v6*)
45	1	1	0	0	1	0	1	1	0	1	(*v1* + *v3* + *v4* + *v6*)
46	1	1	0	0	1	0	1	1	1	0	(*v2* + *v3* + *v4* + *v6*)
47	1	1	0	0	1	0	1	1	1	1	(*v1* + *v2* + *v3* + *v4* + *v6*)
48	1	1	0	0	1	1	0	0	0	0	(*v5* + *v6*)
49	1	1	0	0	1	1	0	0	0	1	(*v1* + *v5* + *v6*)
50	1	1	0	0	1	1	0	0	1	0	(*v2* + *v5* + *v6*)
51	1	1	0	0	1	1	0	0	1	1	(*v1* + *v2* + *v5* + *v6*)
52	1	1	0	0	1	1	0	1	0	0	(*v3* + *v5* + *v6*)
53	1	1	0	0	1	1	0	1	0	1	(*v1* + *v3* + *v5* + *v6*)
54	1	1	0	0	1	1	0	1	1	0	(*v2* + *v3* + *v5* + *v6*)
55	1	1	0	0	1	1	0	1	1	1	(*v1* + *v2* + *v3* + *v5* + *v6*)
56	1	1	0	0	1	1	1	0	0	0	(*v4* + *v5* + *v6*)
57	1	1	0	0	1	1	1	0	0	1	(*v1* + *v4* + *v5* + *v6*)
58	1	1	0	0	1	1	1	0	1	0	(*v2* + *v4* + *v5* + *v6*)
59	1	1	0	0	1	1	1	0	1	1	(*v1* + *v2* + *v4* + *v5* + *v6*)
60	1	1	0	0	1	1	1	1	0	0	(*v3* + *v4* + *v5* + *v6*)
61	1	1	0	0	1	1	1	1	0	1	(*v1* + *v3* + *v4* + *v5* + *v6*)
62	1	1	0	0	1	1	1	1	1	0	(*v2* + *v3* + *v4* + *v5* + *v6*)
63	1	1	0	0	1	1	1	1	1	1	(*v1* + *v2* + *v3* + *v4* + *v5* + *v6*)
Negative half cycle											
0	0	0	0	0	0	0	0	0	0	0	0
1	1	1	0	0	0	0	0	0	0	1	−*v1*
2	1	1	0	0	0	0	0	0	1	0	−*v2*
3	1	1	0	0	0	0	0	0	1	1	−(*v1* + *v2*)
4	1	1	0	0	0	0	0	1	0	0	−(*v3*)
5	1	1	0	0	0	0	0	1	0	1	−(*v1* + *v3*)

(continued)

Table 4 (continued)

Positive half cycle											
6	1	1	0	0	0	0	0	1	1	0	−(v2 + v3)
7	1	1	0	0	0	0	0	1	1	1	−(v1 + v2 + v3)
8	1	1	0	0	0	0	1	0	0	0	−(v4)
9	1	1	0	0	0	0	1	0	0	1	−(v1 + v4)
10	1	1	0	0	0	0	1	0	1	0	−(v2 + v4)
11	1	1	0	0	0	0	1	0	1	1	−(v1 + v2 + v4)
12	1	1	0	0	0	0	1	1	0	0	−(v3 + v4)
13	1	1	0	0	0	0	1	1	0	1	−(v1 + v3 + v4)
14	1	1	0	0	0	0	1	1	1	0	−(v2 + v3 + v4)
15	1	1	0	0	0	0	1	1	1	1	−(v1 + v2 + v3 + v4)
16	1	1	0	0	0	1	0	0	0	0	−(v5)
17	1	1	0	0	0	1	0	0	0	1	−(v1 + v5)
18	1	1	0	0	0	1	0	0	1	0	−(v2 + v5)
19	1	1	0	0	0	1	0	0	1	1	−(v1 + v2 + v5)
20	1	1	0	0	0	1	0	1	0	0	−(v3 + v5)
21	1	1	0	0	0	1	0	1	0	1	−(v1 + v3 + v5)
22	1	1	0	0	0	1	0	1	1	0	−(v2 + v3 + v5)
23	1	1	0	0	0	1	0	1	1	1	−(v1 + v2 + v3 + v5)
24	1	1	0	0	0	1	1	0	0	0	−(v4 + v5)
25	1	1	0	0	0	1	1	0	0	1	−(v1 + v4 + v5)
26	1	1	0	0	0	1	1	0	1	0	−(v2 + v4 + v5)
27	1	1	0	0	0	1	1	0	1	1	−(v1 + v2 + v4 + v5)
28	1	1	0	0	0	1	1	1	0	0	−(v3 + v4 + v5)
29	1	1	0	0	0	1	1	1	0	1	−(v1 + v3 + v4 + v5)
30	1	1	0	0	0	1	1	1	1	0	−(v2 + v3 + v4 + v5)
31	1	1	0	0	0	1	1	1	1	1	−(v1 + v2 + v3 + v4 + v5)
32	1	1	0	0	1	0	0	0	0	0	−(v6)
33	1	1	0	0	1	0	0	0	0	1	−(v1 + v6)
34	1	1	0	0	1	0	0	0	1	0	−(v2 + v6)
35	1	1	0	0	1	0	0	0	1	1	−(v1 + v2 + v6)
36	1	1	0	0	1	0	0	1	0	0	−(v3 + v6)
37	1	1	0	0	1	0	0	1	0	1	−(v1 + v3 + v6)
38	1	1	0	0	1	0	0	1	1	0	−(v2 + v3 + v6)
39	1	1	0	0	1	0	0	1	1	1	−(v1 + v2 + v3 + v6)
40	1	1	0	0	1	0	1	0	0	0	−(v4 + v6)
41	1	1	0	0	1	0	1	0	0	1	−(v1 + v4 + v6)

(continued)

Table 4 (continued)

Positive half cycle											
42	1	1	0	0	1	0	1	0	1	0	$-(v2 + v4 + v6)$
43	1	1	0	0	1	0	1	0	1	1	$-(v1 + v2 + v4 + v6)$
44	1	1	0	0	1	0	1	1	0	0	$-(v3 + v4 + v6)$
45	1	1	0	0	1	0	1	1	0	1	$-(v1 + v3 + v4 + v6)$
46	1	1	0	0	1	0	1	1	1	0	$-(v2 + v3 + v4 + v6)$
47	1	1	0	0	1	0	1	1	1	1	$-(v1 + v2 + v3 + v4 + v6)$
48	1	1	0	0	1	1	0	0	0	0	$-(v5 + v6)$
49	1	1	0	0	1	1	0	0	0	1	$-(v1 + v5 + v6)$
50	1	1	0	0	1	1	0	0	1	0	$-(v2 + v5 + v6)$
51	1	1	0	0	1	1	0	0	1	1	$-(v1 + v2 + v5 + v6)$
52	1	1	0	0	1	1	0	1	0	0	$-(v3 + v5 + v6)$
53	1	1	0	0	1	1	0	1	0	1	$-(v1 + v3 + v5 + v6)$
54	1	1	0	0	1	1	0	1	1	0	$-(v2 + v3 + v5 + v6)$
55	1	1	0	0	1	1	0	1	1	1	$-(v1 + v2 + v3 + v5 + v6)$
56	1	1	0	0	1	1	1	0	0	0	$-(v4 + v5 + v6)$
57	1	1	0	0	1	1	1	0	0	1	$-(v1 + v4 + v5 + v6)$
58	1	1	0	0	1	1	1	0	1	0	$-(v2 + v4 + v5 + v6)$
59	1	1	0	0	1	1	1	0	1	1	$-(v1 + v2 + v4 + v5 + v6)$
60	1	1	0	0	1	1	1	1	0	0	$-(v3 + v4 + v5 + v6)$
61	1	1	0	0	1	1	1	1	0	1	$-(v1 + v3 + v4 + v5 + v6)$
62	1	1	0	0	1	1	1	1	1	0	$-(v2 + v3 + v4 + v5 + v6)$
63	1	1	0	0	1	1	1	1	1	1	$-(v1 + v2 + v3 + v4 + v5 + v6)$

7 Conclusions

The proposed 127-level MLI uses only 10 switches. The performance of this configuration is compared with that of 15-, 31-, and 63-level MLIs having similar topology. The MLIs are compared with respect to the number of switching devices, the number of voltage sources, the number of diodes used, the magnitude of fundamental voltage, and the percentage of total harmonic distortion. This 127-level MLI produces nearly sinusoidal voltage using only 10 IGBT switches. The reduction in the number of IGBT switches reduces the number of gate driving circuits, which in turn reduces

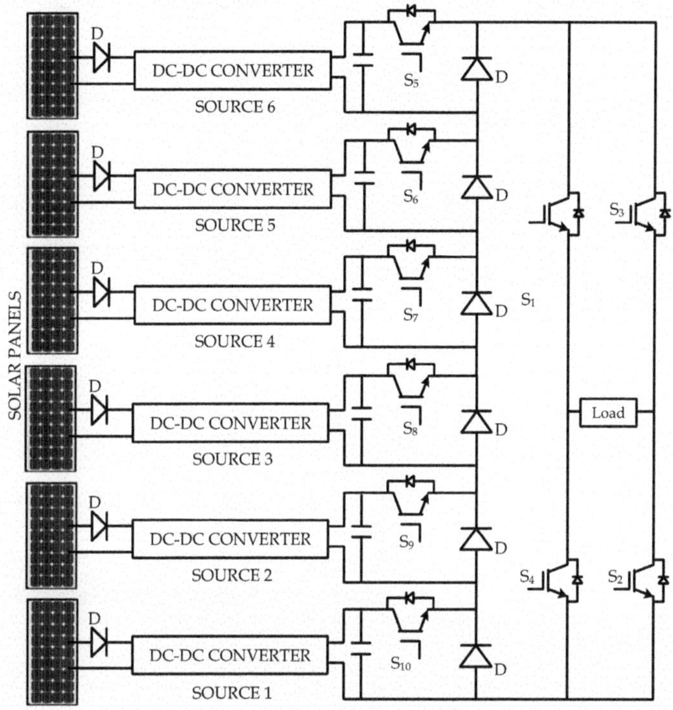

Fig. 4 Circuit diagram of 127-level multilevel inverter

the cost of the system. Harmonics are reduced, percentage of total harmonic distortion is reduced, and distortion factor is reduced. Solar PV cells with DC–DC converters along with battery act as the voltage sources. The voltage stress on individual element is reduced in this topology. The disadvantage of this topology is that the malfunctioning of one of the sources increases the THD in the output waveform. These topologies provide reasonable results with the less number of semiconductor switches. From the comparison between various types of MLIs, the 127-level MLI output waveform is found to have the smallest amount of THD. Power quality in the output waveform is enhanced as the harmonic content is less.

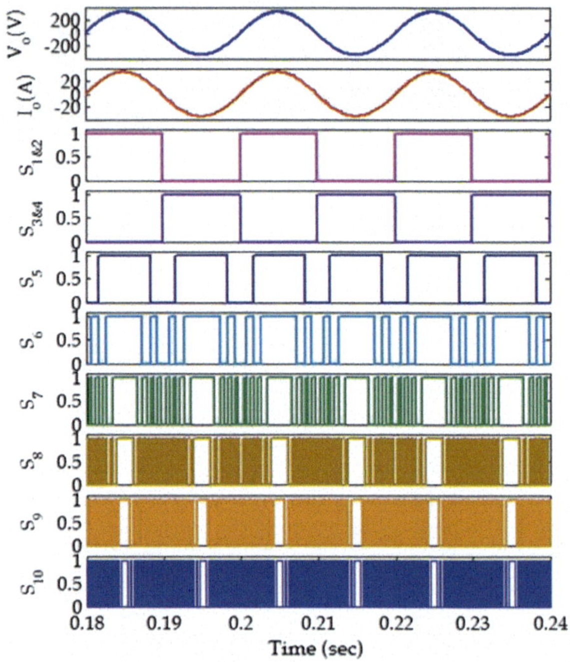

Fig. 5 Output waveforms as per switching pattern for the proposed 127-level multilevel inverter

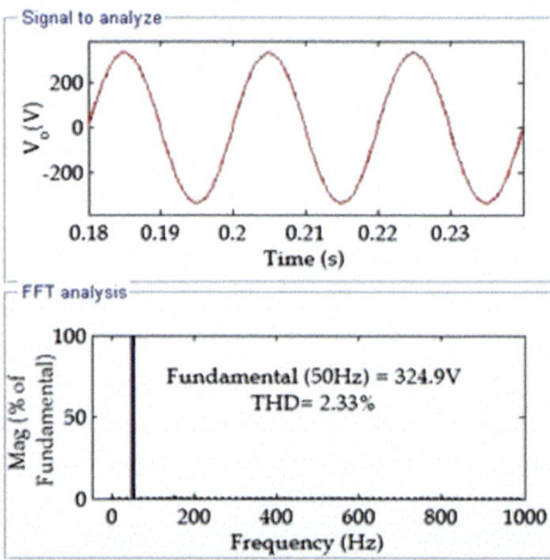

Fig. 6 THD in the output voltage of 127-level MLI

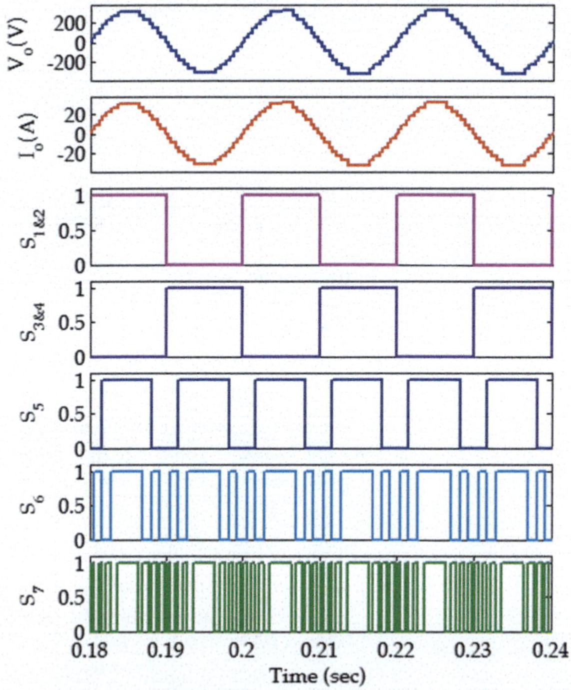

Fig. 7 Output waveforms as per switching pattern for the proposed 15-level multilevel inverter

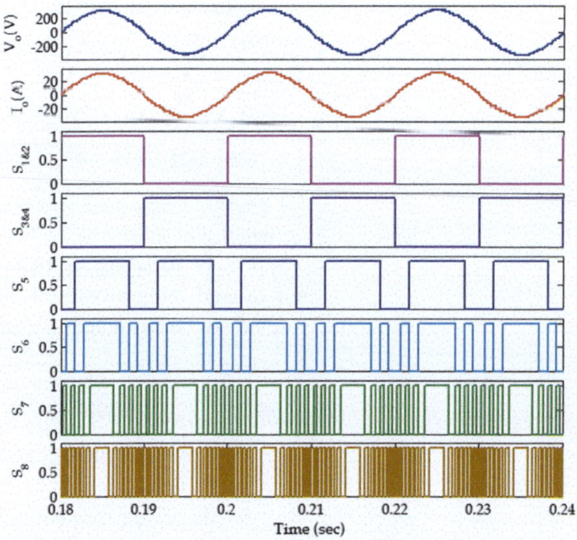

Fig. 8 Output waveforms as per switching pattern for the proposed 31-level MLI

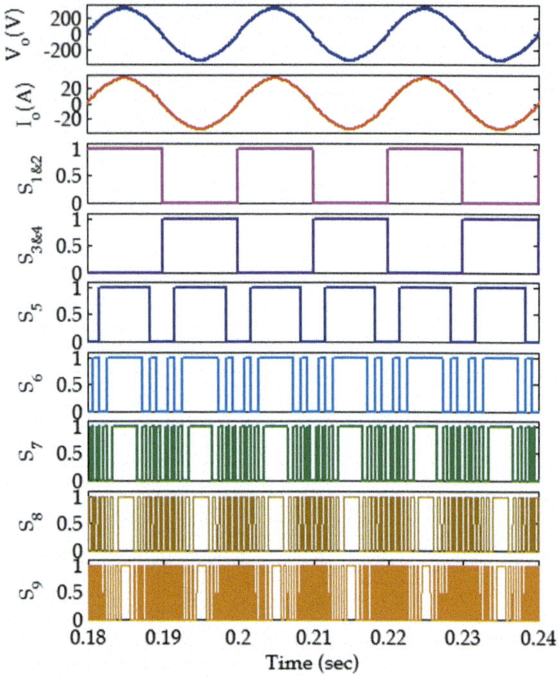

Fig. 9 Waveforms as per switching pattern for the proposed 63-level multilevel inverter

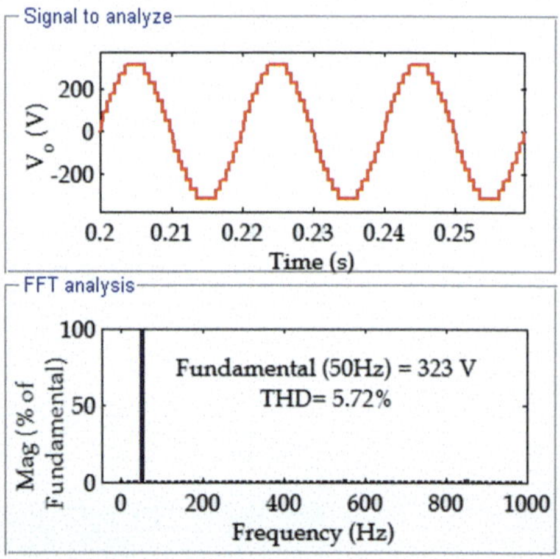

Fig. 10 THD analysis of 15-level MLI

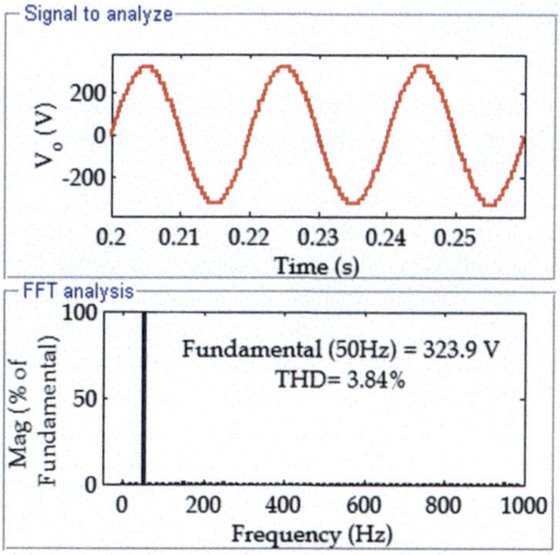

Fig. 11 THD analysis of 31-level MLI

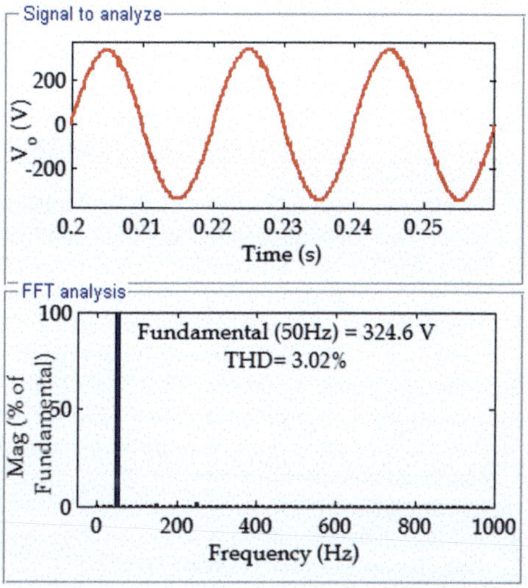

Fig. 12 THD analysis of 63-level MLI

Table 5 Comparison between different level MLIs

No. of levels	15	31	63	127
No. of voltage sources	03	04	05	06
No. of IGBT switches	07	08	09	10
Voltage (Fundamental)	323.1 V	323.91 V	324.61 V	324.91 V
No. of diodes	03	04	05	06
% of THD	5.72	3.84	3.02	2.33

References

1. Babaei E, Jalilzadeh T, Sabahi M, Maalandish M, Alishah RS (2018) High step-up DC–DC converter with reduced voltage stress on devices. IEEE Trans Instr Electr Energy Syst 46:2053–2078
2. Prem P, Sathik J, Sivaraman P, Mathewsaran A, Aleem SHEA (2019) A new asymmetric dual source multilevel inverter topology with reduced power switches. J Chin Instr Eng 42:460–472
3. Kouro S, Malinowski M, Gopakumar K, Pou J, Leopoldo, Franquelo G (2010) Recent advances and industrial applications of multilevel converters. IEEE Trans Indus Electron Syst 57:2553–2580
4. Ali JSM, Alishah RS, Sandeep N, Hosseini SH, Babaei E, Vijayakumar K, Yaragatti UR (2018) A new generalized multilevel converter topology based on cascaded connection of basic units. IEEE Trans J Emerg Sel Topics Power Electron 7:2498–2512
5. Meza C, Negroni JJ, Biel D, Guinjoan F (2008) Energy-balance modeling and discrete control for single-phase grid-connected PV central inverters. IEEE Trans Indus Electron 55(7):2734–2743
6. Oikonomou N, Gutscher C, Karamanakos P, Kieferndorf F, Geyer T (2013) Model predictive pulse pattern control for the five-level active neutral point-clamped inverter. IEEE Transactions on Ind Appl 46:2583–2592
7. Huang J, Corzine K (2006) Extended operation of flying capacitor multilevel inverters. IEEE Trans Power Electron 21:140–147
8. Amini J (2014) An effortless space-vector-based modulation for N-level flying capacitor multilevel inverter with capacitor voltage balancing capability. IEEE Trans Power Electron 57:2633–2642
9. Villalva MG, Gazoli JR, Filho ER (2009) Modeling and circuit-based simulation of photovoltaic arrays. In: Proceedings Brazilian Power Electronics Conference, vol 14, pp 35–45
10. Hasan M, Mekhilef S, Ahmed M (2014) Three-phase hybrid multilevel inverter with less power electronic components using space vector modulation. Power Electron IET 16:1504–1512
11. Mekhilef S, Abdul Kadir MN, Salam Z (2013) Digital control of three phase three-stage hybrid multilevel inverter. IEEE Trans Ind Informat 9:719–727
12. Daher S, Schmid J, Antunes FL (2008) Multi level inverter topologies for stand-alone PV systems. IEEE Trans Ind Electron 55:2703–2712
13. Buticchi G, Barater D, Lorenzani E, Concari C, Franceschini G (2014) A nine-level grid-connected converter topology for single-phase transformerless PV systems. IEEE Trans Ind Electron 61:3951–3960
14. Wu JC, Chou CW (2014) A solar power generation system with a seven-level inverter. IEEE Trans Power Electron 31:2099–2110
15. Gupta KK, Jain S (2014) Comprehensive review of a recently proposed multilevel inverter. Power Electron IET 7:467–479
16. Gupta KK, Ranjan A, Bhatnagar P, Sahu LK, Jain S (2016) Multilevel inverter topologies with reduced device count: a review. IEEE Trans Power Electron 31:135–151
17. Odeh CI (2013) Enhanced three-phase multilevel inverter configuration. Power Electron IET 6:1122–1131
18. Farhadi Kangarlu M, Babaei E (2013) Cross-switched multilevel inverter an innovative topology. IET Power Electron 6:1041–1050
19. Odeh CI (2014) Balancing switching losses in three-phase, five-level pulse width modulation switched voltage source inverter using hybrid modulation techniques. Electro Power Compon Syst 8:1194–1200
20. Ruiz-Caballero A, Ramos-Astudillo RM, Mussa SA, Heldwein ML (2010) Symmetrical hybrid multilevel DC–AC converters with reduced number of insulated DC supplies. IEEE Trans Ind Electron 57:2307–2314
21. Tsai CT, Chen WM (2016) Buck converter with soft-switching cells for PV panel applications. Energies 9(3). https://doi.org/10.3390/en9030148

22. Ando Y, Oku T, Yasuda M, Ushijima K, Murozono M (2017) A Transportable photovoltaic power generation system utilizing a SiC inverter and spherical Si solar cells. Technologies 5(4):18. https://doi.org/10.3390/technologies5020018
23. Amamra SA, Meghriche K, Cherifi A, Francois B (2017) Multilevel inverter topology for renewable energy grid integration. IEEE Trans Ind Electron 64(11):8855–8866. https://doi.org/10.1109/TIE.2016.2645887
24. Gopal Y, Birla D, Lalwani M (2018) Selected harmonic elimination for cascaded multilevel inverter based on photovoltaic with fuzzy logic control maximum power point tracking technique. Technologies 6(3):62. https://doi.org/10.3390/technologies6030062
25. Hota A, Jain S, Agarwal V (2018) An improved three-phase five-level inverter topology with reduced number of switching power devices. IEEE Trans Ind Electron 65(4):3296–3305. https://doi.org/10.1109/TIE.2017.2758722
26. Bassi HM, Salam Z (2019) A new hybrid multilevel inverter topology with reduced switch count and dc voltage sources. Energies 12(6). https://doi.org/10.3390/en12060977
27. Siddique MD, Mekhilef S, Shah NM, Sarwar A, Iqbal A, Memon MA (2019) A new multilevel inverter topology with reduce switch count. IEEE Access 7:58584–58594. https://doi.org/10.1109/ACCESS.2019.2914430
28. Sandeep N, Ali JSM, Yaragatti UR, Vijayakumar K (2019) A self-balancing five-level boosting inverter with reduced components. IEEE Trans Power Electron 34(7):6020–6024. https://doi.org/10.1109/TPEL.2018.2889785
29. Periyanayagam M, Suresh Kumar V, Chokkalingam B, Padmanaban S, Mihet-Popa L, Adedayo Y (2020) A modified high voltage gain quasi-impedance source coupled inductor multilevel inverter for photovoltaic application. Energies 13(4). https://doi.org/10.3390/en13040874
30. Kakar S, Ayob SBM, Iqbal A, Nordin NM, Saad Bin Arif M, Gore S (2021) New asymmetrical modular multilevel inverter topology with reduced number of switches. IEEE Access 9(Mli):27627–27637. https://doi.org/10.1109/ACCESS.2021.3057554
31. Sathik J et al (2021) A multilevel inverter topology using diode half-bridge circuit with reduced power component
32. Rawa M et al (2021) A new multilevel inverter topology with reduced dc sources. Energies 14(15):1–21. https://doi.org/10.3390/en14154709
33. Farahmand MZ, Javadi S, Sadati SMB, Laaksonen H, Shafie-khah M (2021) Optimal operation of solar powered electric vehicle parking lots considering different photovoltaic technologies. Clean Technol 3(2):503–518. https://doi.org/10.3390/cleantechnol3020030
34. Bin Arif MS, Mustafa U, Ayob SBM, Rodriguez J, Nadeem A, Abdelrahem M (2021) Asymmetrical 17-level inverter topology with reduced total standing voltage and device count. IEEE Access 9:69710–69723. https://doi.org/10.1109/ACCESS.2021.3077968
35. Mukundan NC, Kallaveetil V, Suresh S, Pychadathil J (2021) An improved H-bridge multilevel inverter-based. IEEE Trans Industr Electron 57(6):6339–6349

Chapter 11
Solar Photovoltaic System Supported V4G Performance Using IHTQZSC with Constant Switching Frequency Predictive Controller

Anandita Chowdhury and Ramanjaneyulu Alla

Abstract The advancements in energy extraction from renewable energy sources as well as the continuous efforts for the electrification of the automobile sector reduces the issues regarding global warming. The reduction in the cost of photovoltaic modules and energy storage system, accelerate the usage of electric vehicles (EV) and energy extraction from solar power. Power electronic converters are required for charging, discharging, and energy management of EVs. Likewise, power converters are required for grid integration and extraction of maximum power from the photovoltaic systems (PVS). In the present work, improved hybrid three quasi z source converter (IHTQZSC) is utilized for grid integration of PVS in a single-stage system providing higher voltage gain. Penetration of PVS and EVs on large scale into the utility can reduce power system stability, security, and reliability. Meanwhile, EVs are encouraged to charge from PVS reducing burden on the utility. Further, EVs are encouraged to provide grid support during the vehicle idle condition (v4G operation), in addition to vehicle to grid (V2G) and grid to vehicle (G2V) operation. IHTQZSC is employed for supporting V2G, G2V, and V4G operations as well as charging EVs and injecting PV power into the grid. A constant switching frequency predictive controller is intended for the controlling of grid side operation of PVS and EV.

Keywords Electric vehicle · Grid to vehicle (G2V) · Predictive controller · Shoot-through duty ratio · Vehicle to grid (v2G) · Vehicle for grid (V4G)

A. Chowdhury · R. Alla (✉)
Department of Electrical Engineering, Sardar Vallabhbhai National Institute of Technology, Surat 395007, India
e-mail: ramanjaneyulualla@gmail.com; d17el005@eed.svnit.ac.in

A. K. Giri et al. (eds.), *Power Quality: Infrastructures and Control*,
Studies in Infrastructure and Control,
https://doi.org/10.1007/978-981-19-7956-9_11

1 Introduction

Grid-connected solar photovoltaic systems have been substantially increasing in recent years to meet global electrical energy demand. The reduction in the cost of photovoltaic modules and the strong encouragement from local governments accelerate the erection of small and medium-scale photovoltaic systems, reducing global pollution. The power system inertia has been reduced by installing grid-connected photovoltaic systems on a large scale, threatening the power system stability and reliability, and has been explained with the duck curve. The shape of the load demand curve appeared like a duck shown in Fig. 1 published by California independent system operator [1] in 2013.

The excessive deployment of solar power into the grid causes the difference between electrical energy demand and generation in a day. P1, P2, P3, P4, P5, and P6 are net energy demand in megawatts. Each plot in Fig. 1 represents the net load calculated as the difference between forecasted electrical energy demand and energy supplied from the wind, solar, and other renewable energy systems. The net demand is increasing in the morning and evening when there has been no photovoltaic power generation. The photovoltaic power has been rising with the sunlight and is maximum at 12 PM to 3 PM and gradually decreasing. Two significant regions have been identified with the duck curve.

Over Generation (Low energy demand—maximum solar power generation)

The net energy demand has been decreased after 9 AM when the sunlight is available and solar power generation has been started. In mid-day, the net demand is falling and solar power generation has been increasing making the demand curve reach the minimum value. It forms a duck belly. Here the photovoltaic system generates more energy than the required load demand.

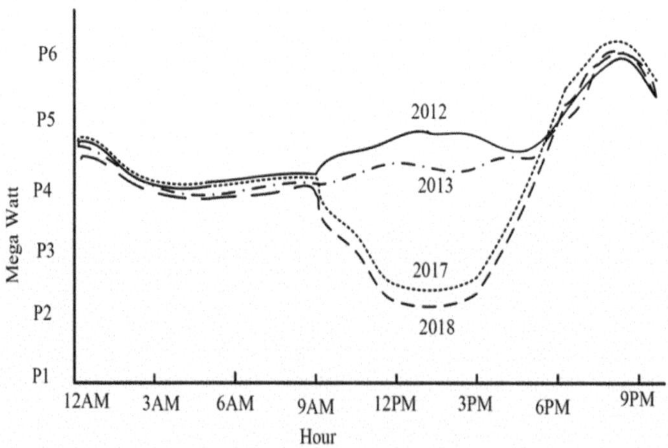

Fig. 1 California Duck curve [1]

High net Demand (Low solar power generation—increase in energy demand)

After 3 PM, the photovoltaic power has been reducing as the sunlight gradually decreases. Simultaneously, the energy demand is increasing. The photovoltaic power has been reaching zero value after 5 PM. The load curve ramp-up close to 6 PM and is continued further. It has been identified as a duck neck in the load curve observed from 3 to 9 PM.

During the duck belly period, a few peak power plants might be shut down to balance energy demand in the grid. The peak power plants are ramped-up at a faster rate to reach the energy demand all along the duck neck period, posing serious power system stability issues. The over-demand region (duck belly) has been controlled by curtailing the energy generation from the photovoltaic system. But, it has been opposing the clean energy generation intention and affecting the economic benefits. Energy storage is one of the alternate solutions for over-generation and over-demand, which will balance the load demand in the system by storing excess renewable energy during the light load period and supplying additional energy into the system during overload demand [2, 3].

Large-scale energy storage systems in the order of megawatt have been discussed and concluded that battery energy storage had been higher energy density and cost-effective among the energy storage technologies [4] and Li-ion battery plays a key role [5]. A review of various energy storage systems in grid-connected renewable energy systems has been discussed [6–8]. The electrification of the automobile industry has been increasing in recent years with the developments in battery energy storage technology and the reduction in the battery cost helps in reducing the consumption of fossil fuels. Energy storage systems (ESS) along with renewable energy sources in grid applications have been discussed. Along with renewable energy sources, ESS is used for energy balance, solving power quality issues, and providing ancillary services to the power grid [9–13]. Most of the time, electric vehicles are in a parking place and are considered a distributed energy source that might be utilized for grid applications.

Electric vehicles with battery ESS are operating to inject energy into the grid when there is a higher net load demand, also profiting with a higher tariff rate. This mode of operation is generally known as vehicle to grid operation (V2G). Electric vehicles can charge energy from the grid during over-generation periods of duck curve or off-peak loads, thereby avoiding higher tariff rates. This mode of operation is generally known as the grid to vehicle operation (G2V). Electric vehicles are also utilized for standalone applications like vehicle to home (V2H). Further, electric vehicles provide ancillary services known as the vehicle for grid operation (V4G). This chapter discusses the power converter configurations incorporating electric vehicles as energy storage systems.

Power converters play a critical role in the conversion, conservation of generated electrical energy, and extraction of energy from renewable energy sources. Power converters are categorized into AC-DC, DC-AC, DC-DC, and AC-AC energy conversion circuits used in many of the applications, a few applications are listed below:

- Variable speed drives,
- Grid integration of renewable sources,
- Extraction of power from renewable sources,
- High voltage AC or DC transmission systems,
- Consumer electronics,
- Energy storage systems,
- Electric vehicle applications,
- AC or DC microgrids,
- Space applications.

This chapter addresses the power converters allowing b-directional power flow used for both grid integration of photovoltaic systems (PVS) and Electric vehicle applications. Energy storage devices, mainly batteries, are used to drive the electric vehicle. DC-DC and DC-AC power converters are required for the grid integration of photovoltaic systems and electric vehicles. DC-DC converters are used to extract maximum energy from the photovoltaic systems and for energy management in electric vehicles by charging and discharging storage devices. DC-AC power converters are used to inject extracted energy from renewable sources into the grid and drive the electric vehicle using induction motors, permanent magnet synchronous motors, etc.

2 Power Converters

Nowadays, power electronic switches can work at vast ranges of voltage, current, and frequency, driving unlimited power converters configurations for industrial applications. Power converters are adapted to renewable energy, micro-grid, and electric drive applications and used for power quality solutions. Power converters are utilized in AC-AC, AC-DC, DC-DC-, and DC-AC energy conversion applications. The list of familiar non-isolated power converters adapted for PVS and EV is shown in Fig. 2. Various impedance source converters (ISC) [14] topologies are developed in continuation to Z source converter [15] and are used for DC-DC and DC-AC energy conversion applications. The voltage source converter (VSC) and current source converter (CSC) are shown in Fig. 3a, b used in DC-AC power conversion systems [16]. A diode bridge rectifier (DBR) is shown in Fig. 3c is used in uni-directional un-controlled AC-DC power conversion systems. DC-DC converters configuration [17–19] with switch on-state and off-state circuits are exhibited in Fig. 4, respectively. G type DC-DC converter in Fig. 4 acts as a A type or B type converter depending on the switching operation of the switches S_1 and S_2.

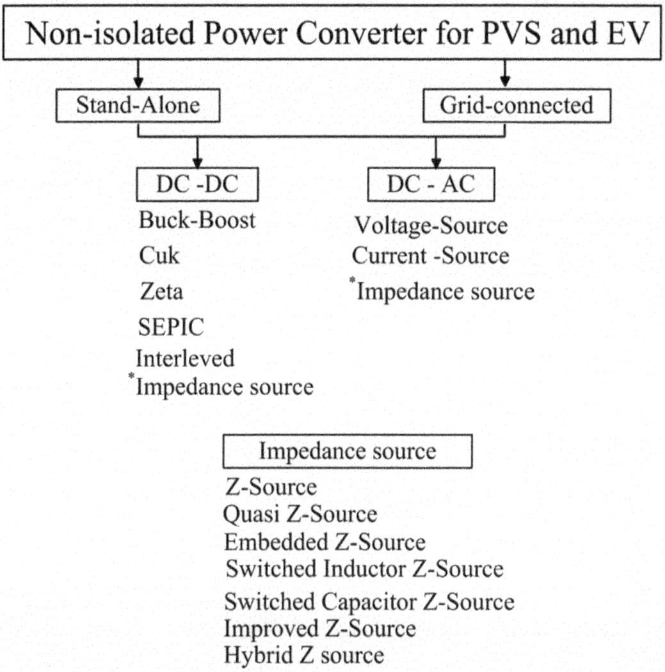

Fig. 2 Types of non-isolated power converters

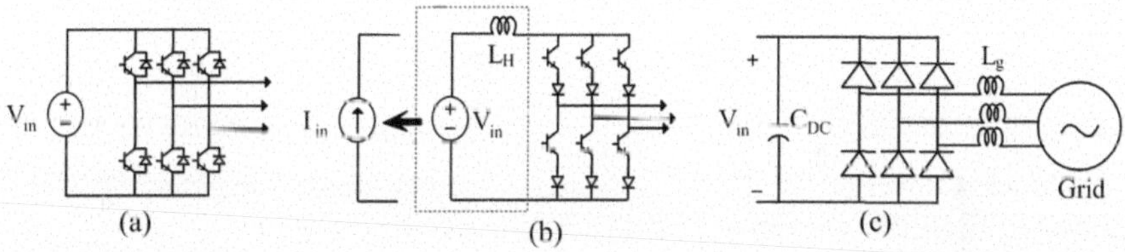

Fig. 3 DC-AC power converter configurations

2.1 Power Converters for Grid-Connected Photovoltaic Systems

The utilization of solar energy in commercial and household loads has been reducing conventional fuel consumption, and hence global pollution has been controlled. Energy generation from renewable sources is an alternative solution for reducing conventional fuel reserves. Figure 5a shows the grid-connected single-stage PVS through a VSC and Fig. 5b shows the two-stage grid-connected PVS through a boost converter along with VSC. The excessive generated power at PVS stored in the battery energy storage system has been shown in Fig. 5c. In place of battery energy storage,

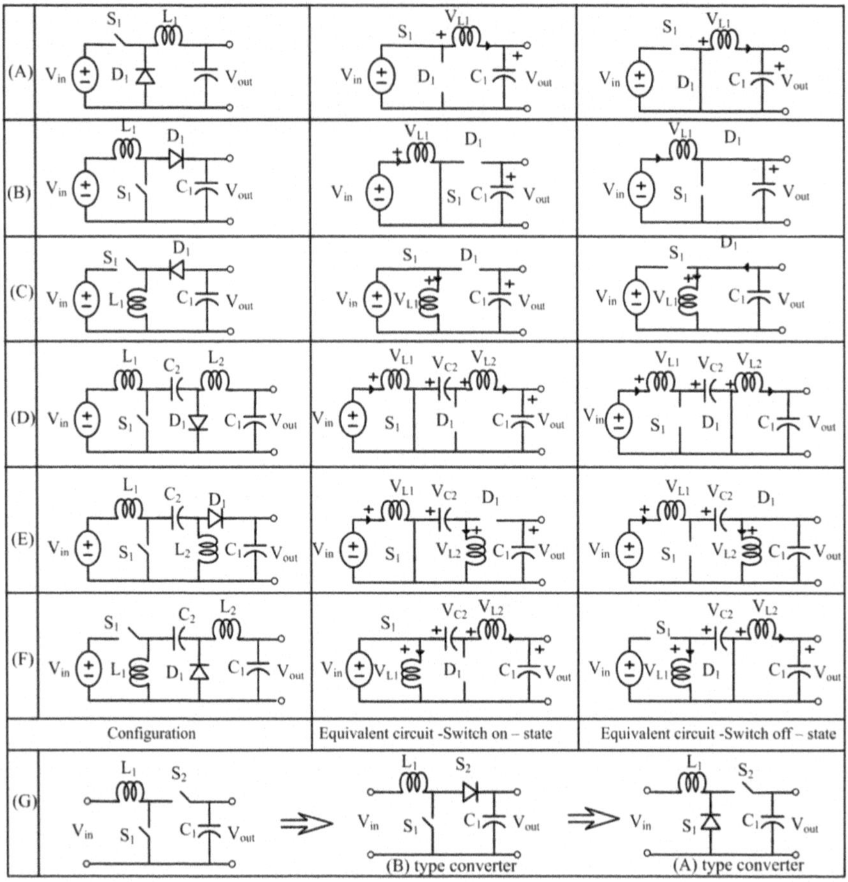

A: Buck Converter; B: Boost converter; C: Buck-Boost converter; D: Cuk Converter; E: SEPIC converter; F: Zeta converter; G: Bi-directional converter

Fig. 4 DC-DC power converter configurations

other energy storage systems like supercapacitors, ultra-capacitors, fly-wheels, and superconducting magnetic storage systems are to be used.

Energy extraction from PVS has been volatile due to changes in solar irradiance, temperature, and other environmental issues. Therefore, maximum power might not be generated from PVS, causing the lesser efficiency of PVS, and thus increasing the cost of the applications working with PVS [20]. Irrespective of environmental issues, the PVS always to be working at the maximum power point. So the efficacy of the PVS has been increasing. The maximum power point tracking has been explained in Fig. 6 with a boost converter. PVS input resistance (Rs) across the PV module has been given in Eq. (1). The value of Rs has been varying from zero (at short circuit) to infinity (at open circuit) depending upon the load connected to PVS. If the load

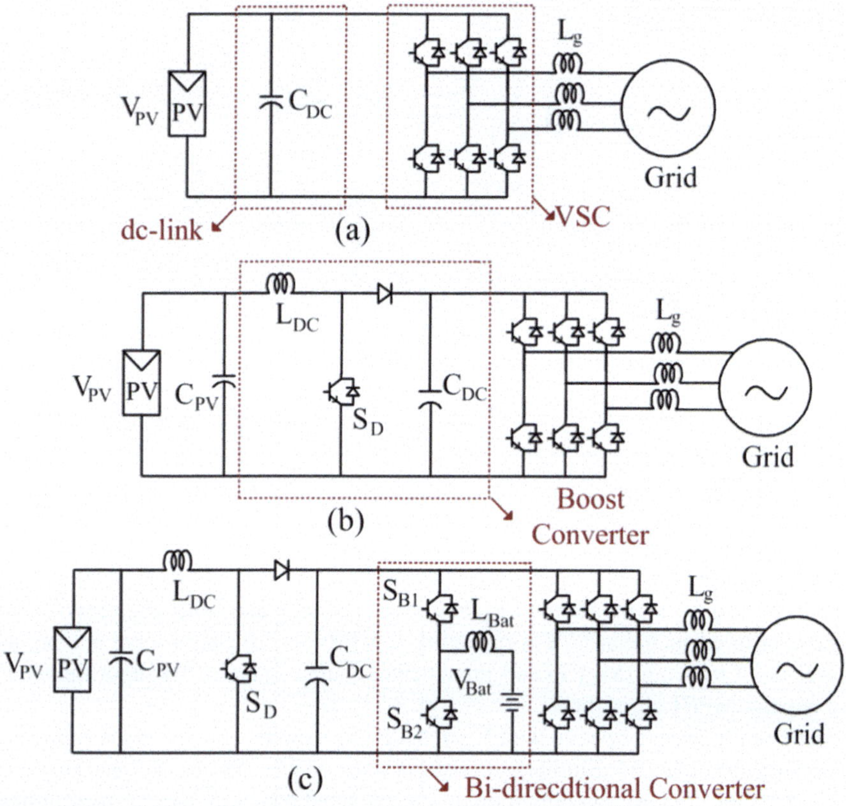

Fig. 5 **a** Single-stage grid-connected PVS, **b** Two-stage connected PVS, and **c** Two-stage grid-connected PVS with ESS

resistance (output resistance) (Ro) equals Rs, then the maximum power has been extracted from PVS.

$$R_S = \frac{V_{PV}}{I_{PV}} \tag{1}$$

For a boost converter, the relation between R_O and R_S is expressed in Eq. (2) at a duty ratio of d

$$R_S = R_O(1 - d)^2. \tag{2}$$

In a boost converter, d = 0 indicates a 100% switch-off state such that there has been no DC voltage step-up. Depending upon the value of Eq. (2), power has been drawn from the PVS. If $R_O = R_A$ or $R_O = R_B$, then the PV power is not at the maximum value observed in Fig. 6 and the PV power has a zero value at PV modules under short circuit (V_{SC}) or the PV modules are open (V_{OC}). Either by changing R_O

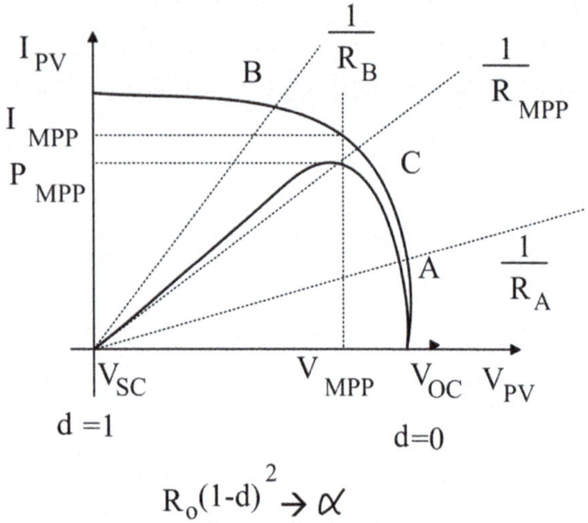

Fig. 6 MPP tracking of PVS with a boost converter

or by changing the d value, PVS has been working at maximum power point (MPP) condition. In general, the value of R_O has been uncertain, and hence duty ratio d has been controlled such that PVS is operating at MPP. Perturb and observe maximum power point tracking (MPPT) and incremental conductance MPPT algorithms are the popular MPPT algorithms.

VSC in Fig. 5a can operate the PVS at MPP and simultaneously injects extracted PV power into the grid. Similarly, boost converter in Fig. 5b can be operating the PVS at MPP by controlling duty ratio d. The extracted power by boost converter has been injected into the grid with the help of VSC. The power generated by the PVS is higher than the VSC rating, then the excessive power has been stored in the ESS using a bi-directional converter operated with the switches S_{B1} and S_{B2} shown in Fig. 5c. Again the ESS has been assisting the power grid by sending stored energy into the grid at the time of less power generation from PVS.

2.2 *Power Converters for Electric Vehicles*

Electric vehicles are one of the alternative solutions for reducing global pollution as EVs reduce conventional fuel consumption. Figure 7a shows the power conversion stages in an EV. EV has been driven with an electric motor controlled electronically with a converter. The induction motor, BLDC motor, permanent magnet synchronous motor, and other motors are designed to operate the EV. Power electronic converters are required for controlling EV speed with sophisticated control algorithms. ESS in the EV generates the required power to run the EV and stores the generated energy back into the ESS during braking. ESS has been charging from the utility through a power conditioning unit. The energy requirement of auxiliary electrical and electronic equipment operated at different voltage levels in the EV is met by the DC-DC converters. Figure 7b shows a buck converter acting as a power conditioning unit. It maintains the grid power factor with well-defined control methods and simultaneously step-down the DC voltage at DBR to a desired voltage at the EV battery. Figure 7c shows a bi-directional DC-DC converter used to control the power in both directions. VSC has been replaced with a DBR such that power has been transferred either from the utility to the EV battery or to the EV battery from the utility.

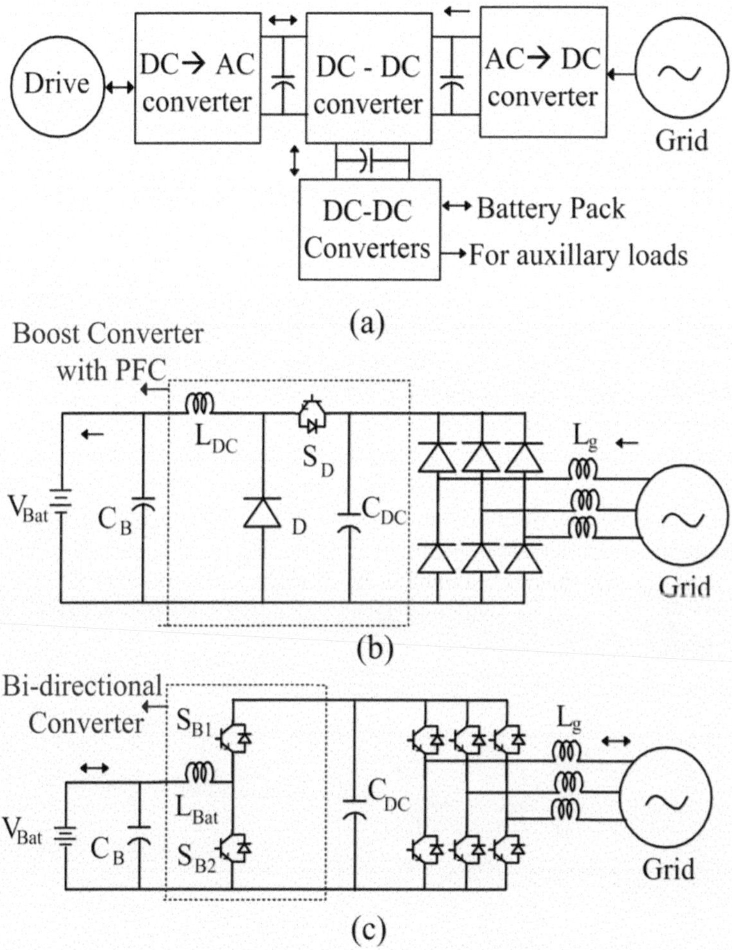

Fig. 7 **a** Power conversion stages in EV, **b** PFC boost converter for battery charging, and **c** Bi-directional converter for battery charging

3 Impedance Source Converters

Z source inverter is a basic impedance source network, developed in 2003 by Peng [15] to overcome the shoot-through related problems in Voltage source converters and open circuit problems in current source converters (CSC). The Z source converter (ZSC) consists of two inductors and two capacitors of lattice structure connect between the input DC source and converter along with a diode shown in Fig. 8. ZSC can work as either VSC or CSC. The source may be a PV cell or Fuel cell or rectifier or a DC source or a possible combination of these. If the input source is a DC voltage source, it acts as a VSC. It acts as CSC when the input DC voltage source is connected in series with the large inductor. If the value of inductances in the lattice network is too small and tends to zero, the lattice network seems to be two capacitors connected in parallel and resembles a VSC. Therefore, the worst-case capacitor requirement for a Z source converter is nothing but the capacitor requirements in traditional VSC. Likewise, the capacitance values in the lattice network are small and tend to be zero; the impedance source network appeared to be two inductors in series with an input DC source and becomes a standard CSC. Therefore, the worst-case inductor requirement for a Z source converter is the inductance required for a traditional CSC. Obviously, low-value and smaller-sized inductors and capacitors are sufficient for the design of a Z source converter with additional filtering and energy storage by the inductors and capacitors compared with the existing converters. Comparisons among VSC, CSC, and ZSC are showed in Table 1.

The main objective of the Z source converter is capable to vary the output voltage in between zero to infinity during an unpredictable input voltage. Generally, an inverter switching states are defined in 6 active and 2 zero vectors and should not allow the shoot through on any leg. But ZSC has nine switching vectors, eight switching vectors from traditional converters, and one more zero vector named as shoot-through state (STS). Shoot-through state is obtained in seven ways: a zero state occurs due to the triggering of top and bottom switching devices on any one leg or any two legs or on all three legs at the same instant.

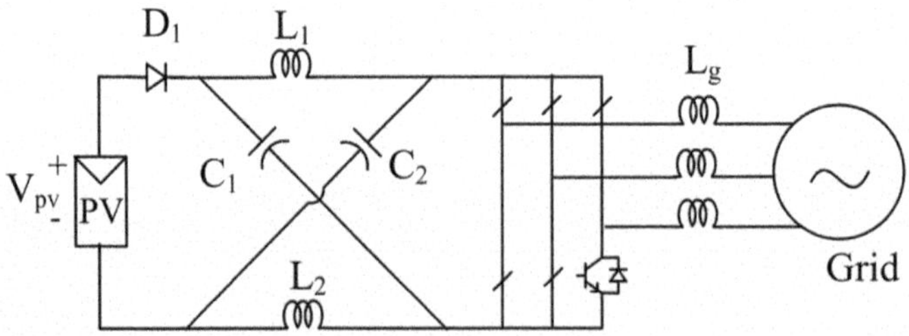

Fig. 8 General Topology of Z source inverter [15]

Table 1 Comparision of VSC, CSC, and Z Source converter

VSC	CSC	ZSC
Capacitors are employed in the input DC link to represent a voltage source	An inductor in series with the source represents a constant current source	The capacitors and inductors in the impedance network on the DC side works as either a current or voltage source
Avoid Miss-firing switches on any leg. As it causes shoot through and does damage to switching devices	It can accept the misfiring of switches	In ZSI, overcome the problems related to missing-firing of the switches
Waveform distortion is more for the inclusion of dead time in switching states	Switching losses are high due to the inclusion of overlap time between the switching states to avoid open state	There is no need to provide additional dead time or overlap time among the switches
It is a voltage buck or current boost converter	It is a voltage boost or current buck converter	It can act as either buck or boost converter
The impact of EMI noise can influence the converter operation	The effect of EMI noise can affect the converter operation	It is less sensitive to the EM1 noise

3.1 *Working of ZSC*

The operating principle of a Z source inverter is explained in two operating modes:

- Shoot-through state mode.
- Non-Shoot-through state mode.

Including two zero states (000 and 111) and six active states (100,110,010,011,001 and 101) total of eight switching states are considered for the non-shoot-through state mode (nSTS). The additional ninth state can be considered for shoot-through operation of the Z source converter. All PWM control methods implemented for VSC can be implemented in a new topology. Note that the shoot-through zero state will produce a zero voltage across the load terminals as like as open states. The shoot-through zero state is inserted in the positions of traditional open zero states in the gating pulses by any of the pulse width modulation techniques. It does not alter the normal working of the inverter. One drawback of the Z source converter is the practical gain, which is limited and is not up to the theoretical values due to the interdependence of shoot-through duty ratio (d_{ST}) and M given by Eq. (3).

The two inductance and capacitance values are assumed to be equal for the symmetry of the circuit $L_1 = L_2 = L$ and $C_1 = C_2 = C$. The STS equivalent circuit of ZSC during an interval of T_{ST} with a switching period of T is shown in Fig. 9. The diode is reverse biased and the inverter legs are short-circuited. Voltage equations in STS are expressed in Eqs. (4) and (5). The nSTS equivalent circuit of ZSC is represented in Fig. 10 during an interval of T_{nST} with the inverter operated in the traditional eight switching states. The inverter terminals are replaced with a current source

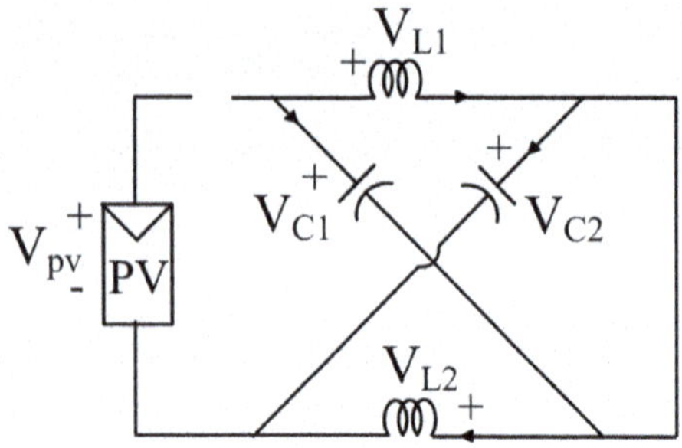

Fig. 9 Equivalent circuit of ZSC in STS [15]

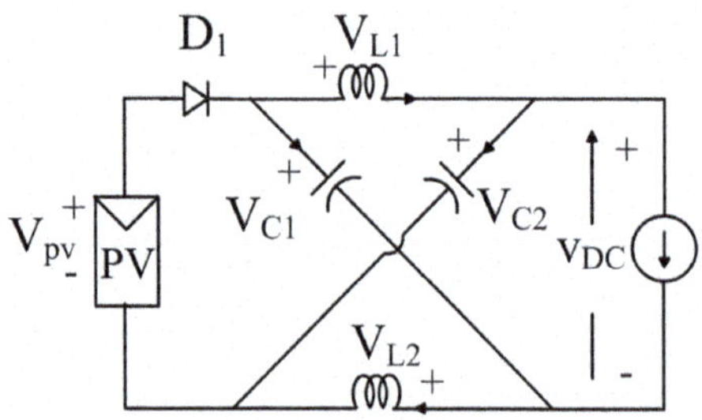

Fig. 10 Equivalent circuit of ZSC in nSTS [15]

having a terminal voltage equivalent to DC-link voltage V_{DC}. Voltage equations in STS are expressed in Eqs. (5) and (6).

$$M \leq 1 - d_{ST} \tag{3}$$

$$v_{L1} = v_{C1} \tag{4}$$

$$v_{L2} = v_{C2} \tag{5}$$

$$v_{L1} = v_{PV} - v_{C2} \tag{6}$$

$$v_{L2} = v_{PV} - v_{C1} \tag{7}$$

$$v_{DC} = -v_{PV} + v_{C1} + v_{C2} \tag{8}$$

According to the V-s balance principle, from Eqs. (4) and (8)

$$V_{C1} = V_{C2} = \frac{1 - d_{ST}}{1 - 2d_{ST}} V_{PV} \tag{9}$$

$$V_{DC} = \frac{1}{1 - 2d_{ST}} V_{PV} \tag{10}$$

$$V_{DC} = B V_{PV} \tag{11}$$

$$B = \frac{1}{1 - 2d_{ST}} \tag{12}$$

$$v_{peak} = M B \frac{V_{PV}}{2} \tag{13}$$

Here, B is the voltage boost factor. The relation between the peak voltage of AC supply across the load terminals or grid and the input DC voltage is expressed in Eq. (13).

Example 1 For a three-phase 400 V supply on AC side of the converter with an input DC voltage of 400 V, find out the required DC-link voltage and shoot-through duty ratio.

Sol:

$V_{Peak} = 326.56$ V

$V_{PV} = 400$ V

$d_{ST} = 0.279$

$M = 0.721$

$B = 2.26$

$V_{DC} = 905$ V

If the system requires high gain, then d_{ST} value has to be increased, which causes high voltage stress across the switching devices. The modulation index (M) decreases and causes high harmonic distortion in the output waveform. The Z source inverter can give a voltage gain from zero to infinity theoretically, but it does not provide the required voltage gain under practical conditions. The limitations of voltage-fed ZSI are listed as:

- It may be operated at discontinuous conduction mode (DCM), to meet the high boost voltage gain influence of high ripple in inductor current.
- There is no common ground in between the source and load causes EMI (Electromagnetic interference) noise.
- High starting inrush current will damage the switching devices.

To improve the operation of the Z source network, several topologies are developed to meet specified needs. The inclusion of various switching devices in combination with inductor and capacitor elements at a proper location can generate a variety of impedance source converters (ISC) to improve the performance of the basic Z source converter. Some of the topologies are developed for soft start capability, providing a common DC ground, and maintaining the converter under CCM. The impedance source ideology has opened up a new research area in various fields related to power electronics. Many new modifications can be brought in impedance source converters such that each topology is designed to meet a particular application and exhibits its unique features. Modern manufacturing trends in power electronic devices like silicon carbide (SiC) and gallium nitride (GaN) devices, will maximize the performance of Z source converters. The impedance source networks can be categorized based on whether coupled coil or transformer is placed in the impedance network or not. Again the Non transformer/uncoupled coil-based converters can be of various types. Some of them are given below:

- Quasi Z source Converters.
- Improved Z source Converter.
- Switched Inductance Z source converter.
- Embedded Z source Converter.
- Hybrid quasi Z source converter…etc.

3.2 Quasi Z Source Converter (QZSC)

The Quasi Z source converter (QZSC) [21] can be developed to obtain a common DC rail. Also, obtains a continuous current from the DC source, which enhances renewable sources' efficiency. These characteristics are not found in the original Z source converter. The quasi Z source converter configuration shown in Fig. 11 inherits the features of the Z source converter. The present topology reduces the voltage stress on one of the capacitors compared with the Z source inverter and runs under CCM (continuous source current). Like a Z source converter, the quasi Z source inverter also runs in two modes: Shoot-through and non-shoot-through states. It develops the voltage boost gain like the Z source inverter at reduced voltage stress across one capacitor. The input source current of the quasi Z Source Inverter is continuous due to the input inductor L1 in series with the DC source, significantly reducing the ripple current in inductors. It has a common DC ground between the converter and driving source, which mitigates EMI problems to some extent.

3.3 Improved Z Source Converter

Despite the aforesaid merits, the traditional Z source inverter topology also shows the following pitfalls:

- High voltage rating capacitors are needed in the Z source converter to be at a voltage not less than the input voltage, resulting in larger volume and cost expensive to the system.

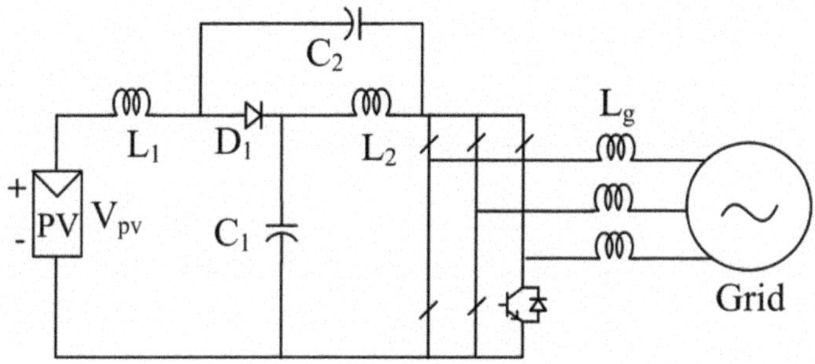

Fig. 11 Circuit configuration of Quasi Z source Converter [21]

- High inrush current and the voltage surges due to resonance due to energy stored elements appearing in traditional Z source converter at start-up may destroy the switching devices.

To get rid of the obstacles in basic Z source inverter, an improved Z source inverter topology [22] has been developed, which is shown in Fig. 12. The proposed topology shall diminish the voltage stress on capacitors, provide a soft start capability by smooth variation of shoot-through zero states, and suppress resonance. The DC-link voltage boost gain of the proposed topology is the same as traditional ZSC.

$$V_{DC} = \frac{1 + 3d_{ST}}{1 - 3d_{ST}} V_{PV} \quad (14)$$

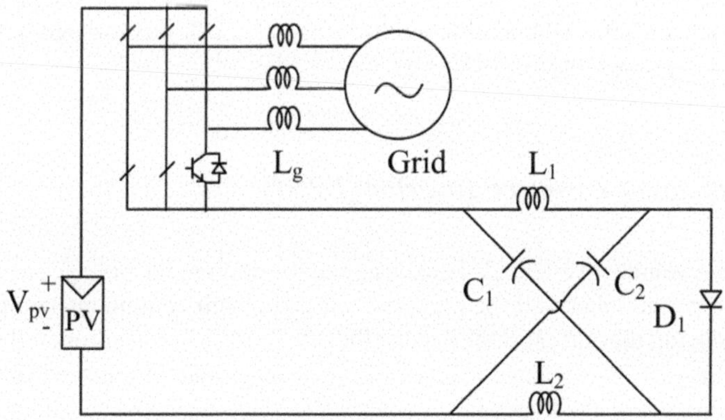

Fig. 12 Improved Z source Inverter Topology [22]

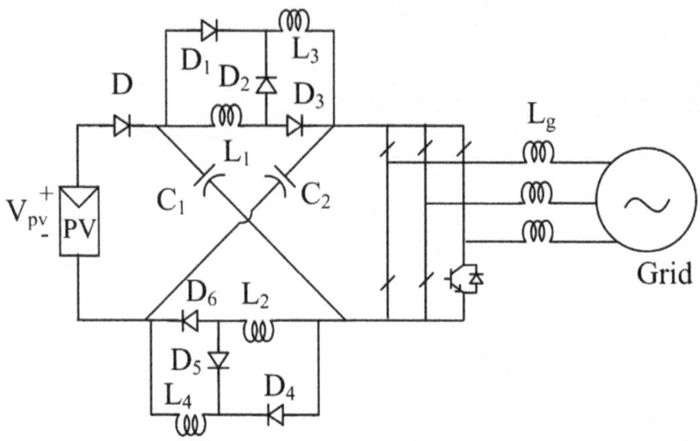

Fig. 13 Switched Inductor Z source Inverter topology [23]

3.4 Switched Inductor Z Source Converter (SL ZSI)

To obtain a higher gain and inherit all the features of traditional ZSC, a new topology known as Switched Inductor Z source Converter [23] is developed with the same lattice impedance network shown in Fig. 13. In contrast with the Z Source converter, the switched inductor Z source inverter remarkably increases the voltage boost gain at the converter terminals. A narrow short shoot-through zero state is sufficient to attain high boost voltage gain, which is instrumental for enhancing the output power quality. Obviously, it can also meet the low AC voltages with the conventional voltage buck inversion ability. The SL ZSI can also be employed in various AC-DC, DC-AC, DC-DC, and AC-AC power conversion applications at higher voltage gain, which will be worthwhile for engineering applications. The DC-link voltage V_{DC} of the converter is expressed in Eq. (14).

3.5 Embedded Z Source Inverter (EZSI)

Embedded Z source inverters [24] are designed to suppress the voltage stress across the capacitors and inherit all the properties of ZSC, with continuous input current, thereby reducing the current ripples in the inductor current. In the proposed structure shown in Fig. 14, two voltage sources of equal magnitude are placed in series to the two inductors of the impedance network to reduce the capacitor voltage stress, instead of using one DC voltage source with high magnitude. Renewable energy sources particularly solar panels are used as input DC voltage sources.

The essential contrast between the discussed topology and the conventional ZSC is that the input source voltage can be bifurcated as two halves and located in series with the inductors to attain CCM (continuous current mode) shown in Fig. 14. The

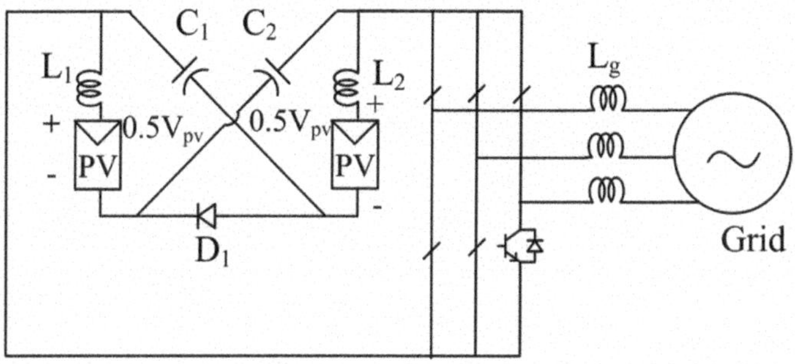

Fig. 14 Embedded Z source network Topology [24]

converter can give the same voltage gain as ZSC but with reduced voltage stress across the capacitance. The capacitors are having voltage stress equal to half of the boost DC-link voltage value compared with ZSC. Also, the proposed EZSI can work in CCM.

3.6 Improved Hybrid Three Quasi Z Source Converter (IHTQZSC)

With the knowledge about a basic Z source converter and quasi Z source converter, a new set of DC-DC hybrid Z source boost converters [25, 26] are designed to take care of the better utilization of renewable sources, particularly photovoltaic applications. The high magnified boost voltage DC-AC converters are needed to uplift the low source voltages to application needs. The converter is designed to upgrade the voltage boost ability by merging quasi Z source converters of two configurations. The newly designed impedance source converters are termed improved hybrid three quasi Z source converters. It inherits the advantages of Z source networks with stronger boost voltage caliber implemented to reach other AC-AC, DC-AC, and AC-DC power conversions. L_3, L_4, C_5, and C_6 form one QZSC configuration-1. Similarly, L_1, L_2, C_1, and C_2 form the second QZSC configuration-2. Configuration-1 and configuration-2 with the D_2, C_4, and C_3 forms QSZC configuration-3. Configuration-3 has been different from configurations-1 and 2. This means IHTQZSC has two kinds of three QZSC. Hence, the term Three and hybrid is given. The DC source is connected in series with the inductor to achieve CCM operation; accordingly, the term improved is mentioned in IHTQZSC and shown in Fig. 15.

All the diodes are off in STS and the converter legs are shorted. The STS equivalent circuit of IHTQZSC is shown in Fig. 16. The nSTS equivalent circuit of IHTQZSC is shown in Fig. 17. In nSTS, all the diodes are on and the converter is represented with a current source with a DC-link voltage of V_{DC}. The voltage and current relations in STS of IHTQZSC are given in Eqs. (15)–(24). The voltage and current relations in nSTS of IHTQZSC are given in Eqs. (25)–(38). Consider $R_{LZ}$and R_{CZ} are equivalent

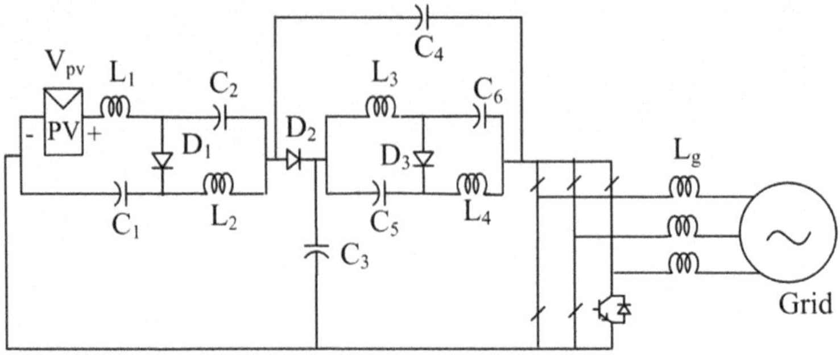

Fig. 15 Improved Hybrid Three Quasi Z source converter topology [26]

series resistance (ESR) of inductors and capacitors, respectively. The value of current at the current source in Fig. 17 is represented with i_{Con}. V_L and V_C are the voltage across the particular inductor and capacitor, respectively. i_L and i_C are current flowing through the specific inductor and capacitor, respectively.

$$V_{L1} = -R_{LZ}i_{L1} + R_{CZ}i_{C2} + R_{CZ}i_{C4} + V_{C2} + V_{C4} + V_{PV} \tag{15}$$

$$V_{L2} = -R_{LZ}i_{L2} + R_{CZ}i_{C1} + R_{CZ}i_{C4} + V_{C1} + V_{C4} \tag{16}$$

$$V_{L3} = -R_{LZ}i_{L3} + R_{CZ}i_{C3} + R_{CZ}i_{C6} + V_{C3} + V_{C6} \tag{17}$$

$$V_{L4} = -R_{LZ}i_{L4} + R_{CZ}i_{C3} + R_{CZ}i_{C5} + V_{C3} + V_{C5} \tag{18}$$

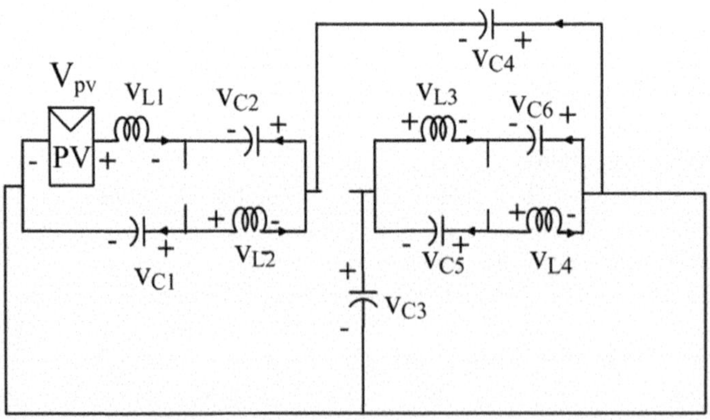

Fig. 16 STS operation of IHTQZSC [26]

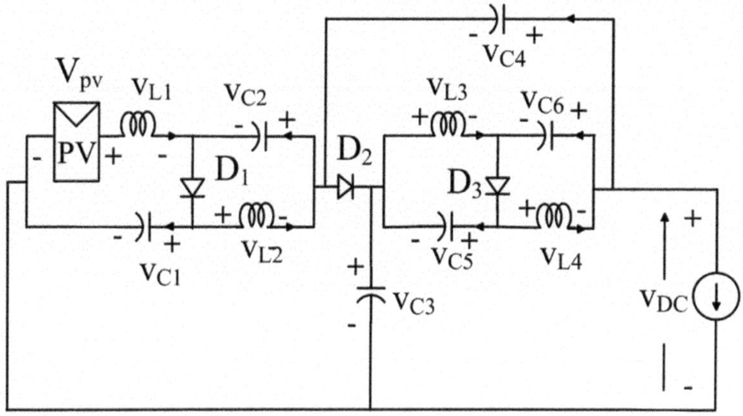

Fig. 17 nSTS operation of IHTQZSC [26]

$$i_{C1} = -i_{L2} \tag{19}$$

$$i_{C2} = -i_{L1} \tag{20}$$

$$i_{C3} = -(i_{L3} + i_{L4}) \tag{21}$$

$$i_{C4} = -(i_{L1} + i_{L2}) \tag{22}$$

$$i_{C5} = -i_{L3} \tag{23}$$

$$i_{C6} = -i_{L4} \tag{24}$$

$$V_{L1} + R_{CZ}i_{C1} = -R_{LZ}i_{L1} - V_{C1} + V_{PV} \tag{25}$$

$$V_{L2} + R_{CZ}i_{C2} = -R_{LZ}i_{L2} - V_{C2} \tag{26}$$

$$V_{L3} + R_{CZ}i_{C5} = -R_{LZ}i_{L3} - V_{C5} \tag{27}$$

$$V_{L4} + R_{CZ}i_{C6} = -R_{LZ}i_{L4} - V_{C6} \tag{28}$$

$$V_{C1} + V_{C2} = V_{C3} + R_{CZ}i_{C1} + R_{CZ}i_{C2} + R_{CZ}i_{C3} \tag{29}$$

$$V_{C5} + V_{C6} = V_{C4} + R_{CZ}i_{C4} + R_{CZ}i_{C5} + R_{CZ}i_{C6} \tag{30}$$

$$i_{C1} - i_{C2} = i_{L1} - i_{L2} \tag{31}$$

$$i_{C5} - i_{C6} = i_{L3} - i_{L4} \tag{32}$$

$$i_{C2} + i_{C3} - i_{C4} - i_{C5} = i_{L2} - i_{L3} \tag{33}$$

$$i_{C1} + i_{C3} = i_{L1} - i_{Con} \tag{34}$$

$$i_{C4} + i_{C5} = i_{L3} - i_{Con} \tag{35}$$

$$i_{C2} + i_{C3} = i_{L2} - i_{Con} \tag{36}$$

$$i_{C4} + i_{C6} = i_{L4} - i_{Con} \tag{37}$$

The voltage distribution among the capacitors is expressed in Eqs. (38)–(41) for an ideal network and DC-link voltage is in Eq. (42)

$$V_{C1} = \frac{1 - 3d_{ST}}{(1 - 4d_{ST})} V_{PV} \tag{38}$$

$$V_{C2} = V_{C5} = V_{C6} = \frac{d_{ST}}{(1 - 4d_{ST})} V_{PV} \tag{39}$$

$$V_{C3} = \frac{1 - 2d_{ST}}{(1 - 4d_{ST})} V_{PV} \tag{40}$$

$$V_{C4} = \frac{2d_{ST}}{(1 - 4d_{ST})} V_{PV} \tag{41}$$

3.7 *Simple Boost Control (SBC) Pulse Width Modulation*

Various pulse width modulation (PWM) control techniques are implemented in conventional inverter circuits to meet the application needs of consumers. Conventional inverter circuits have six active vectors (switching states 100, 110, 010, 011, 001, and 101) and two zero vectors (switching states 000 and 111), and any phase leg of VSC couldn't be short-circuited which causes damage to the switching devices and circuit components. The impedance source converters utilize the prohibited shoot-through zero state in the traditional inverter switching sequence, thereby providing the boost voltage. Hence, impedance source converters perform single-stage voltage buck-boost action. The shoot-through zero states are formed by simultaneously switching the top and bottom devices of the same phase leg. There are seven possible ways to generate the shoot-through zero state. They are switching on both the top and bottom devices of any one-phase leg or the combination of any two-phase legs or all the three-phase legs of the three-phase voltage source inverter.

Based on the way of generating shoot-through zero state and inserted in the available sinusoidal PWM methods, the PWM methods applicable to impedance source converters [27] are classified as:

- Simple boost control (SBC) method.
- Maximum boost control (MBC) method.
- Maximum constant boost control (MCBC) method.

The insertion of shoot-through zero state has not caused any changes to the normal switching operation. It will be inserted at the time duration of zero switching vector which provides a zero voltage across load terminals.

$$VDC = \frac{1}{(1 - 4d_{ST})} V_{PV} \tag{42}$$

$$B = \frac{1}{(1 - 4d_{ST})} \tag{43}$$

Simple Boost Control Method

The simple boost control (SBC) method [27] is the initial PWM technique developed along with the introduction of a Z source inverter. In this method, the three modulating signals are set against with triangular carrier signal as it is in the conventional sinusoidal pulse width modulation (SPWM) method. The additional shoot-through state is provided by comparing the magnitude of the carrier signal with positive and negative DC voltages (+DC and –DC signal) in the upper half cycle and lower half cycle, respectively, without disturbing the regular operation of the converter circuit. The magnitudes of DC voltage values are equal to $1-d_{ST}$ value. Figure 18 elaborates on the pulses generation from SPWM and SBC PWM.

Three modulating signals m_a, m_b, and m_c and a carrier signal c_r is chosen for SPWM. Whenever modulating signal magnitude is higher than the carrier signal, the top switch is on, else the bottom switch is on for a three-phase converter. For the demonstration, phase a modulating waveform and corresponding switching pulses S_a for the top switch and S_{ac} for the bottom switch are chosen. Input dc voltage has been boosted to a value of V_{DC} at DC-link of the converter by proper insertion of STS in SPWM pulses observed in Fig. 18. Shoot-through switching pulses are generated whenever +DC signal is higher than the c_r signal or –DC signal is lower than the c_r signal. As a result, the shoot-through switching pulses are inserted twice the switching frequency of the carrier signal. During STS, S_a and S_{ac} are turned on simultaneously as shown in Fig. 18. Similarly, switching pulses are generated for the other two phases.

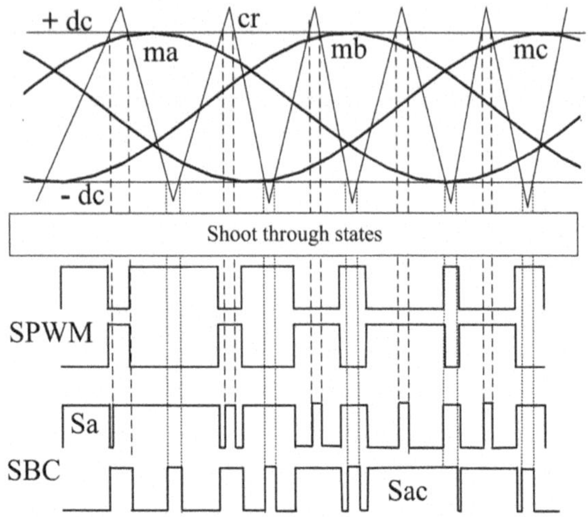

Fig. 18 Switching Pulses from SBC method [27]

4 Controlling of IHTQZSC in Grid-Connected Application

Most of the control methods developed for VSC are to be implemented on impedance source converters for a proper response on grid-side parameters. For validation of the control methods, the system parameters are given in Table 2. V_{in} is the input DC voltage, v_{abc} is grid voltage, i_{abc} is grid current, i_{L1} is inductor current, V_{C3} is the voltage across the capacitor C_3 and V_{DC} is the DC-link voltage. In this chapter, three control methods are discussed with IHTQZSC:

- PI controller
- MPC
- CSF-PC

Example 2 Design the IHTQZSC inductance value at 10 kW output power for a three-phase 400 V supply on AC side of the converter with an input DC voltage of 400 V.

Table 2 System parameters—grid connected IHTQZC

Parameter	Value
Inductance	1 mH
Capacitance	470 μf
ESR of Inductor	0.1 Ω
ESR of Capacitor	0.1 Ω
Switching Frequency	10 kHz
Grid Voltage (line)	220 V
Interfacing Inductance	10 mH

Sol.

$V_{peak} = 326.56$ V

From Eq. (13), $d_{ST} = 0.114$

$V_{DC} = 735.3$ V

For a lossless converter $P_{ou}t = P_{in}$

Input current $I_{in} = 10{,}000/400 = 25$A

The current stresses across the inductors are maximum in STS. Hence, STS voltage relation is considered for L and C calculation.

From Eq. (15), with ideal elements, consider the converter switching frequency as 10 kHz. Then the STS switching frequency becomes two times of converter switching frequency for a SBC PWM.

Allow 20% ripple in the inductor current IL1 = 0.743 mH.

All inductance values are equal in a IHTQZSC.

4.1 *PI Controller*

The system block diagram implemented with PI controller for grid-integrated IHTQZSC has been shown in Fig. 19. All the control methods used for general voltage source converter in grid integration application are applicable to AC side control. The grid voltage e_g is measured to extract the grid angle which is used for extraction of grid current as dq components. The dq component of grid currents is compared with reference grid current components. Error in grid current components is controlled by PI control and generating modulation index (m).

On the DC side, a dual-loop control method [29] is applied for generating shoot-through duty ratio (d_{ST}). Interaction of d_{ST} and m by simple boost control method providing required boost voltage on DC side as well as the converter has been grid integrated. The DC link voltage of the converter varies from a zero value to a peak value. Hence, it has been controlled with the use of capacitor voltages. The average DC link voltage and all capacitor voltages are a function of shoot-through ratio. Here indirect control of DC-link voltage can be done in the outer loop, by controlling capacitor voltage V_{C3} and V_{C4} as shown in Fig. 19. Capacitor voltage control regulates V_{DC} value. Apart from controlling DC link voltage, inductor current ripples are also regulated with an inner current loop. The outcome of the DC side control is d_{ST}.

Response of the IHTQZSC is shown in Fig. 20 at a step change in V_{in} from 200 to 140 V at the instant 1 s. The converter has been injecting i_{abc} of 5A into the grid. Current i_{L1} has been changed to 13.5A from 9.5A, V_{C3} has been changed from 307 to 286 V and V_{DC} has been changed to 432 V from 413 V with the change in V_{in}. The converter has taken 0.2 s to restore back to steady-state response with input voltage disturbance. The disturbance in V_{in} affects i_{abc} for a few cycles to reach a steady state. Figure 21 visualize IHTQZSC response at a step change in i_{abc} from

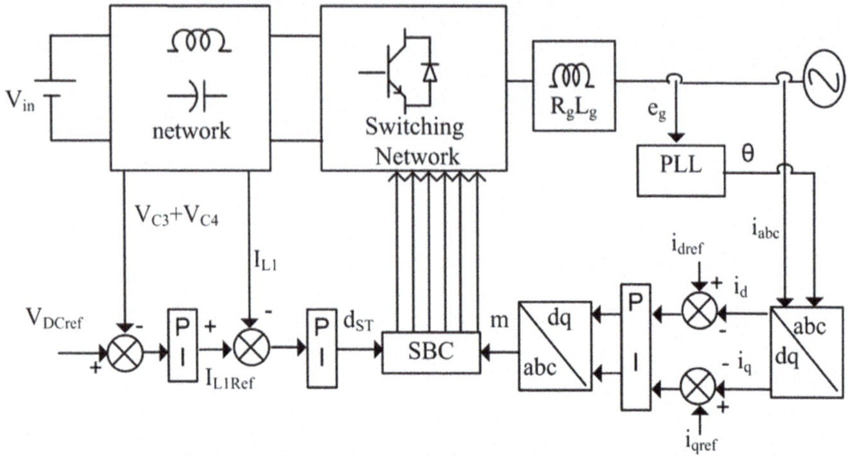

Fig. 19 Grid tied IHTQZSC with PI Control [28]

5A to 2.5A at the instant 2 s at V_{in} of 140 V. The current i_{L1} has been changed to 6.85A from 13.5A. V_{in} is maintained at 140 V, hence V_{C3} and V_{DC} are continuing at 286 V and 432 V at a steady state. The converter has taken 0.1 s to restore back to steady-state response with i_{abc} variation.

4.2 *Model Predictive Controller for Grid-Connected Converter*

Finite control set model predictive control (MPC) [30] is an online optimizing control method also called receding (or moving) horizon MPC. MPC has been forseen the future value of the variables or functions and conducts control action accordingly with well-defined objective function or cost function expressed in Eq. (44). The cost function is defined in terms of error between the control variables and their reference values. So that MPC has been minimizing the cost function value at every predicted state. MPC is a multi-objective optimization control. Error (ε) has been calculated for each switching state and is given to the cost function. With the proper selection of weighting factors (W), the system has achieved a good steady-state response. MPC is a non-linear control method providing an excellent dynamic response.

$$C(\varepsilon) = w_1\,|\varepsilon_1| + w_2|\varepsilon_2| + \ldots + w_n|\varepsilon_n| \tag{44}$$

In the field of power electronics, digital control techniques predictive control has been rigorously researched for almost three decades. Finite control set MPC (FCS-MPC) was proposed nearly a decade ago for a two-level voltage source converter

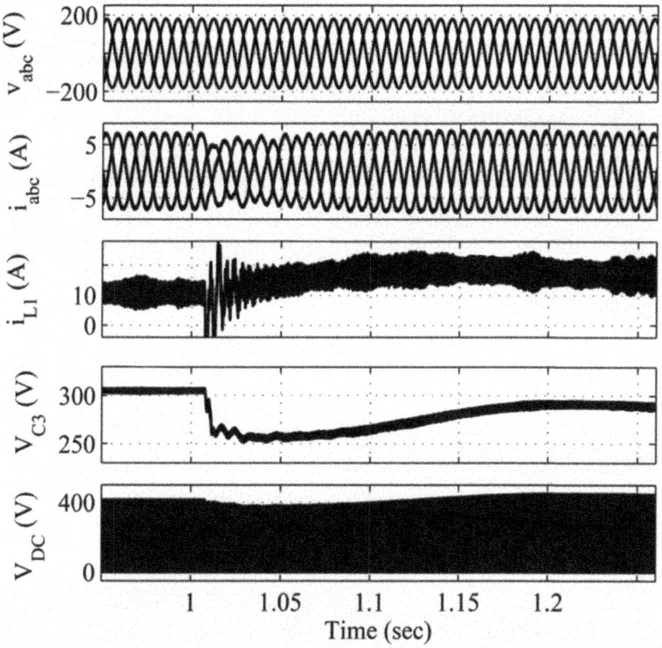

Fig. 20 Grid tied IHTQZSC with PI Control with input voltage variation [28]

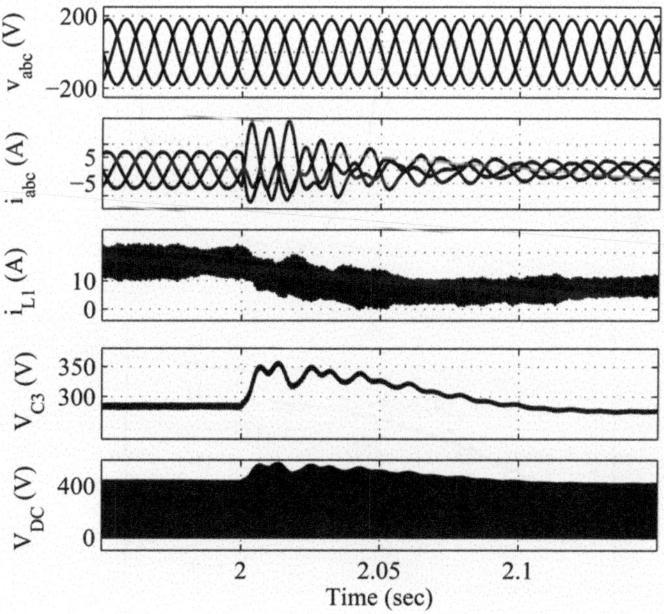

Fig. 21 Grid tied IHTQZSC with PI Control with grid current variation [28]

(2L-VSC). MPC has been theoretically analyzed and experimentally tested for a wide range of low, medium, and high power converter topologies and power conversion applications.

Model predictive control has been implemented by discretizing all control variables at a sampling time of Ts. MPC has provided an optimal switching vector to the switching networks (converters). Figure 22 explains the working of MPC. The reference current i_x has been discretized at the Ts sampling rate. For each switching state, it calculates the predicted current i_α values at each sample and compares them with its reference value. The switching state that provides the lowest error between the actual and predicted current values will be the most appropriate switching vector and is given to the converter. The process has continued for the following switching states. Switching vectors 4 and 7 are the optimal switching states at the instant Ts and 2Ts, respectively.

Figure 23 depicts the block diagram for implementing MPC in the current converter. The predictive controller receives grid currents i_g (representing i_{abc}) and grid voltages e_g (representing v_{abc}) as feedback signals. Inductor currents and capacitor voltages are also supplied to the predictive controller as control parameters on

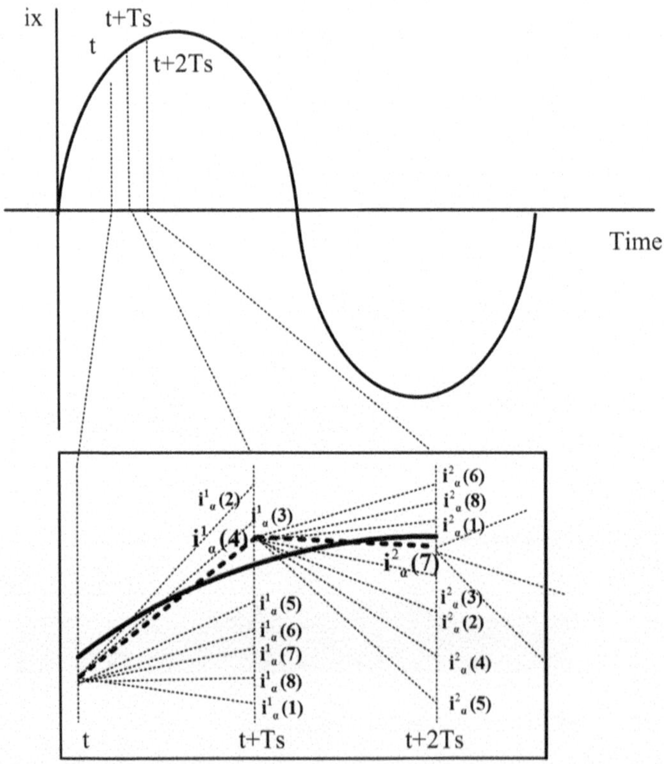

Fig. 22 Working of MPC [30]

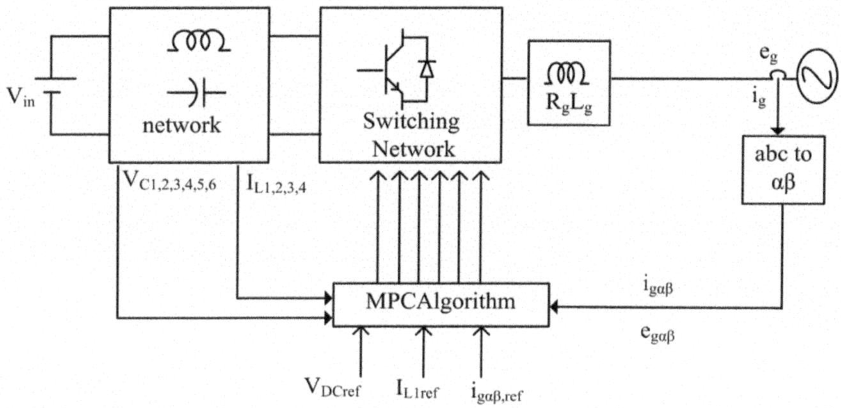

Fig. 23 Grid tied IHTQZSC with MPC Control [28]

the dc side (feedback). The anticipated control parameter values are sent into the cost function to select the best switching vector and feed it to the converter. Differential equations representing the mathematical model of the converter on the grid side and on the DC side of the converter are required for implementing MPC.

Grid Side Parameter Prediction

The converter voltage v_{Con} is described as a space vector expressed in Eq. (45) as a function of converter switching states. Table 3 list all the switching states. Two conventional zero states of the VSC due to either all top switches of legs are on are off indicated with V_o. V_7 is the shoot-through zero state that is allowed in ISC and is formed by turning on each switch on any one or multiple legs. The remaining are active states. S_a, S_b, and S_c indicate the state of the top switches and Sac, Sbc, and Scc indicate the state of the lower switches of phases a, b, and c of the converter.

$$v_{\text{con}}(k+1) = \frac{2}{3} V_{DC}(S_a + a\, S_b + a^2\, S_c) \tag{45}$$

Converter voltage o grid side at (k + 1)th instant is v_{Con} (k + 1) and a is the operator given as $a = -\frac{1}{2} + j\frac{\sqrt{3}}{2}$.

The converter has been grid-connected with an interfacing filter inductor. R_g and L_g are the per phase net inductance and resistance of the grid. The voltage equation on the grid side is given in (46). A three-phase current of i_g is injected by the converter into the grid.

$$v_{Con}(k) = R_g\, i_g(k) + L_g\, \frac{d\, i_g}{dt} + e_g(k) \tag{46}$$

$$\frac{di}{dt} = \frac{i(k+1) - i(k)}{T_s} \tag{47}$$

Table 3 Switching vector for a 2L—VSC

Switching vector	Sa	Sb	Sc	Sac	Sbc	Scc
Vo	0	0	0	1	1	1
	1	1	1	0	0	0
V1	0	0	1	1	1	0
V2	0	1	1	1	0	0
V3	0	1	0	1	0	1
V4	1	1	0	0	0	1
V5	1	0	0	0	1	1
V6	1	0	1	0	1	0
V7	1	1	1	1	1	1

$$\frac{d\,v_C}{dt} = \frac{v_C(k+1) - v_C(k)}{T_s} \tag{48}$$

i(k), and $v_C(k)$ are the current instant current and capacitor voltage values. The control variables in the discrete model are expressed with Euler's forward approximation method at a sampling time of T_S in Eqs. (47) and (48). $i_g(k+1)$, $i_{L1}(k+1)$, $v_{C3}(k+1)$, and $v_{C4}(k+1)$ are the predicted values at k + 1 instant. From Eqs. (46) and (47), the predicted future value of grid current at the (k + 1)th instant is given in Eq. (49)

$$i_g(k+1) = (1 - \frac{R_g\,T_S}{L_g})\,i_g(k) + \frac{T_S}{L_g}\big(v_{Con}(k) - e_g(k)\big) \tag{49}$$

$$\begin{aligned} i_{L1}(k+1) &= \left\{1 - \frac{(2R_{CZ} + R_{LZ})T_S}{L}\right\} i_{L1}(k) \\ &\quad + \{i_{L2}(k) + v_{C2}(k) + v_{C4}(k) + V_{in}\}\frac{T_S}{L} \end{aligned} \tag{50}$$

$$v_{C3}(k+1) = v_{C3}(k) - (i_{L3} + i_{L4})\frac{T_S}{C} \tag{51}$$

Dc Side Parameter Prediction

As previously stated, the converter operates in STS and nSTS modes, their future capacitor voltage and inductor current estimations are given for these modes. STS predicted voltage and currents are expressed in Eqs. (50)–(52) based on the Eqs. (15), (21), and (22). nSTS predicted voltage and currents are expressed in Eqs. (53)–(55) based on the Eqs. (25)–(37). The predictive control method identifies the best switching vector by optimizing the cost function at each instant of time. This optimal switching vector is delivered to the converter in order to meet the desired system response. The switching states are listed in Table 3.

$$v_{C4}(k+1) = v_{C4}(k) - (i_{L1} + i_{L2})\frac{T_S}{C} \tag{52}$$

$$i_{L1}(k+1) = \left\{1 - \frac{(\frac{2}{3}R_{CZ} + R_{LZ})T_S}{L}\right\} i_{L1}(k) + \left\{\frac{1}{3}R_{CZ}i_{L2}(k) - \frac{2}{3}v_{C1} + \frac{1}{3}v_{C2}(k) - \frac{1}{3}v_{C3}(k) + V_{in} + \frac{1}{3}R_{CZ}i_{Con}\right\}\frac{T_S}{L} \tag{53}$$

$$v_{C3}(k+1) = \left\{\frac{1}{3}i_{L1}(k) + \frac{1}{3}i_{L2}(k) + \frac{1}{3R_{CZ}}v_{C1}(k) + \frac{1}{3R_{CZ}}v_{C2}(k) - \frac{2}{3}i_{Con}\right\}\frac{T_S}{C} + \left\{1 - \frac{T_S}{3R_{CZ}C}\right\} v_{C3}(k) \tag{54}$$

$$v_{C4}(k+1) = \frac{1}{3}(i_{L3}(k) + i_{L4}(k)) + \left(1 - \frac{T_S}{3R_{CZ}C}\right)v_{C4}(k) + \frac{T_S}{3R_{CZ}C}(v_{C5}(k) + v_{C6}(k)) - \frac{2T_S}{3C}i_{Con}(k) \tag{55}$$

$$v_{C4}(k+1) = \frac{T_S}{3C}(i_{L3}(k) + i_{L4}(k)) + (1 - \frac{T_S}{3R_{CZ}C})v_{C4}(k) + \frac{T_S}{3R_{CZ}}(v_{C5}(k) + v_{C6}(k)) - \frac{2T_S}{3C}i_{Con} \tag{56}$$

$$i_{Con} = S_a\, i_a(k) + S_b\, i_b(k) + S_c\, i_c(k) \tag{57}$$

$$C_i = w_i\left\{\left|i_{g\alpha ref}(k+1) - i_{g\alpha}(k+1)\right| + \left|i_{g\beta ref}(k+1) - i_{g\beta}(k+1)\right|\right\} \tag{58}$$

$$C_L = w_L\{|i_{L1ref}(k+1) - i_{L1}(k+1)|\} \tag{59}$$

$$C_C = w_c\left\{\left|V_{DCref}(k+1) - (v_{C3}(k+1) + v_{C4}(k+1))\right|\right\} \tag{60}$$

The future anticipated grid current $i_g(k + 1)$ and respective reference grid current $i_{gref}(k + 1)$ is expressed in α β components as $i_{g\alpha}(k + 1)$, $i_{g\beta}(k + 1)$, $i_{g\alpha ref}(k + 1)$, and $i_{g\beta ref}(k + 1)$. With the weighting factor W_i, the absolute error between the future projected and the corresponding grid current reference value is included in the cost function in Eq. (58). Likewise, the absolute error between the estimated future inductor current capacitor voltages, respectively, $i_{L1}(k + 1)$ and $v_{C3}(k + 1) + v_{C4}(k + 1)$ and the associated references $I_{L1ref}(k + 1)$ and $V_{DCref}(k + 1)$ are incorporated in cost function with weighting factors w_L and w_C, respectively. The cost function describing inductor current is expressed in Eq. (59). The capacitor voltage cost function is represented in Eq. (60)

$$C\,(x) = C_i + C_L + C_C \tag{61}$$

Figure 24 depicts the flow chart of MPC implementation on grid-connected IHTQZC. Cost function $C(x)$ in Eq. (61) is evaluated for every switching vector in a sampling period T_S. $C(x)$ value due to switching vectors is compared among them and the switching vector that minimizes $C(x)$ will be the optimum switching vector given to the converter switches. The estimated future parameters are extremely nearer or the same as their reference value when the cost function has been minimized.

Figure 25 shows the response of the IHTQZSC while a step change in V_{in} from 200 to 140 V at the instant 1 s. The converter has been injecting i_{abc} of 5A into the grid. Current i_{L1} has been changed to 13.5 A from 9.5A, V_{C3} has been change from 307 to 286 V and V_{DC} has been changed to 432 V from 413Vwith the change in V_{in}. Compared to Fig. 20, the converter exhibiting faster dynamic response with input voltage disturbance with MPC control. The disturbance in V_{in} does not affect i_{abc} response with MPC when compared with PI controller. Figure 25 shows the response of the IHTQZSC while a step change in i_{abc} from 5A to 2.5A at the instant 2 s at V_{in} of 140 V. Current i_{L1} has been changed to 6.85A from 13.5A. V_{in} is maintained at 140 V hence V_{C3} and V_{DC} are continuing at 286 V and 432 V at steady state.

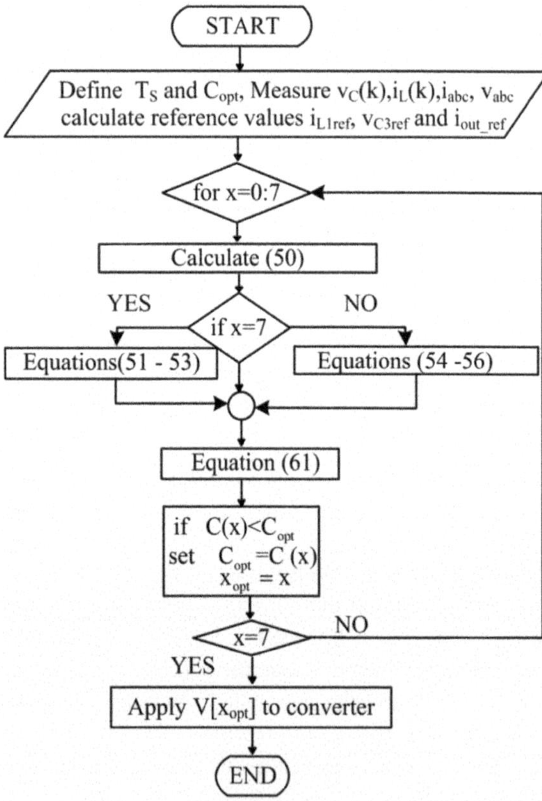

Fig. 24 Flow chart for implementing MPC for grid-connected IHTQZSC

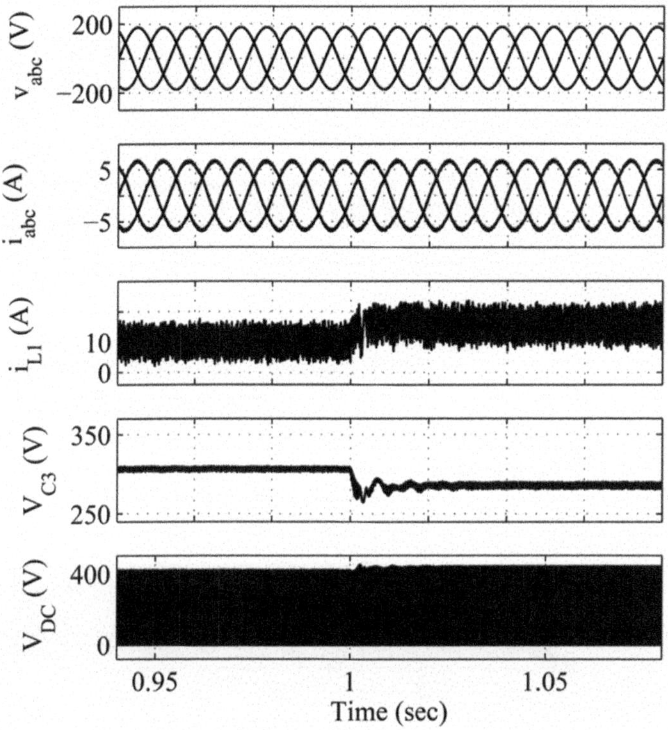

Fig. 25 Response of grid-connected IHTQZSC with at step change in input voltage

Tracking of i_{L1} is also exhibiting better response than the PI controller response by observing Figs. 26 and 27.

4.3 Grid Tied ImHTQZSI with CSF—PC

MPC has provided variable switching frequency makes the difficulties in filter design. The time instants of the switching state are decided by the MPC algorithm causing higher ripples in ISC. MPC implementation with long prediction horizons provides faster and smoother dynamic performance. But the long prediction horizon MPC compels a more significant number of computations on controlled parameters posing a huge computation burden on the digital controller. Direct MPC implementation on single-phase grid-connected converters provides higher THD in current due to fewer switching states. The burden on digital processors has been enormously increased if the number of switching states are increased depending upon the converter topology. To avoid these difficulties, constant switching frequency (CSF) predictive control (PC) has been implemented.

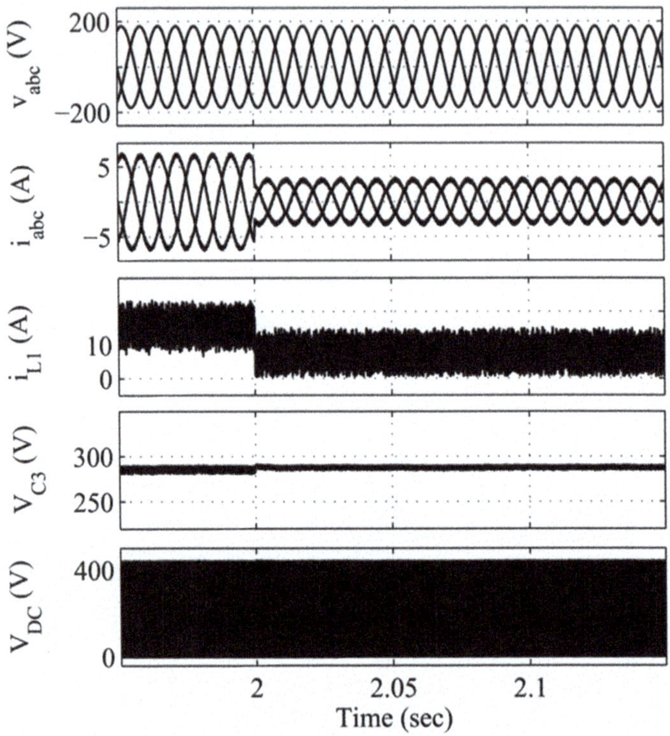

Fig. 26 Response of grid-connected IHTQZSC with at step change grid current

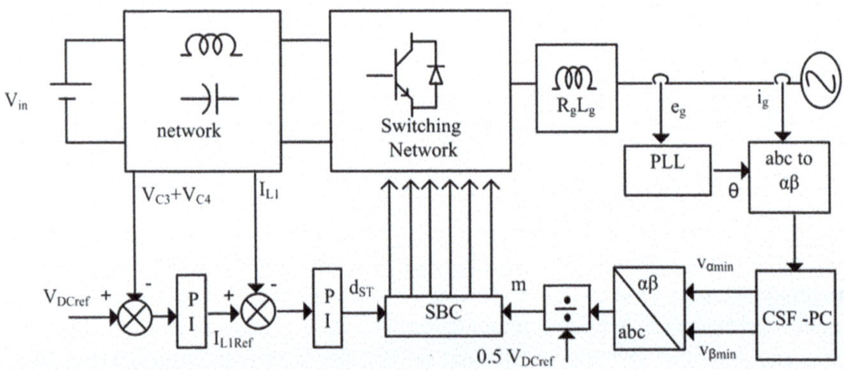

Fig. 27 Grid tied IHTQZSC with CSF-PC Control [28]

CSF-PC has been applied for IMHTQZSC for grid integrated applications shown in Fig. 24. PI controller has been intended for DC-link voltage regulation and the grid current can be controlled with CSF-PC [28] method and generate $v_{\alpha min}$ and $v_{\beta min}$ and converted them into abc reference frame and given to modulator. As the modulator has been used in the control method, it provides a constant switching frequency. Equation (58) is an absolute cost function evaluating error in the predicted current values. Here the quadratic cost function is considered [30] for CSF-PC expressed in Eq. (62). The partial derivative of these individual cost functions with respect to the converter voltage v_{Con} is carried out to find out minimum voltage vectors given in Eqs. (65) and (66) to minimize the current error.

$$C_i = \left|i_{g\alpha ref}(k+1) - i_{g\alpha}(k+1)\right|^2 + \left|i_{g\beta ref}(k+1) - i_{g\beta}(k+1)\right|^2 \tag{62}$$

$$C_I = \left|i_{g\alpha ref}(k+1) - i_{g\alpha}(k+1)\right|^2 \tag{63}$$

$$C_{II} = \left|i_{g\beta ref}(k+1) - i_{g\beta}(k+1)\right|^2 \tag{64}$$

$$v_{\alpha\,\min} = -\frac{L_g}{T_s}\left\{\left(1 - \frac{R_g T_s}{L_g}\right)i_{g\alpha}(k) + \frac{T_s}{L_g}e_{g\alpha}(k) - i_{g\alpha ref}(k+1)\right\} \tag{65}$$

$$v_{\beta\,\min} = -\frac{L_g}{T_s}\left\{\left(1 - \frac{R_g T_s}{L_g}\right)i_{g\beta}(k) + \frac{T_s}{L_g}e_{g\beta}(k) - i_{g\beta ref}(k+1)\right\} \tag{66}$$

The discrete model on the grid side of the converter generates optimized voltage vectors $v_{\alpha min}$ and $v_{\beta min}$ which are transformed into abc modulating signals like the sinusoidal PWM method. The CSF-MPC does not use any weighting factors. Compared with the MPC method, the number of feedback signals and the computations are reduced for generating optimized voltage vectors.

Figure 28 shows the response of the IHTQZSC while a step change in V_{in} from 200 to 140 V at the instant 1 s with CSF-PC. The converter has been injecting i_{abc} of 5A into the grid. Current i_{L1} has been changed to 13.5A from 9.5A, V_{C3} has been changed from 307 to 286 V, and V_{DC} has been changed to 432 V from 413 V with the change in V_{in}. Compared with Fig. 20, the converter exhibits a faster dynamic response with input voltage disturbance with CSF-PC but a slower dynamic response when compared with MPC control. Figure 29 shows the response of the IHTQZSC while a step change in i_{abc} from 5A to 2.5A at the instant 2 s at V_{in} of 140 V. The disturbance in V_{in} affect i_{abc} response with CSF-PC compared with MPC, but exhibits faster dynamic response compared to PI controller. The current i_{L1} has been changed to 6.85A from 13.5A. V_{in} is maintained at 140 V, hence V_{C3} and V_{DC} are continuing at 286 V and 432 V at a steady state. Tracking of I_{L1} is also exhibiting better response than the PI controller response by observing Figs. 21 and 29.

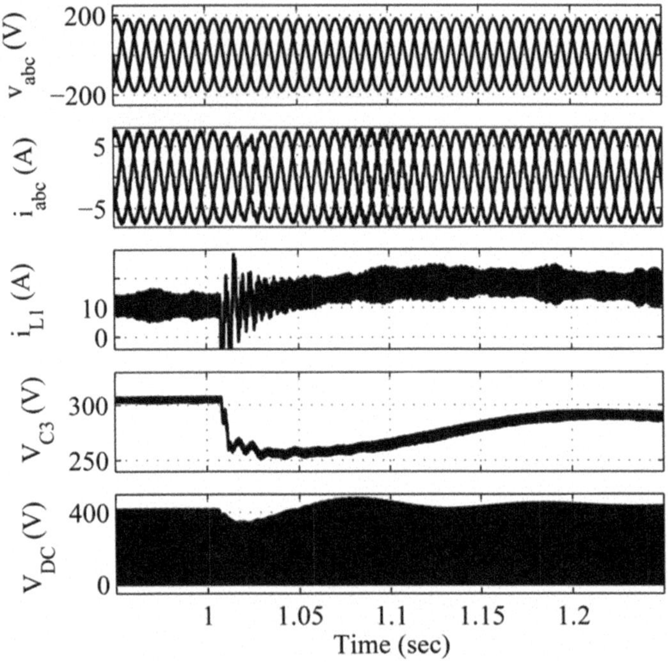

Fig. 28 Response of grid-connected IHTQZSC with at step change in input voltage

5 PVS Supported V2G and V4G System

According to the duck curve, the PV power has been injected in the daytime and reaches zero during the evening hours. The rest of the time, the architecture of the photovoltaic grid-connected system has been not utilized. Meanwhile, the power demand on the grid has been increased after evening hours of the day to mid-night. Most of the time, vehicles are parked in the homes. Hence, the vehicles with energy sources are connected to the grid, injecting power into the grid whenever required, and providing ancillary services. Also, the vehicle is used to compensate harmonics raises due to the non-linear load in the homes. Figure 5c configurations can be used for connecting PVS as well as EVs together with the utility. Battery-connected QZSC [31] shown in Fig. 30 is also used for connecting PVS and Battery storage system together with the grid. Here grid-connected PVS with IHTQZSC architecture has been considered for demonstrating the grid support by EV. During night time solar power is unable. The architecture of PVS is not utilized at these hours. Hence, the input DC source terminals at photovoltaic modules changeover to EV.

System configuration of the IHTQZSC for vehicle application has been shown in Fig. 31. On the grid side at the point of common coupling (PCC) current injected by the vehicle i_V, current drawn by the home appliances i_h and grid current i_g are observed. In general, EVs are drawing power from the grid (G2V) for charging battery

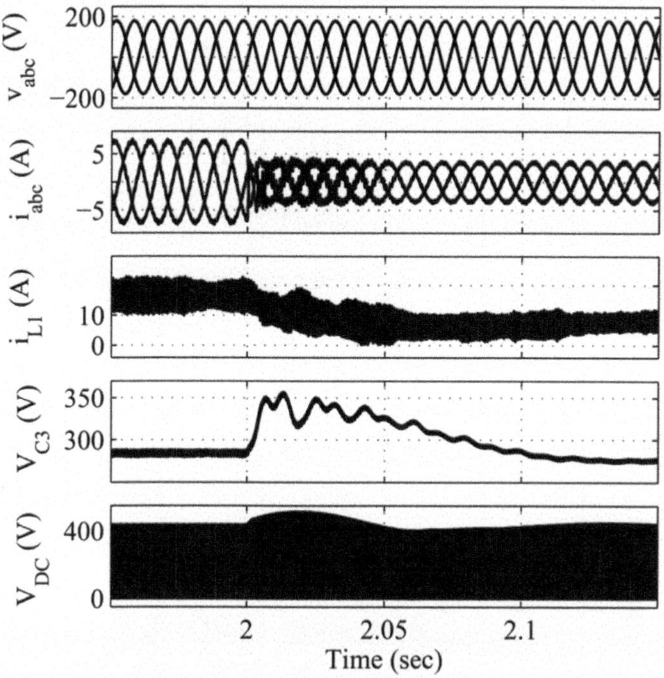

Fig. 29 Response of grid-connected IHTQZSC with at step change in grid current

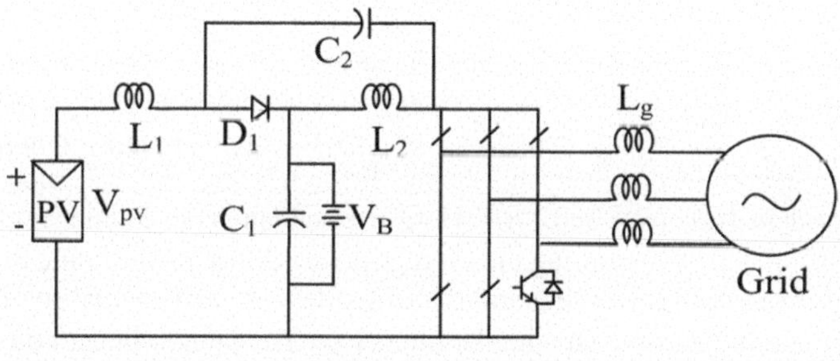

Fig. 30 Battery-connected QZSC in Grid-integrated PVS [31]

energy storage system. In general, G2V operation has been carried out at unity power factor by controlling the converter switching pulses. In this chapter, V2G and V4G operations of the EV are explained.

$$i_{V\alpha ref} = \frac{2}{3}\left(\frac{v_{g\alpha}P_{EV} + v_{g\beta}Q_{EV}}{v_{g\alpha}^2 + v_{g\beta}^2}\right) \tag{67}$$

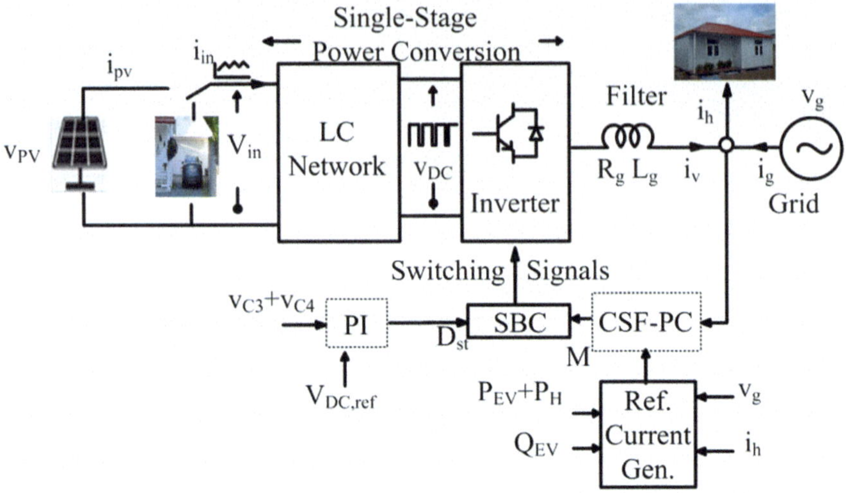

Fig. 31 System Configuration of the grid-connected PV system with IHTQZSC

$$i_{V\beta ref} = \frac{2}{3}\left(\frac{-v_{g\alpha} Q_{EV} + v_{g\beta} P_{EV}}{v_{g\alpha}^2 + v_{g\beta}^2}\right) \tag{68}$$

The reference active power has been chosen for charging and discharging of the vehicle accordingly, the reference grid current has been calculated and is expressed in Eqs. (66) and (67) [32]. Here P_{EV} and Q_{EV} are active and reactive powers consumed by the vehicle In G2V operation. P_{EV} and Q_{EV} are positive values and negative values for V2G operation. The value of Q_{EV} is changing depending upon the reactive power consumed and generated. For vehicle for grid (V4G) application for compensating the non-linear current drawn by the home, Eqs. (67) and (68) are modified and are expressed in Eqs. (69) and (70). Here the fundamental component of home current is calculated by measuring the average power drawn by the home P_H at PCC. Accordingly, reference current equations are modified. A fundamental component of load current might be extracted with the filters [33]. Actual non-linear home current is subtracted from the fundamental component of home current in addition to the current injected by the EV into the grid.

$$i_{V\alpha ref} = \frac{2}{3}\left(\frac{v_{g\alpha}(P_{EV} + P_H) + v_{g\beta} Q_{EV}}{v_{g\alpha}^2 + v_{g\beta}^2}\right) - i_{h\alpha} \tag{69}$$

V2G operation of EV connected in place of the PVS terminals is shown in Fig. 32. Voltage and current of phase a have been considered for exhibiting the V2G operation. EV is injecting power into the grid at the unity power factor. At the instant of 1 s, the home load is connected to the PCC consuming reactive power from the grid. The value of Q_{EV} is set to the value equal to the reactive power consumed by the home. The value of Q_{EV} might be the different value defined according to the system operator [32–35]. Though the home loads consume reactive power, an equal quantity of reactive power has been supplied by the vehicle observed in i_V response which is leading in nature. The net current at PCC has been reduced and it is at the unity power factor observed in i_g response. The phasor sum of i_v and i_h is equal to i_g. After 1.1 s, the home loads are disconnected, then i_v and i_g are equal up to 1.2 s in Fig. 31. In between 1.2 s to 1.3 s, home loads are supplying reactive power into the PCC. Here EV consumes reactive power at PCC and makes i_g at unity power factor. IHTQZSC has been boosting the input dc voltage V_{in} at 325 V to DC-link voltage v_{DC} at 760 V by maintaining capacitor v_{C3} at 542 V.

$$i_{V\beta ref} = \frac{2}{3}\left(\frac{-v_{g\alpha}Q_{EV} + v_{g\beta}(P_{EV} + P_H)}{v_{g\alpha}^2 + v_{g\beta}^2}\right) - i_{h\beta} \tag{70}$$

The system parameters are given in Table 4 to observe the converter response in V4G application. Up to 0.3 s, the vehicle has been injecting a peak current of 10A into the grid, thus i_g is following i_V. After 0.3 s, a non-linear load is connected to the grid, hence distorted current response i_h is observed in Fig. 33. The controller has

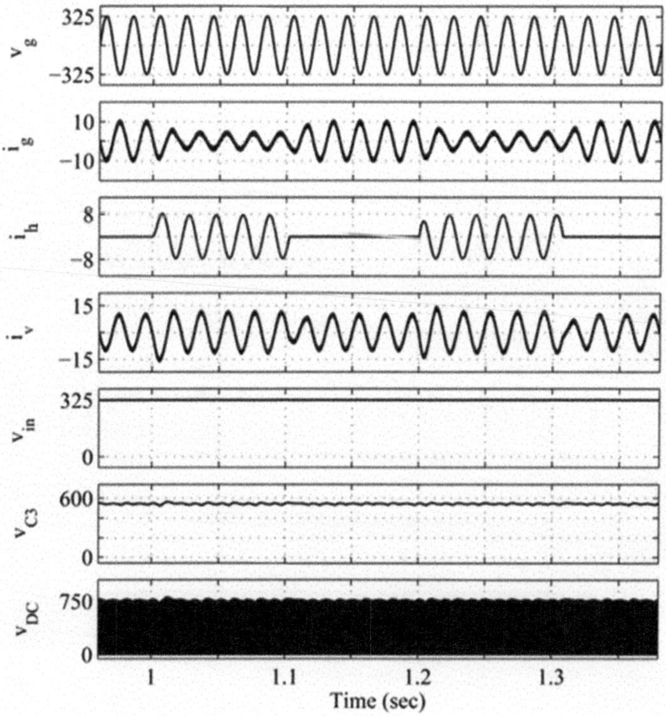

Fig. 32 V2G response of IHTQZSC

Table 4 System parameters—V4G

Parameter	Value
Grid Voltage	230 V
Grid Frequency	50 Hz
Filter Inductance	10 mH
Switching Frequency	10 kHz
Converter Rating	5 kVA
IHTQZSC inductance	1 mH
IHTQZSC Capacitance	470 μF
Sampling time	20 μs
Non-linear load	Diode rectifier with resistance 50 Ω and inductance of 100 mH

been calculating the reference current according to Eqs. (69) and (70) injecting both injected and compensation currents i_V. Here, between 0.3 s to 0.5 s, V2G operation along with V4G operation are performed. The resultant grid current i_g is sinusoidal in nature even though a non-linear load is connected from home. After 0.5 s, the vehicle does not inject power into the grid, but it compensates the non-linear lad current generated with the home load. The response of i_h, i_g, and i_v are identified in Fig. 32. After 0.5 s, the non-linear home load is disconnected and EV is injecting power with V2G operation. The DC-link voltage v_{DC} and the capacitor C3 voltagev_{C3} are also observed in Fig. 32.

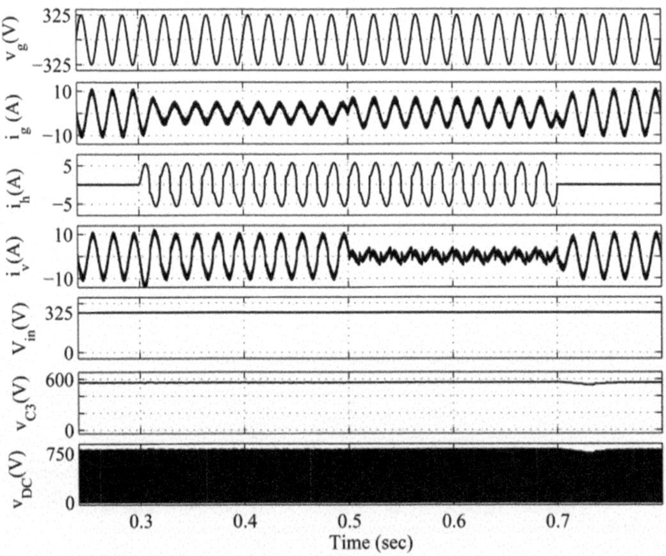

Fig. 33 V4G response of IHTQZSC

6 Summary

The phenomenon of the California duck curve is explained. The necessity of an energy storage system for the stability of the grid during a large number of renewable sources' penetration into the grid has been discussed. EVs are having energy storage capacity through batteries. The charging mechanism of EVs and energy extraction from renewable sources required power converters. Power converters configurations used in PVS and EV are discussed. Impedance source converter types and the operating principle have been discussed. SBC method is providing PWM signals for the IHTQZSC for providing higher voltage boost gain along with grid integration of PVS and EV has been studied. Control methods for impedance source converters are examined in grid-connected applications. The concept of MPC is addressed. CSF-PC is working at less number of computations and sensing signals for grid integration of IHTQZSC. V2G and V4G operation of the EV with the use of IHTQZSC is addressed.

References

1. Denholm P, O'Connell M, Gregory Brinkman, and Jennie Jorgenson. Overgeneration from solar energy in California: a field guide to the duck chart. National Renew Energy Laboratory. https://www.nrel.gov/docs/fy16osti/65023.pdf
2. Su H-Y et al (2021) Developing an optimal scheduling of Taiwan power system with highly penetrated renewable energy resources and pumped hydro storages. IEEE Trans Indus Appl 57(3):1973–86
3. Majzoobi A, Khodaei A (2017) Application of Microgrids in supporting distribution grid flexibility. IEEE Trans Power Syst 32(5):3660–3669
4. Hameer S, van Niekerk JL (2015) A review of large-scale electrical energy storage. Int J Energy Res 39:1179–1195
5. Luo X, Wang J, Dooner M (2015) Jonathan Clarke "overview of current development in electrical energy storage technologies and the application potential in power system operation". Appl Energy 137:511–536
6. Dekka A, Ghaffari R, Venkatesh B,. Wu B (2015) A survey on energy storage technologies in power systems. IEEE electrical power and energy conference (EPEC), pp 105–111
7. Ould Amrouche S et al (2016) Overview of energy storage in renewable energy systems. Int J Hydrogen Energy 41:20914–20927
8. Chen H et al (2009) Progress in electrical energy storage system: a critical review. Prog Nat Sci 19:291–312
9. Alsharif A et al (2021) A comprehensive review of energy management strategy in Vehicle-to-Grid. Sustain Energy Technol Assess 47:101439
10. Sujitha N, Krithiga S (2017) RES based EV battery charging system: a review. Renew Sustain Energy Rev 75:978–988
11. Gomes ISF, Perez Y, Suomalainen E (2020) Coupling small batteries and PV generation: a review. Renew Sustain Energy Rev 126:109835
12. Vazquez S et al (2010) Energy storage systems for transport and grid applications. IEEE Trans Indus Electron 57(12)
13. Mwasilu F et al (2014) Electric vehicles and smart grid interaction: a review on vehicle to grid and renewable energy sources integration. Renew Sustain Energy Rev 34:501–516

14. Siwakoti YP, Peng FZ, Blaabjerg F, Loh PC, Town GE (2015) Impedance-source networks for electric power conversion part I: a topological review. IEEE Trans Power Electron 30(2):699–716. https://doi.org/10.1109/TPEL.2014.2313746
15. Peng FZ (2003) Z-source inverter. IEEE Trans Indus Appl 39(2):504–510. https://doi.org/10.1109/TIA.2003.808920
16. Vázquez N, López JV (2018) 11—Inverters. Rashid MH (eds), Power electronics handbook (Fourth Edition), Butterworth-Heinemann, pp 289–338
17. Gorji SA et al (2019) Topologies and control schemes of Bidirectional DCDC power converters: an overview. IEEE Access
18. Czarkowski D (2018) 10—DC-DC converters. Rashid MH (eds), power electronics handbook (Fourth Edition), Butterworth-Heinemann, pp 275–288
19. Reshma Gopi R, Sreejith S (2018) Converter topologies in photovoltaic applications—a review. Renew Sustain Energy Rev 4:1–14
20. Reshma Gopi R,. Sreejith R (2018) Converter topologies in photovoltaic applications–a review. Renew Sustain Energy Rev 94:1–14
21. Anderson J, Peng FZ (2008) Four quasi-Z-Source inverters. IEEE Power Electron Specialists Conf 2008:2743–2749. https://doi.org/10.1109/PESC.2008.4592360
22. Tang Y, Xie S, Zhang C, Xu Z (2009) Improved Z-source inverter with reduced Z-source capacitor voltage stress and soft-start capability. IEEE Trans Power Electron 24(2):409–415. https://doi.org/10.1109/TPEL.2008.2006173
23. Zhu M, Yu K, Luo FL (2010) Switched Inductor Z-Source Inverter. IEEE Trans Power Electron 25(8):2150–2158. https://doi.org/10.1109/TPEL.2010.2046676
24. Loh PC, Gao F, Blaabjerg F (2010) Embedded EZ-source inverters. IEEE Trans Indus Appl 46(1):256–267. https://doi.org/10.1109/TIA.2009.2036508
25. Shen H, Zhang B, Qiu D (2017) Hybrid Z-source boost DC–DC converters. IEEE Trans Industr Electron 64(1):310–319. https://doi.org/10.1109/TIE.2016.2607688
26. Alla R, Chowdhury A (2018) Model predictive controller for improved hybrid three quasi Z source inverter for DG applications. 2018 IEEE international conference on power electronics, drives and energy systems (PEDES), pp 1–6. https://doi.org/10.1109/PEDES.2018.8707495
27. Siwakoti YP, Peng FZ, Blaabjerg F, Loh PC, Town GE, Yang S (2015) Impedance-source networks for electric power conversion Part II: review of control and modulation techniques. IEEE Trans Power Electron 30(4):1887–1906. https://doi.org/10.1109/TPEL.2014.2329859
28. Alla R, Chowdhury A (2020) CSF-PC based grid tied improved hybrid three quasi Z source converter. COMPEL Int J Comput Math Electr Electron Eng 39(4):927–942
29. Ellabban O, Van Mierlo J, Lataire P (2012) A DSP-based dual-loop peak DC-link voltage control strategy of the Z-source inverter. IEEE Trans Power Electron 27(9):4088–4097. https://doi.org/10.1109/TPEL.2012.2189588
30. Rodriguez J, Cortes P (2012) Predictive control of power converters and electrical drives. Publication USA, A John Wiley and Sons
31. Sun D, Ge B, Bi D, Peng FZ (2013) Analysis and control of quasi-Z source inverter with battery for grid-connected PV system. Int J Electr Power Energy Syst 46:234–240
32. Monteiro G, Pinto JG, Afonso JL (2016) Operation modes for the electric vehicle in smart grids and smart homes: present and proposed modes. IEEE Trans Veh Technol 65(3):1007–1020. https://doi.org/10.1109/TVT.2015.2481005
33. Singh B, Chandra A, Al-Haddad K (2014) Power quality: problems and mitigation techniques. Wiley, Hoboken, NJ, USA
34. Sarkar MNI, Meegahapola LG, Datta M (2018) Reactive power management in renewable rich power grids: a review of grid-codes, renewable generators, support devices, control strategies and optimization algorithms. IEEE Access 6:41458–41489. https://doi.org/10.1109/ACCESS.2018.2838563
35. Chao P et al (2019) A unified modeling method of photovoltaic generation systems under balanced and unbalanced voltage dips. IEEE Trans Sustain Energy 10(4):1764–1774. https://doi.org/10.1109/TSTE.2018.2870617

Chapter 12
Application of Ant Lion Optimization Technique for Microgrid Infrastructure

More Raju, Haritha Dhanlalji Parmar, Ankur Singh Rana, and Shailendra Kumar

Abstract This paper presents frequency regulation in a microgrid consisting of wind turbine generator (WTG), photovoltaic (PV), and fuel cell (FC) along with the energy storage system such as flywheel energy storage (FESS) and battery energy storage (BESS). Here, the Ant lion optimization technique (ALOT) is applied to tune the parameters of the proportional–integral–derivative (PID) controller to minimize the frequency deviation in the microgrid system under different operating conditions, namely with and without energy storage elements. The comparative analysis reflects the improvements in frequency deviation in microgrids using the proposed ALOT-optimized PID controllers in the presence of an energy storage system. The different case studies like variations in the propagation delay and increment in load perturbation are performed on the system. The case study on propagation delay provides a more oscillatory response in frequency. The increment in load perturbation shows that the system response deteriorates slightly. The cost function of the system is found minimum when the energy storage system is incorporated.

Keywords Ant Lion Optimization technique · Energy storage · Frequency deviation · Microgrid system · PID controller

M. Raju (✉) · S. Kumar
Department of Electrical Engineering, Maulana Azad National Institute of Technology Bhopal, Bhopal 462003, India
e-mail: rajunitt1@gmail.com

H. D. Parmar
MVSR Engineering College, Nadergul 501510, Hyderabad, India

A. S. Rana
Department of Electrical and Electronics Engineering, National Institute of Technology Tiruchirappalli, Tiruchirappalli 620015, Tamil Nadu, India

A. K. Giri et al. (eds.), *Power Quality: Infrastructures and Control*,
Studies in Infrastructure and Control,
https://doi.org/10.1007/978-981-19-7956-9_12

Nomenclature

BESS	Battery Energy Storage System
ESE	Energy storage elements
FESS	Flywheel Energy Storage System
ALOT	Ant lion optimization technique
MG	Micro Grid
FC	Fuel Cell
WTG	Wind turbine generator
PV	Photovoltaic
PID	Proportional Integral Derivable Controller
PEM	Proton Exchange Membrane

1 Introduction

The traditional electrical power network which consists of three main parts namely generation, transmission, and distribution systems, has changed dramatically as a result of significant advancements in electrical power generating and storage technology. Conventionally, the centrally produced electricity is supplied to all customers via the transmission system in which the power flow is unidirectional, with generation occurring only on the generation side and consumption occurring only on the distribution side of the electrical power system. Decentralized generation facilities, such as distributed generation comprising of wind turbine generators (WTG), fuel cells (FC), and photovoltaic (PV) panels are used in modern power systems along with battery energy storage system (BESS). As a microgrid (MG) is placed near consumers, it offers the advantage of reduction in transmission stage power loss when compared to the conventional power system.

Numerous sustainable resources, like wind, photovoltaic (PV), power device (FC), and diesel engine generator (DEG), as well as energy storage systems like flywheel energy (FESS) and battery energy (BESS), are arranged in a microgrid [1]. An examination by A. Anil and N. Srikanth was carried out and the impact of superconducting magnetic energy storage (SMES) on MG load frequency control was explored using the fuzzy-PI control strategy [2].

The performance of microgrid infrastructure has been significantly improved with the assistance of energy storage devices, as they add inertia to the grid [3]. The parameters are optimized not only in PID controllers, but also in adaptive fractional-order fuzzy control systems. Many optimization algorithms, such as particle swarm optimization (PSO), artificial bee colony (ABC) algorithm, Harris hawks optimization (HHO), and strategies such as centralized model predictive control (MPC), sliding mode control (SMC), and internal model approach (IMA), are used to improve the frequency that results in the better dynamic response of system [4–8].

The energy storage devices are prominent in providing improved dynamic response when intermittent power sources like wind turbines are installed in the isolated microgrid [9]. The genetic algorithm-tuned PID controller was found superior over the Ziegler–Nicholas tuned PID controller for the frequency regulation of the standalone ac microgrid system [10]. The author in [11] has investigated the time delay stability region for fuel cell-based microgrid system.

By generating one or more random solutions for a given issue, evolutionary algorithms search for the global optimum in a search space. These techniques provide various inherent benefits, including issue independence, derivation independence, avoidance of local optima, and simplicity. Genetic Algorithms (GA), Particle Swarm Optimization (PSO), Ant Colony Optimization (ACO), Differential Evolution (DE), Evolutionary Programming (EP), Gray Wolf Optimizer (GWO), Artificial Bee Colony (ABC) calculation, Firefly Algorithm (FA), Cuckoo Search (CS) calculation, Cuckoo Optimization Algorithm (COA), Gravitational Search Algorithm are well-known techniques that provide suitable parameters of controllers for a microgrid. As an alternative to tackling optimization issues, the present research introduces the Ant Lion Optimizer technique (ALOT), a revolutionary nature-inspired method. The ALOT is based on the intelligent behavior of antlions in nature when hunting ants [12].

The successful implementation of ALOT was performed by the authors in [13, 14]. In [13], the ALOT is applied to solve the conventional power system frequency regulation. Mouassa et al. [14] have shown the superiority of ALOT in solving the problem of reactive power dispatch. The advantages of ALOT in solving different real-world engineering problems motivated the authors to apply it in optimizing the controller parameters for enhancing the frequency regulation in microgrid system.

Based on the aforementioned discussion, the main contributions of the present research are to:

(i) Study the impact of energy storage system on frequency regulation in presence of ALOT based PID controller.
(ii) Observe the microgrid system dynamic behavior with different magnitudes of propagation delay.
(iii) Compare the oscillations in the event of increased perturbation level.
(iv) Carry out the comparative analysis of the cost function.

2 Microgrid Modeling

Wind turbine generators (WTG), fuel cells (FC), photovoltaic (PV) system, battery energy storage system (BESS), and flywheel energy storage system (FESS) are all part of the AC Microgrid system explored in this study. Figure 1 depicts the mathematical model using first-order transfer functions, of the subsystems that make up the microgrid [11].

The fuel cell block model comprises three sections: a PEM FC, a DC to AC converter, and an interconnection gadget. The following are the transfer functions:

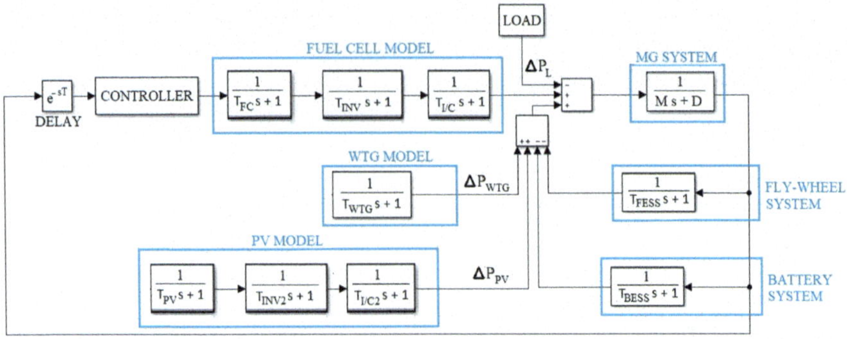

Fig. 1 The mathematical models of the subsystems constituting the microgrid

$$G_{fc} = \frac{1}{T_{fc}s + 1}, \ G_{inv} = \frac{1}{T_{inv}s + 1}, \ G_{i/c} = \frac{1}{T_{i/c}s + 1} \tag{1}$$

Energy storage systems are critical for supplying the needed energy to energy-producing subsystems in a timely manner in order to maintain system stability. To provide the necessary time to charge the battery cells, the time constant in BESS is restricted to a few seconds. FESS stores extra energy as kinetic energy, that releases back into the system whenever required by the system. The BESS and FESS transfer functions are written as follows:

$$G_{bess} = \frac{1}{T_{bess}s + 1}, \ G_{fess} = \frac{1}{T_{fess}s + 1} \tag{2}$$

The PID controller is utilized to control the fuel cell in response to variations in system load and power caused by WTG and PV systems. A PV system is made up of three components: a photovoltaic panel, a linking device, and an inverter.

$$G_{pv} = \frac{1}{T_{pv}s + 1}, \ G_{inv2} = \frac{1}{T_{inv2}s + 1}, \ G_{i/c2} = \frac{1}{T_{i/c2}s + 1} \tag{3}$$

The first-order transfer function of the Wind Turbine Generator can be expressed as below.

$$G_{wtg} = \frac{1}{T_{wtg}s + 1} \tag{4}$$

The microgrid transfer function is described below.

$$G_{sys} = \frac{1}{Ms + D} \tag{5}$$

Table 1 The Parameters of the proposed Microgrid System [11]

Parameter symbol	value	Parameter symbol	value
T_{fc}	0.26 s	T_{bess}	0.1 s
$T_{inv} = T_{inv2}$	0.04 s	T_{fess}	0.1 s
$T_{i/c} = T_{i/c2}$	0.004 s	M	0.1667
T_{pv}	1.8 s	D	0.015
T_{wtg}	1.5 s		

Here, M and D refer to the equivalent inertia constant and damping constant of the hybrid power system, respectively.

The most adaptable and extensively used feedback controllers are PID controllers, which have acquired popularity due to their high dynamic performance and durable properties. Proportional gain (K_P), integral gain (K_I), and derivative gain (K_D) are the three gains of the PID controller. The system model is subjected to 1% unit of load disturbance (PL) (Table 1).

3 ALO Technique-Based PID Controller

3.1 Ant Lion Optimizer Technique

The Ant Lion Optimizer technique (ALOT) [12], is a new meta-heuristic algorithm that numerically mimics insect antlion communications in nature. The execution of a random walk of ants, building traps, catching insects in traps, gathering prey, and once again fabricating traps are the steps involved in ALOT. The flowchart of ALOT is mentioned in Fig. 2. For optimization of controller parameters, search agents (population size (n)) of 5 and the iteration count as 100 are considered in this study subjected to optimization of the cost function with squared integral error method described as (6).

$$J_{ISE} = \int_{O}^{T} \Delta f^2 dt \tag{6}$$

4 Results and Discussions

Different values for K_P, K_I, K_D, and N (optimal values) are obtained when the controller is adjusted using the abovementioned ALOT. The studies are divided into 3 case studies, (i) with and without energy storage elements (ESE) such as BESS

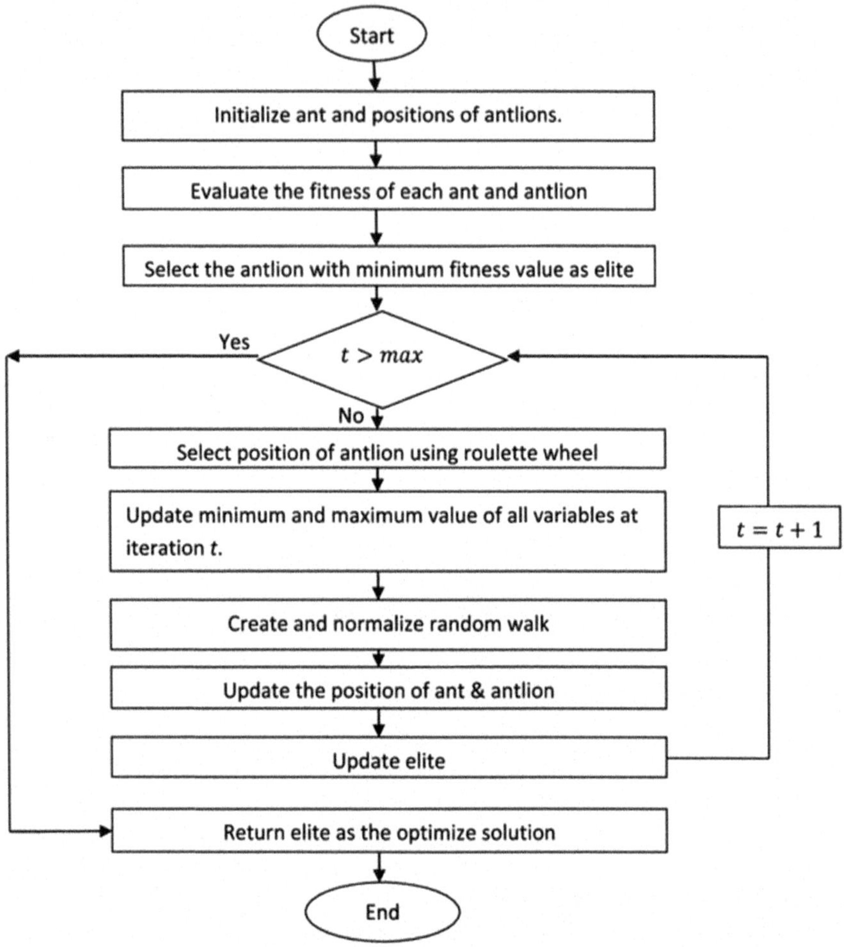

Fig. 2 Flow chart diagram of the ant lion optimization technique

and FESS, (ii) with different propagation delays, and (iii) with different magnitudes of load Perturbations.

4.1 Case Study I

4.1.1 Microgrid System Response Using the Proposed ALOT Optimized PID Controllers with Different Energy Storage Elements (ESE)

To study the impacts of different energy storages like BESS and FESS present in this system, a study on the microgrid system has been done where the controller is tuned using ALOT-based optimization under different energy storage elements (ESE) (Table 2).

When the system is examined without any energy storage elements in the microgrid, then the system response has much oscillation, but as the time goes on, the system reaches a steady value. The rise time of the system is 2.5 s and the settling time is 12 s (Fig. 3).

When the system has a battery energy storage element in the microgrid, having the controller values tuned using ALOT, then the system response is much better

Table 2 Controller parametric values under various energy storage elements (ESE)

S. no	Systems with/without ESE	K_P	K_I	K_D	N	J
1	System without ESS	0.42956	0.6140	0.17966	51.963	4.6576
2	System with BESS	0.60693	1	0.07989	99.283	0.8177
3	System with BESS and FESS	1	0.9999	0.02847	96.401	0.5327

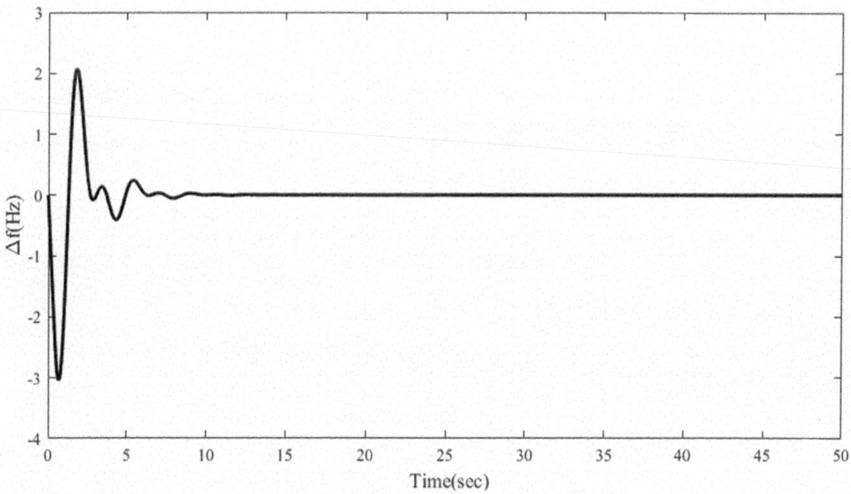

Fig. 3 The dynamic response of the system when energy storage elements are not present in the system

compared to the response obtained without an energy storage element (ESE). The rise time of the system is 1.5 s and the settling time is 7 s (Fig. 4).

When both battery and flywheel energy storage elements are present in the micro-grid, having the controller values tuned using ALOT, then the system response is much improved and has very less settling and rise time. The rise time of the system is 1.2 s and the settling time is 5.7 s (Fig. 5).

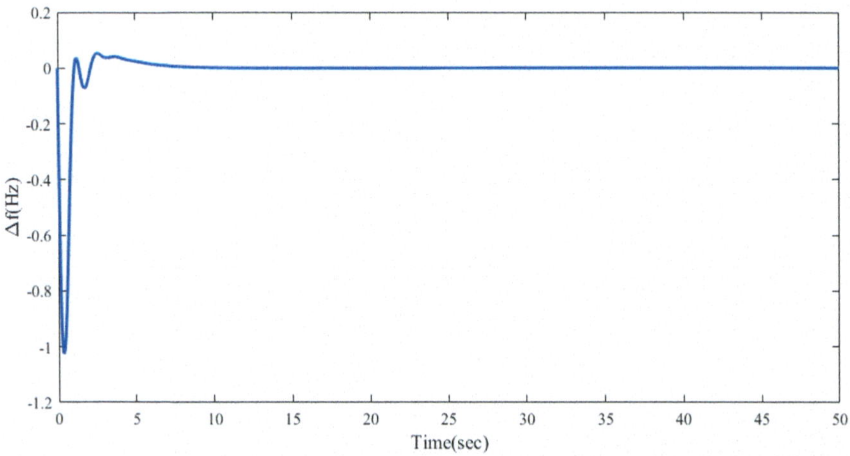

Fig. 4 The dynamic response of the system, when Battery energy storage is present in the system

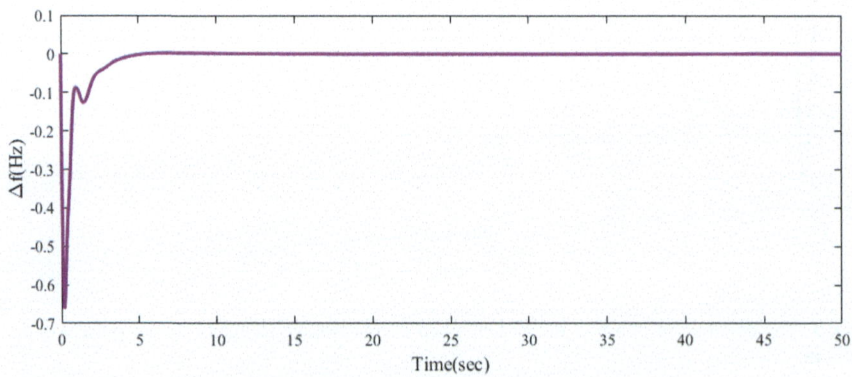

Fig. 5 The dynamic response of the system when BESS and FESS are present in the system

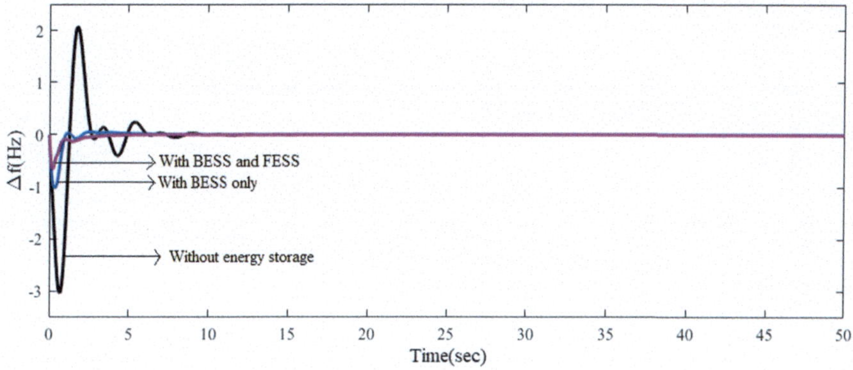

Fig. 6 The dynamic responses for various systems with energy storage

4.1.2 Comparison Between the Responses of the System

When a comparative analysis is done on the system responses with and without different energy storage elements (ESEs), then it can be observed that a system with both energy storages has a better and more robust response than without/with either of the energy storage elements (ESEs) (Fig. 6).

4.2 Case Study II

4.2.1 Microgrid System Response Using the Proposed ALOT Optimized PID Controllers with Different Propagation Delays

To study the impacts of propagation delay on the system, a study on the microgrid system has been done where the controller is tuned using ALOT-based optimization considering different values for propagation delays.

The various parametric values of PID controllers after the optimization for various cases have been shown below.

When the system is examined with the energy storage elements (ESEs) for various values of the propagation delays in the microgrid, then with PID parameters the system oscillations are shown below in Fig. 7.

As the propagation delay in the system is increased, the system cost function also increases (Table 3) and disturbances in the system also increase (Fig. 7). The system with low propagation delay has better performance having a rise time of 1.5 s and settling time of 5 s.

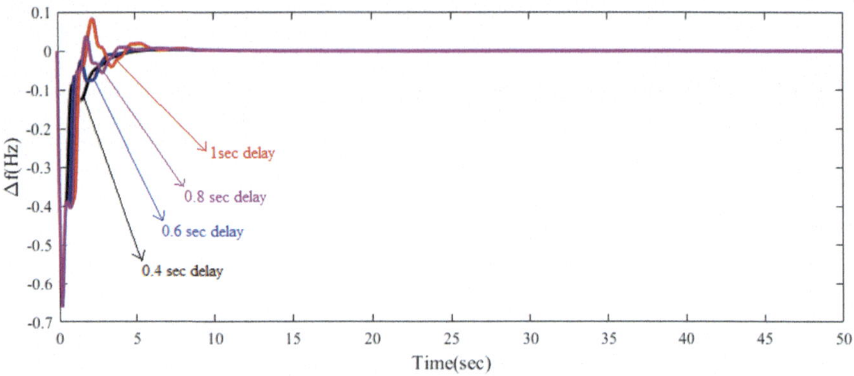

Fig. 7 The dynamic responses for systems with different propagation delays

Table 3 Controller parametric values under different values of propagation delays

S. no	System delay	P	I	D	N	J
1	0.4 s	1	0.9999	0.02847	96.401	0.53271
2	0.6 s	1	1	0.03658	73.1988	0.54526
3	0.8 s	1	1	0.10699	75.944	0.57967
4	1 s	0.94906	0.98102	0.11471	68.7564	0.64498

4.3 *Case StudyIII*

4.3.1 Microgrid System Response Using the Proposed ALOT Optimized PID Controllers with Different Load Perturbations

To study the impacts of load perturbations on the system, a study on the microgrid system has been done where the controller is tuned using ALOT-based optimization.

The various values of PID controllers after the optimization for various perturbations have been shown below.

When the system is examined having the energy storage elements (ESEs) for various load perturbations in the microgrid (2% and 3%), the system responses has been shown below.

In Fig. 8, as the load on the system is increased, the system produced more oscillations with the surge in cost values (Table 4).

4.4 *Comparative Analysis of J*

A comparison of cost value (J) is performed here for above discussed different cases. It is observed from the figures (Fig. 9a–c) that when the system has no energy storage

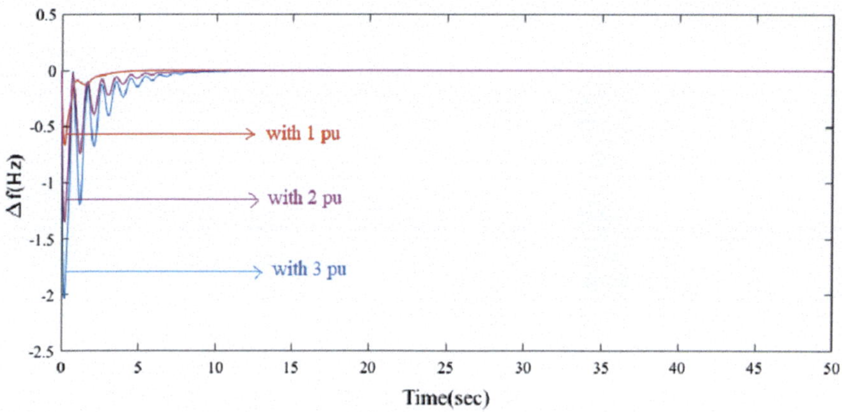

Fig. 8 The dynamic responses for systems with different load perturbations

Table 4 Controller parametric values under different values of load perturbations

S. no	SLP value	P	I	D	N	J
1	1 pu (%)	1	0.9999	0.02847	96.401	0.53271
2	2 pu (%)	0.955061	1	0.26015	97.7832	1.5003
3	3 pu (%)	0.996526	1	0.272396	100	2.5004

elements (ESEs) then the system responses have more oscillations and the value of cost function is also high. But as the system is included with energy storage elements ESE, the dynamic response of the system improves effectively and the value of the cost function also decreases. The system with BESS and FESS has the best response among all the compared systems.

5 Conclusions

The influence of energy storage devices, such as BESS and FESS, on microgrid frequency regulation, is addressed in this work. The PID controller, which is adjusted by ALOT, is utilized in this study. The research demonstrates that including energy storage features improves not just limited to the system dynamics, but also associated cost values. Different case studies are carried out, such as differences in propagation latency and perturbations. The dynamic response got more oscillatory as the propagation delay was increased. The perturbation variation demonstrated that as the load in the system increases, the dynamic response becomes more oscillatory. As a result, the suggested ALO-tuned PID controller may be employed in a variety of smart grid applications for effective frequency regulation.

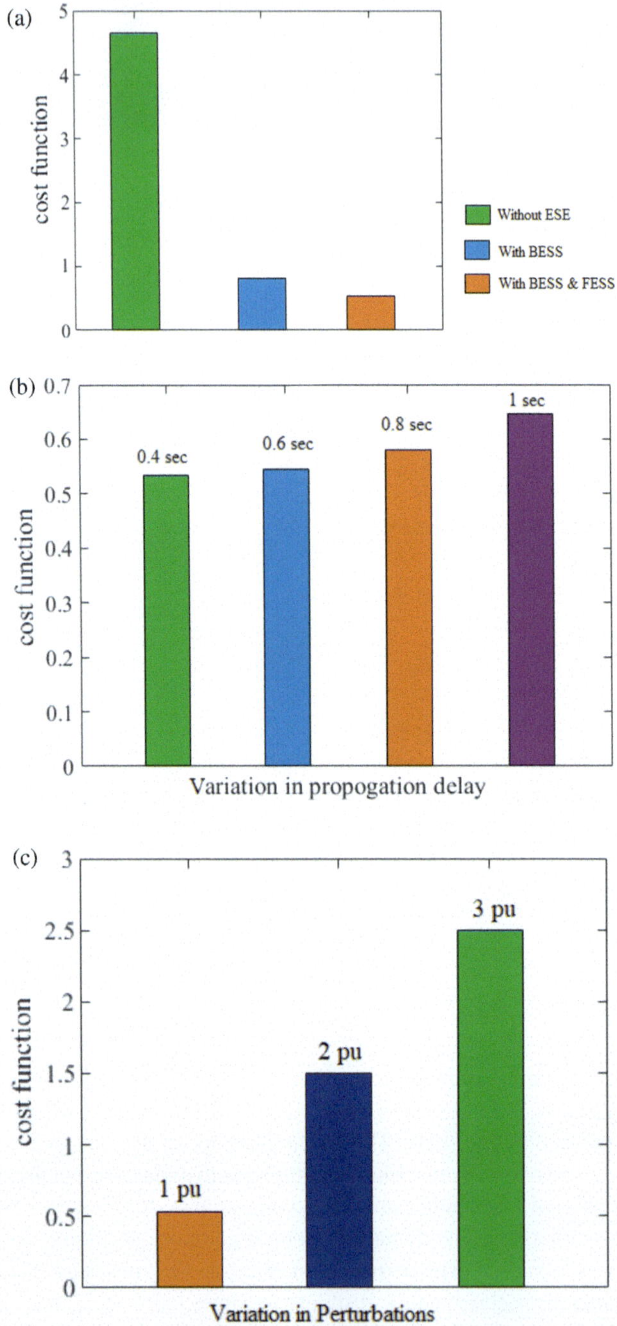

Fig. 9 The Cost function comparison for different cases. **a** With and without ESEs. **b** With different propagation delays. **c** Different load perturbations

References

1. Ray PK, Mohanty A (2019) A Robust firefly–swarm hybrid optimization for frequency control in wind/PV/FCbased microgrid. Appl Soft Comput 85:105823
2. Annamraju A, Nandiraju S (2019) Load frequency control of an autonomous microgrid using robust fuzzy PI controller. 8th international conference on power systems (ICPS)2019, pp 1–6
3. Mishra S, Nayak PC, Prusty UC, Prusty RC (2020) Frequency regulation of an islanded microgrid integrated by virtual inertia control. 2020 IEEE international symposium on sustainable energy
4. Mohammadzadeh A, Kayacan E (2020) A novel fractional-order type-2 fuzzy control method for online frequency regulation in AC microgrid. Eng Appl Artif Intell 90:103483
5. Sahoo BP, Panda S (2020) load frequency control of solar photovoltaic/wind/biogas/biodiesel generator based isolated microgrid using harris hawks optimization. First international conference on power, control and computing technologies (ICPC2T), pp 188–193
6. Kayalvizhi S, Vinod Kumar DM (2017) Load frequency control of an isolated micro grid using fuzzy adaptive model predictive control. IEEE Access 5:16241–16251
7. Patel V, Guha D, Purwar S (2019) Frequency regulation of an islanded microgrid using integral sliding mode control. 8th international conference on power systems (ICPS), pp 1–6
8. Jeya Veronica A, Senthil Kumar N (2017) Load frequency controller design for microgrid using internal model control approach. Int J Renew Energy Res 7(2):778–786
9. Nandar CSA (2013) Robust PI control of smart controllable load for frequency stabilization of microgrid power system. Renew Energy 56:16–23
10. Singh A, Sathans GA (2016) Optimized PID controller for frequency regulation in standalone AC microgrid 7th India international conference on power electronics (IICPE), pp 1–5
11. Mahmut Temel €Ozdemir (2020) The effects of the FOPI controller and time delay on stability region of the fuel cell microgrid. Int J Hydrogen Energy 45(60):35064–35072
12. Mirjalili S, The antlion optimizer (2015) Adv Eng Softw 83:80–98
13. Raju M, Saikia LC, Sinha N (2016) Automatic generation control of a multi-area system using ant lion optimizer algorithm based PID plus second order derivative controller. Electr Power Energy Syst 80:52–63
14. Mouassa S, Bouktir T (2017) Ahmed. Salhi, Ant lion optimizer for solving optimal reactive power dispatch problem in power systems. Eng Sci Technol Int J 20:885–895

Chapter 13
Protection of High Voltage Transmission Lines Connected to Large-Scale Wind Farms: A Review

Sauvik Biswas, Paresh Kumar Nayak, and Bijaya Ketan Panigrahi

Abstract In recent years, wind energy technology (WET) has been a promising alternative to meet the growing power demand worldwide. This technology is gaining popularity due to its cleanliness and economic attractiveness. Among the various turbine generators (TGs) in WET, the Type-3 doubly-fed induction generator (DFIG) and Type-4 permanent magnet synchronous generator (PMSG) is the most used and technically attractive TGs in large-scale wind power generation. The wind farm consists of several such TGs integrated into the main utility grid and the generated large power is injected through high voltage (HV) long transmission lines (TLs). Therefore, an adequate distance relaying is provided in the long TL for an uninterrupted power supply to the main utility grid. However, the impedance of the TL is greatly affected both during fault and normal conditions due to the nonlinear wind speed versus generated power characteristics, complex control action, and low voltage ride-through capability of the TGs. It confounds the operation of the traditional fixed impedance-based distance relay protecting the TL during fault. These motivate to develop the advanced relaying schemes to provide accurate protection to such TLs. For this purpose, numerous research works have been published over the last decade. The present paper discusses the various impacts of grid-integrated large-scale wind farms on distance relaying-based transmission line protection schemes first. Further, it provides a comprehensive chronological review of the relaying schemes available in the literature to mitigate the protection problems of such wind farm integrated crucial transmission lines. This study may be found useful for future research and to gain insight into the scope of protection of HV-TLs connected to large-scale wind farms for further development.

S. Biswas · B. K. Panigrahi
Indian Institute of Technology Delhi, Delhi, India
e-mail: sauvik@ee.iitd.ac.in

B. K. Panigrahi
e-mail: Bijaya.Ketan.Panigrahi@ee.iitd.ac.in

P. K. Nayak (✉)
Indian Institute of Technology (ISM) Dhanbad, Dhanbad, India
e-mail: paresh@iitism.ac.in

A. K. Giri et al. (eds.), *Power Quality: Infrastructures and Control*,
Studies in Infrastructure and Control,
https://doi.org/10.1007/978-981-19-7956-9_13

1 Introduction

In the last decade, a considerable growth has been observed in the development and installation of large-capacity wind turbine generators (WTGs) across the world, including in India [1, 2]. The DFIG and PMSG are currently the most accepted WTG due to their various technical and economic advantages [3]. In large-scale wind farms, output powers of individual WTGs are collected at the local substation through intricate collector lines and further transmitted to the power system through TLs. Due to the growing integration of WTGs into the power system, the latest grid codes necessitate these WTGs should remain connected to the grid in the presence of grid voltage sags caused due to grid faults, which is stated as low voltage ride-through (LVRT) capability [4, 5] and should supply reactive power to ensure system stability.

Currently, various compensating devices like fixed series capacitors (FSC), thyristor-controlled series capacitors (TCSC), static synchronous compensators (STATCOM), and unified power flow controllers (UPFC) are installed in the connecting TLs to the grid [6–8]. These devices have the capability of improving power transfer through the TLs [9, 10]. Despite of several benefits of these WTG and compensating devices, the control schemes of DFIG or PMSG-WTG and compensating devices vary the transient as well as the steady-state behavior of currents and voltages and, thereby, vary the apparent impedances seen by the distance relays. These may cause maloperation of the traditional distance relaying scheme.

Substantial research work has been carried out in the past decade with an aim to provide newer advanced protection solutions to the wind farm-connected HV-TLs or to enhance the protection performance of traditional distance relaying scheme when used for protection such crucial TLs. The number of papers published related to the enhanced protection of large-scale wind farm-connected TLs in various reputed journals, book chapters, and conference proceedings from 2007 to till date is shown in Fig. 1. The increasing trend in the number of published papers in the figure justifies the importance of the topic of research.

2 Impact of Grid-Integrated Large-Scale Wind Farm on Distance Protection

To study the impact of grid-connected large-scale wind farms on distance relaying-based protection scheme, in the present study, the 2-bus test power system, as shown in Fig. 2 is considered. As shown in the figure, the wind farm consists of a number of DFIG or PMSG-WTGs connected to the grid TL through intricate collector lines (CLs). Further, to evacuate a large amount of wind-generated power to the utility grid, the interconnected grid TLs are usually compensated with either FSC or TCSC or STATCOM or UPFC. In this study, the control strategies of the grid-connected DFIG-WTGs or PMSG-WTGs are modeled to satisfy the LVRT requirement of the

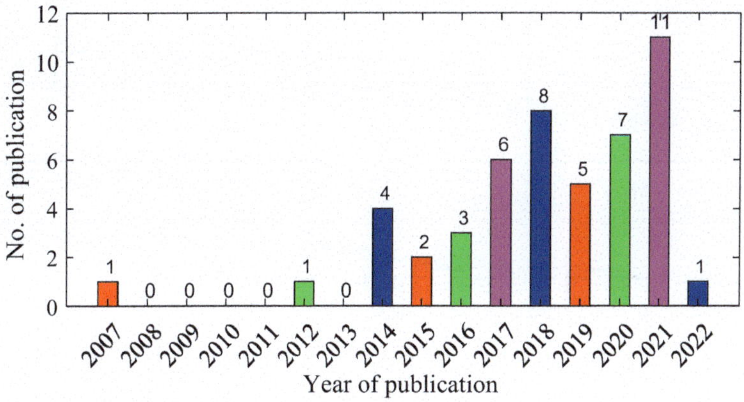

Fig. 1 Number of publications with years

Indian grid code [11]. As per the grid code, the WTGs should remain connected during grid disturbances and provide reactive power support to improve the voltage profile at the connected bus.

The equivalent circuit of the uncompensated TL model in Fig. 2 during pre-fault and post-fault is shown in Fig. 3a and b, respectively [12, 13]. The voltage phasors, current phasors, and source impedances of the wind farm and the grid are denoted by E_{WT}, E_G, I_W, I_G, and Z_W, Z_G, respectively. The voltages at Bus-W and Bus-G are denoted by V_W and V_G, respectively. The impedances of TL are denoted as Z_L. The superscripts '*pre*' and '*f*' indicate that the parameters correspond to pre-fault and

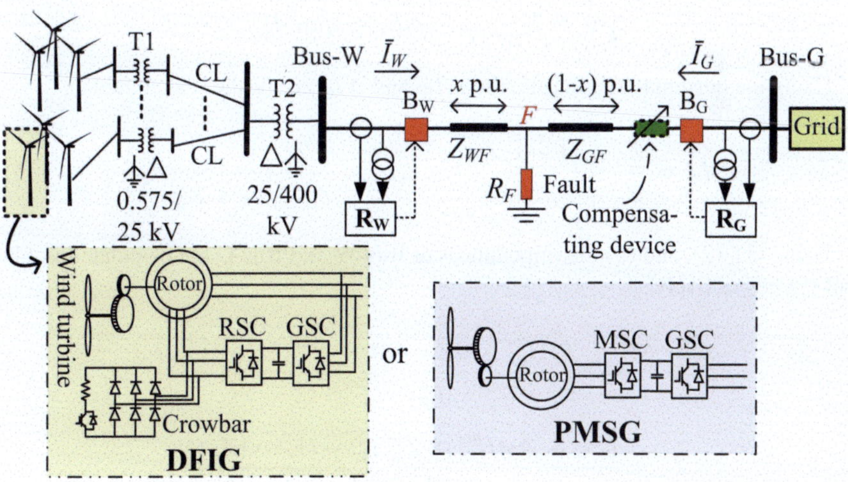

Fig. 2 Test system of the wind farm-integrated TL

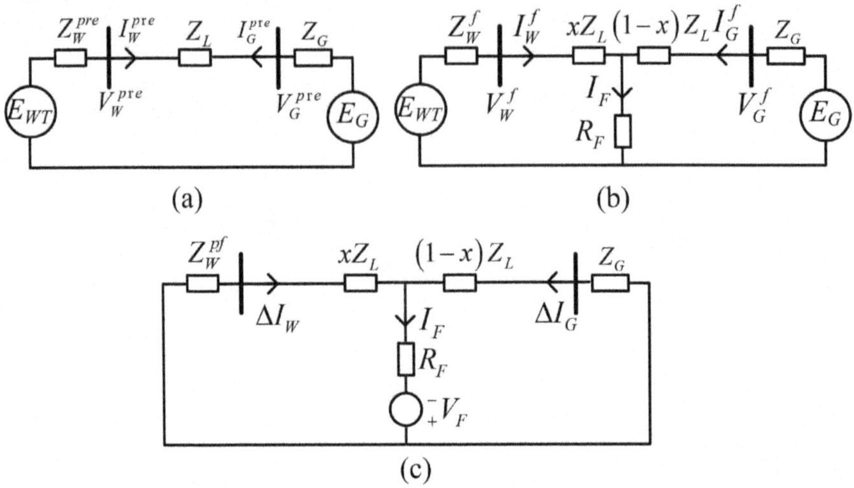

Fig. 3 Equivalent circuit of the test system during, **a** pre-fault, **b** post-fault, and **c** pure fault

post-fault, respectively. A fault with resistance R_F is created at x p.u. the location from Bus-W and the fault current through R_F is I_F.

Using Kirchhoff's current law in Fig. 3b, the fault current is given as

$$I_F = I_W^f + I_G^f \tag{1}$$

The incremental currents in Fig. 3c are considered as

$$\Delta I_W = I_W^f - I_W^{pre}, \Delta I_G = I_G^f - I_G^{pre} \tag{2}$$

Using the current division rule, the ΔI_W and ΔI_G are further calculated as

$$\Delta I_W = \frac{Z_G + (1-x)Z_L}{Z_G + Z_L + Z_W^{pf}} I_F, \ \Delta I_G = \frac{Z_W^{pf} + xZ_L}{Z_G + Z_L + Z_W^{pf}} I_F \tag{3}$$

Using (1), (2), and (3) the impedances at Bus-W and Bus-G are calculated in (4) and (5), respectively.

$$Z_W^{pf} = \frac{(Z_G + Z_L)I_W^{pre} + [Z_G + (1-x)Z_L]I_G^f - xZ_L I_W^f}{I_W^f - I_W^{pre}} \tag{4}$$

$$Z_G = \frac{(Z_W + Z_L)I_G^{pre} + Z_W^{pf} I_W^f - (1-x)Z_L I_G^f}{I_G^f - I_G^{pre}} \tag{5}$$

Considering compensating devices of impedance Z_C in the TL, the Eqs. (6) and (7) can be modified as.

$$Z_W^{pf} = \frac{(Z_G + Z_L + Z_C)I_W^{pre} + [Z_G + (1-x)Z_L + Z_C]I_G^f - xZ_L I_W^f}{I_W^f - I_W^{pre}} \quad (6)$$

$$Z_G = \frac{(Z_W + Z_L)I_G^{pre} + Z_W^{pf} I_W^f - [(1-x)Z_L + Z_C]I_G^f}{I_G^f - I_G^{pre}} \quad (7)$$

From the above equations, it is clearly evident that the impedances Z_W^{pf} and Z_G are dependent upon the impedances of WTG. The control scheme of grid side converter (GSC), rotor side converter (RSC) in case of DFIG or machine side converter (MSC) in case of PMSG, crowbar circuit and compensating devices vary the impedances Z_W^{pf} and Z_G based on the severity of the faults. Apart from the fundamental components, other components such as decaying DC, forced AC comportments are generated in the fault current signal imposing difficulty in the extraction of fundamental components correctly just after the initiation of the fault [14]. The approximate short circuit current [15] of DFIG calculated using circuit analysis from Fig. 4 is in (8).

$$I_{S-DFIG} = \frac{V_1(1 - k_s k_r)e^{j\omega_s t} - V_2 e^{-j\omega_s t}}{j\omega_s L'_S} + \frac{V_{pre}(V_1 - V_2)}{j\omega_s L'_S} e^{-t/T_s} - k_s k_r \frac{V_{pre} - V_1}{j\omega_s L'_S} e^{j\omega_r t} e^{-t/T_r} \quad (8)$$

$$T_s = \frac{L'_S}{R_S} = \frac{L_S - \frac{L_m^2}{L_r}}{R_S} \quad (9)$$

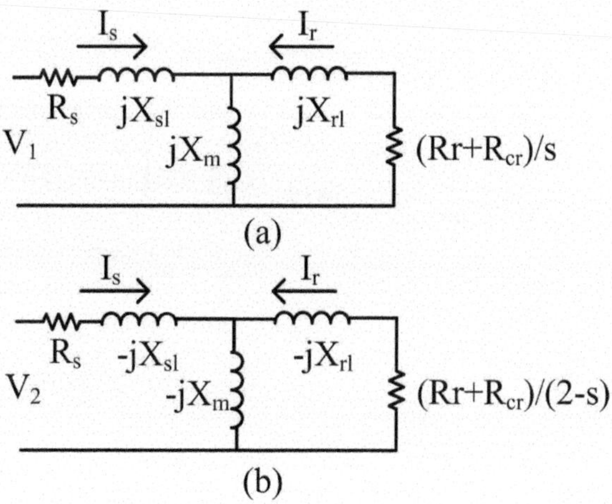

Fig. 4 Positive and negative-sequence circuits of a DFIG with rotor resistance (R_r) connected to a crowbar resistance (R_{cr})

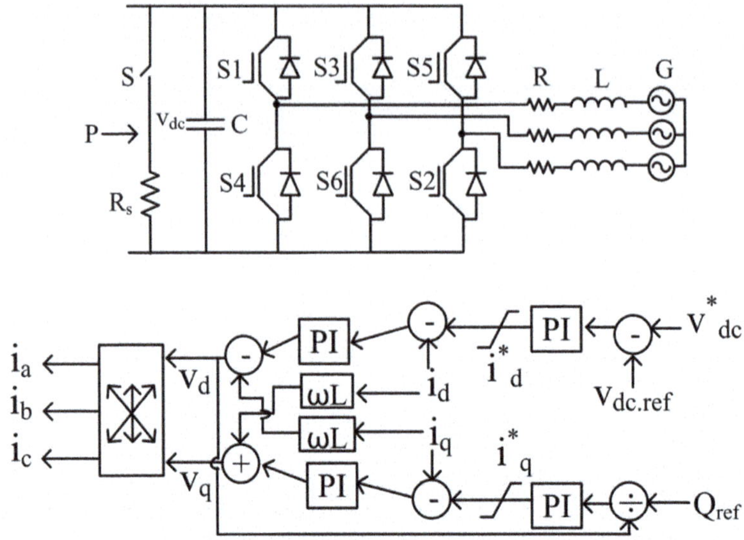

Fig. 5 Circuit diagram and control logic of PMSG-wind farms

$$T_r = \frac{L'_r}{R_r} = \frac{L_r - \frac{L_m^2}{L_S}}{R_r} \tag{10}$$

In case of PMSG-WTG, the approximate short circuit current [16] from Fig. 5 is in (11).

$$\begin{aligned} I_{S-PMSG} &= \frac{2P}{3v_d}\cos(\omega t + \theta) - i_q^* \sin(\omega t + \theta) \\ &+ \left[i_d - \frac{2P}{3v_d} \right] \frac{1}{\sqrt{1-\xi^2}} e^{-\xi \omega_n t} \sin(\omega_d t + \gamma)\cos(\omega t + \theta) \\ &+ i_q^* e^{-\frac{kip}{2L}} \frac{2\sqrt{Lk_{ii}}}{\sqrt{4Lk_{ii} - k_{ip}^2}} \sin\left(\frac{\sqrt{4Lk_{ii} - k_{ip}^2}}{2L} t + \gamma \right) \sin(\omega t + \theta) \end{aligned} \tag{11}$$

$$\omega_d = \omega_n \sqrt{1-\xi^2} \tag{12}$$

$$\gamma = \tan^{-1} \frac{\sqrt{1-\xi^2}}{\xi} \tag{13}$$

Therefore, the impedance calculated at distance relays R_W and R_G at both ends of the TL (Fig. 2) may alter from the actual cases and may fail to provide reliable protection to the TL. It motivates researchers to investigate the traditional distance relaying under the operation of WTG and compensating devices in the TLs and to propose an accurate and reliable scheme to protect such HV-TLs.

2.1 *Performance of WTG Side Relay*

To understand the operation during SLG fault, the sequence diagram during this condition is shown in Fig. 6. Due to LVRT requirement of the DFIG, the positive and negative-sequence source impedances of the WTGs, i.e., Z_{1WT} and Z_{2WT} are controlled to limit the positive- and negative-sequence fault currents I_{1W} and I_{2W}, respectively. The power electronics interfaced WTGs usually do not supply zero-sequence current during ground faults in the grid TL. Only the star-grounded high voltage winding of the transformer T2 provides path for zero-sequence current to flow during ground faults in the grid TL. Consequently, the magnitude of zero-sequence current I_{0W} is expected significantly high compared to I_{1W} and I_{2W}. The unusual variation of sequence currents during SLG fault can affect the performance of the wind farm side distance relay R_W.

To test the performance of the distance relay protecting a wind farm-connected TL, the system shown in Fig. 2 is simulated in MATLAB/Simulink. In this study, a total of 105 MW DFIG-WTGs (70 Nos. × 1.5 MW each) based wind farm is connected to the grid transmission line. Length of the transmission line is 320 km. The positive- and zero-sequence impedance of the line are 0.033 + j0.332 Ω/km and 0.329 + j1.1137 Ω/km, respectively. Performance of the wind farm side distance relay R_W is tested for different faults. The reach of zone 1 of R_W covers up to 80% of the TL.

Test result of two typical faults is demonstrated here. First, a bg-fault is created at 90% length of the TL from R_W (outside of zone 1) at 1.5 s with fault resistance

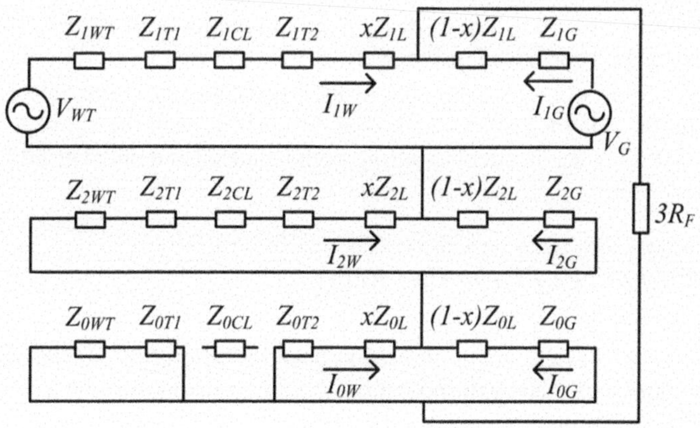

Fig. 6 Sequence network connection diagram during a SLG fault

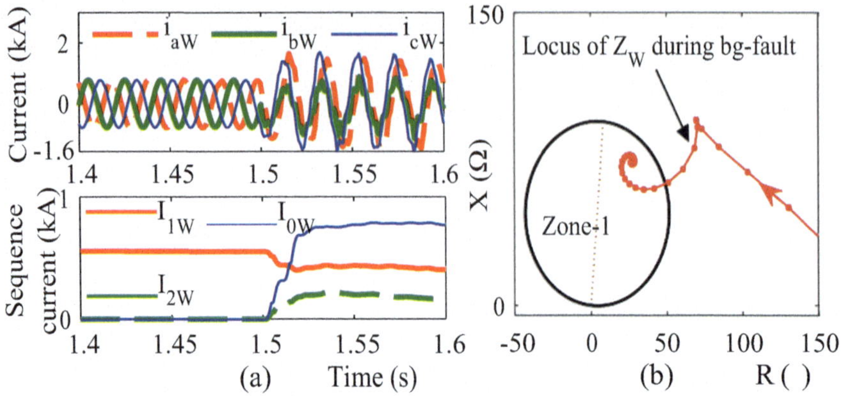

Fig. 7 Effect on relay R_W during a bg-fault, **a** 3-phase current, and **b** impedance plot (reach of zone 1 setting of R_W cover up to 80% length of the TL)

$r_f = 10\ \Omega$, $v = 15$ m/s and the results are provided in Fig. 7. It is clearly evident from Fig. 7a that the current magnitude of the faulted phase (phase-b) is coming even less compared to the healthy phases (phase-c and phase-a). This is mainly due to the control action and weak power supply system of the DFIG-based wind farm. It is also clear from the figure that as the wind farm side acts as a week power system with star-grounded transformer, the zero-sequence current I_{0W} is larger than the positive- and negative-sequence currents (I_{1W} and I_{2W}). As a result, the locus of impedance Z_W ($= V_{bW}/(I_{bW} + k_0 I_{0W})$, where V_{bW} and I_{bW} are the voltage and current phasors of phase-b, k_0 is the residual compensation factor and I_{0W} is the zero-sequence current) measured at R_W enters into its zone 1 characteristic as seen in Fig. 7b making R_W overreach.

Performance of RW is also evaluated for a 3-phase ground fault with $r_f = 5\ \Omega$ initiated at 1 s at 70% line length (within zone 1 of R_W) for $v = 5$ m/s. The ratios of currents (wind to grid side), i.e., β ($= |I_W/I_G|$) and the locus of Z_W ($= (Z_{WF} + r_f) + r_f/\beta$) seen at R_W is provided in Fig. 8. As $\beta < < 1$ during fault (Fig. 8a), the Z_W is magnified falsely even a small value of fault resistance r_f is present in the fault path, making the relay R_W under each (Fig. 8b). Thus, from the above result, it is confirmed that relay R_W is affected during both balanced and unbalanced faults.

2.2 *Performance of Grid Side Relay*

The compensating devices are connected at bus-G location (Fig. 2). Therefore, the relay RG is mainly affected by the operation of these compensating devices. Here, the TL is compensated with different devices like FSC, TCSC, STATCOM, and UPFC. Schematic of these compensating devices with their control circuit are provided in Figs. 9, 10, and 11, respectively.

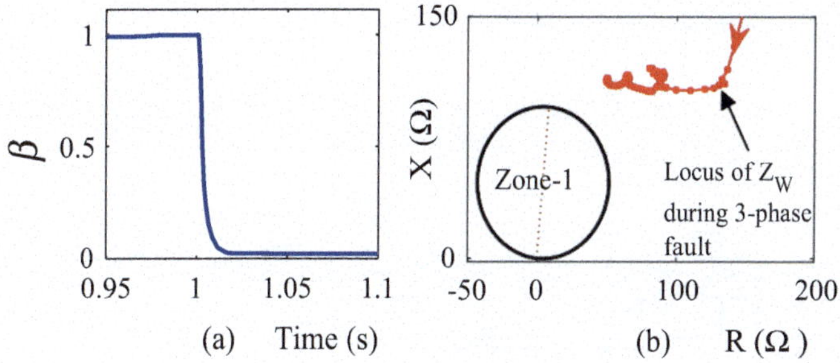

Fig. 8 Effect on relay R_W during the occurrence of a 3-phase fault, **a** β, and **b** impedance locus at relay R_W

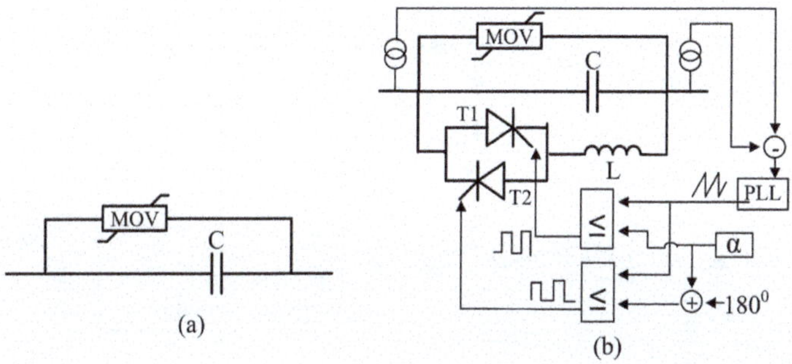

Fig. 9 Single line diagram and control scheme of connected compensating devices in the network in Fig. 2, **a** FSC and **b** TCSC

Taking FSC into consideration, a 3-phase fault created at 288 km (90%-line length) away from R*G* while v = 10 m/s, and 40% and 70% FSC compensation levels and the results are provided in Fig. 12. It is evident from Fig. 12a that the presence of FSC compensation in the fault loop causing relay R_G overreach. Another ag-fault is created keeping the fault distance and wind speed the same. The characteristics of the relay R_G is shown in Fig. 12b. In this fault case also, R*G* is facing overreaching problem.

The TCSC (Fig. 9b) is basically a 30–40% compensated FSC connected parallel with thyristor control reactor *L*. The TCSC is controlled by changing the value of α and it operates in capacitive mode during $150^0 \leq \alpha \leq 180^0$ or inductive mode during $90^0 \geq \alpha \geq 140^0$. The metal oxide varistor (MOV) is installed across the capacitor *C* to protect it from over-voltage. The variation of impedance of TCSC with α is shown in Fig. 13a. During fault, when the voltage across MOV exceeds the preset value $v_{OP\text{-}MOV}$, MOV operates and bypasses the fault current through it and protects *C* from over-voltage which is demonstrated in Fig. 13b and c, respectively. This nonlinear operation of MOV will impose difficulty in the functioning of distance relay.

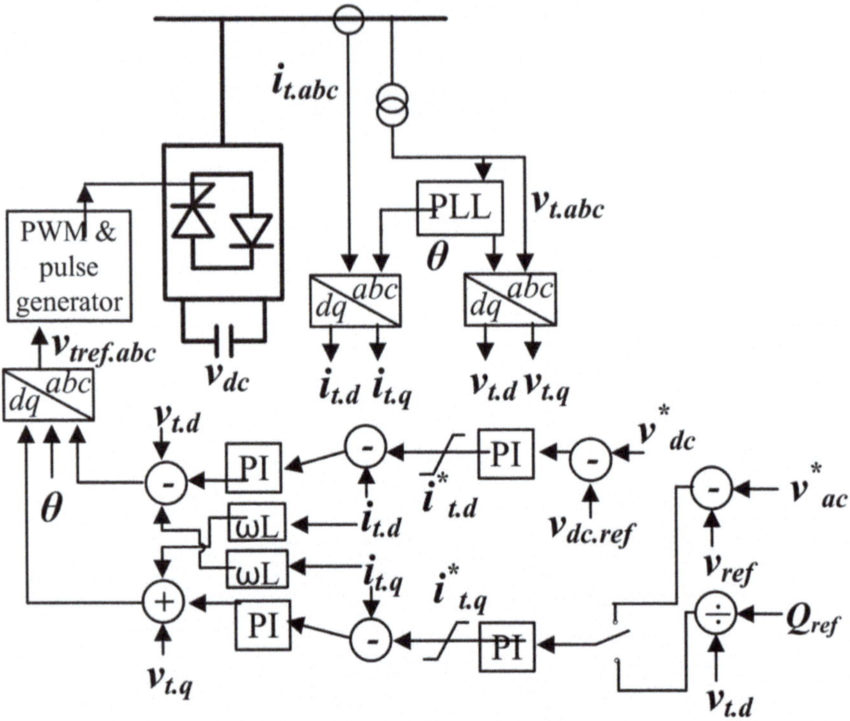

Fig. 10 Single line diagram and control scheme of connected STATCOM in the network in Fig. 2

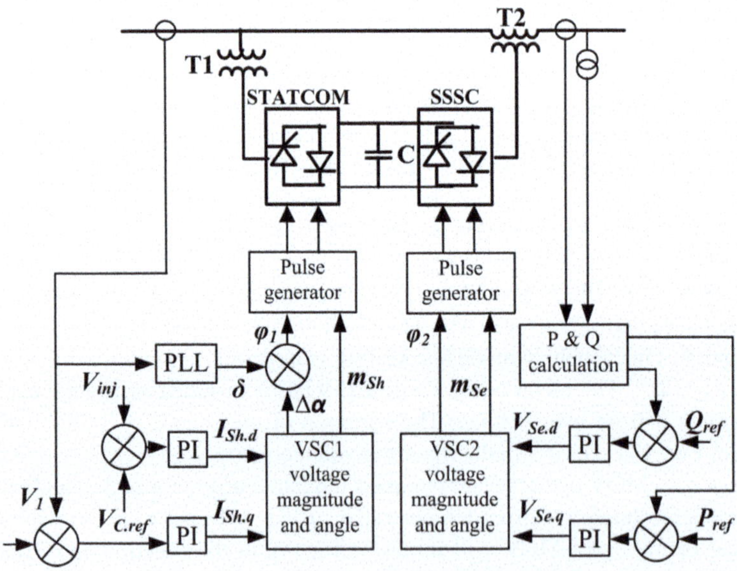

Fig. 11 Single line diagram and control scheme of connected UPFC in the network in Fig. 2

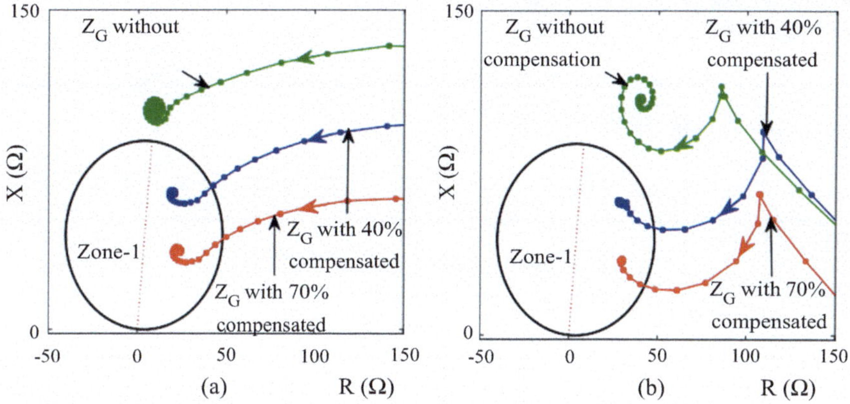

Fig. 12 Demonstration of the impacts of balanced and unbalanced ground faults on relay R_G during, **a** abc-fault, and **b** ag-fault

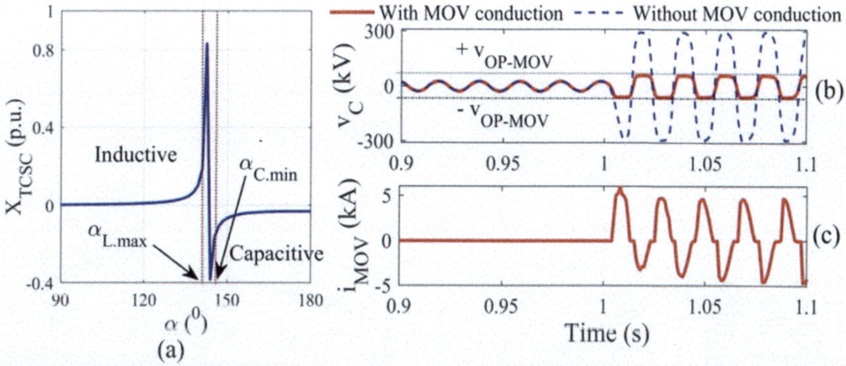

Fig. 13 Performance of the TCSC, **a** α versus X_{TCSC} during steady-state, **b** voltage across the capacitor and **c** current through MOV

An ag-fault ($r_f = 20\ \Omega$) is created at 70% of line length (within zone 1) from R*G* during inductive modes of operation of the TCSC ($\alpha = 120^0$ and 140^0). Another 3-phase fault is created at 90% of line length (outside zone 1) from R_G during capacitive modes of operation of the TCSC ($\alpha = 160^0$ and 180^0). During both the fault cases, wind speed is kept at 5 m/s. The loci of the impedances seen by relay R_G for the two fault cases are plotted in R-X plane as shown in Fig. 14a and b, respectively. As observed in Fig. 14a, though the fault has occurred inside zone 1 of R_G, the locus of Z_G is coming outside zone 1 due to inductive operating modes of the TCSC. Thus, making relay R_G underreach. Similarly, in the second case (Fig. 14b), due to the capacitive operating mode of the TCSC, through the fault has occurred outside of zone 1, the locus Z_G enters into zone 1. Thus, making a relay R_G overreach.

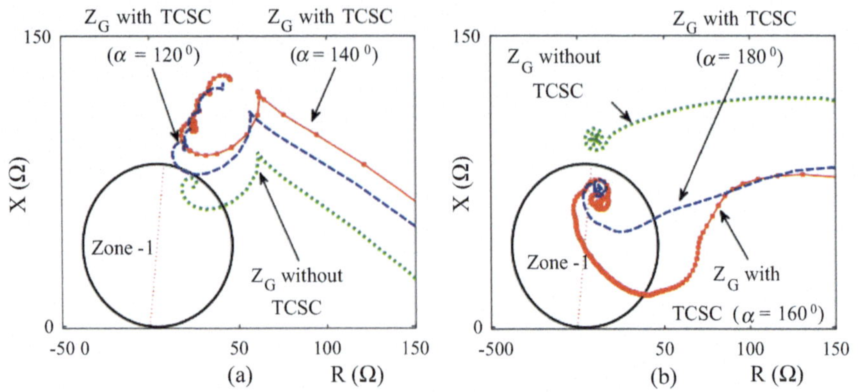

Fig. 14 Demonstration of, **a** underreaching and **b** overreaching problem faced by distance relay R_G

Similarly, the distance relay may provide unusual operation while STATCOM and UPFC is in the network as shown in Figs. 15 and 16, respectively.

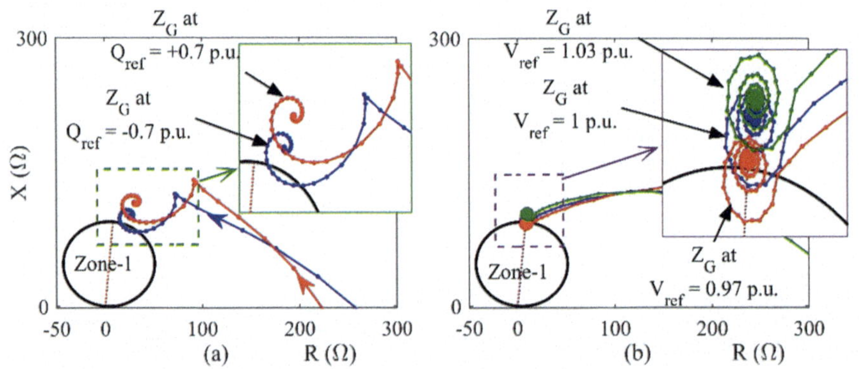

Fig. 15 Performance during three-phase fault for: **a** VARCM (-0.7 p.u. $\leq Q_{\text{ref}} \leq +0.7$ p.u.) and **b** VRM (0.97 p.u. $\leq V_{\text{ref}} \leq 1.03$ p.u.)

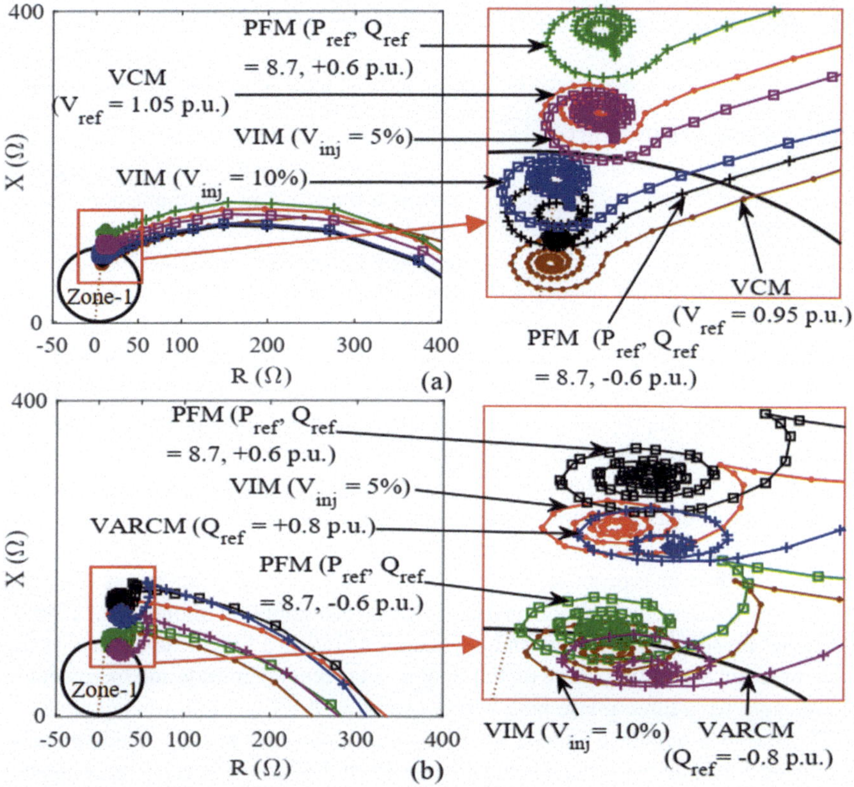

Fig. 16 Effect on relay R_G during the occurrence of, **a** a 3-phase fault at 70% length of the line from R_G, and **b** an ag-fault at 90% length of the line from R_G

3 State of the Art on Fault Detection, Classification, and Location Estimation on High Voltage TLs Connected to Large-Scale Wind Farms

In this section, a brief and concise review of the different papers published in various journals, book chapters and conference proceedings related to the detection, classification, and location estimation of faults in wind farm integrated HV-TLs with or without compensation is provided.

3.1 Relaying Algorithms to Protect DFIG-Wind Farm Integrated TLs

In [17–19], the adaptive trip boundary setting-based relaying approach using voltage and current from the wind side is proposed for protecting DFIG-wind farm integrated TL. In [20], the modified permissive overreach transfer trip communication-based balance fault isolation is proposed utilizing voltage and current from both ends at a sampling frequency of 3.84 kHz to protect DFIG-wind farm integrated TL. In [21], the equivalent sequence impedance-based adaptive scheme is proposed for zone-2 of distance relay for detecting and isolating ground faults in DFIG-wind farm integrated TL using the voltage and current from both ends of the protected line. In [22], a wavelet transforms, Clarke transform and correlation coefficient operated fault detection scheme is proposed for DFIG-wind farm integrated TL. In that work, the authors have utilized the voltage and current measurements at a sampling frequency of 2.4 kHz from the wind farm side for performing the fault detection task.

In [23], the discrimination between forward and reverse fault using correlation coefficient and discrimination between internal and external fault using RL-differential equation is proposed for protecting DFIG-wind farm integrated TL. Here, the voltage and current from both ends are utilized at a sampling frequency of 12 kHz. Here, the current phasors at both ends at a sampling rate of 1 kHz are used. In [24], the fault direction estimation based on the change in current magnitude and increment in phase angle for a three-phase fault is proposed for DFIG-wind farm integrated TL. In [25], the zero-sequence impedance using voltage and current at the wind farm side is utilized for fault detection. However, a delay in the farm side relay as compared to grid side relay is provided to avoid the abrupt effect in impedance by the wind farm. In [26], it is expressed that the frequency of short circuit current of DFIG varies between 35 and 65 Hz. The Prony algorithm is employed to extract these frequency components and taking the ratio with the fundamental component is considered as an index for fault detection for DFIG-wind farm integrated TL.

In [27], the Walsh-Hadamard transform-based fault detection for DFIG-wind farm integrated double-circuited TL is proposed. In [28], a Bayesian optimized support vector machine-based approach in a differential protection scheme is reported for protecting DFIG-wind farm integrated TL. Here, the upper confidence bound, the maximum probability of improvement, and the expected improvement of current are utilized as the features for fault detection. The fault detection accuracy is found to be 97.5%. This method is validated in FPGA DE2-115 board equipped with Cyclone IV-E device (EP4CE115F29C7). In [29], a current differential protection algorithm-based fault detection scheme is presented for protecting DFIG-wind farm integrated TL. Here, particle swarm optimization is used to set an optimum threshold for the differential current.

In [30], the traveling wave approach based on rotational invariant techniques for fault location is presented for protecting DFIG-wind farm integrated TL. Here, the Clarke transform of voltages from both ends is extracted at a sampling frequency of 1000 kHz and then reconstructed. The squared errors of the original and reconstructed wave from both ends are taken for this purpose. Here, the fault location is found to be less than 1%. In [31], the Pearson correlation coefficient of S-transform data and random forest-based fault classification and location scheme is presented for protecting DFIG-wind farm integrated TL. Here, the currents at both ends are extracted at a sampling frequency of 10 kHz for this purpose. The fault classification and location accuracies are 99.5 and 99.75%, respectively.

3.2 *Relaying Algorithms to Protect DFIG-Wind Farm Integrated Compensated TLs*

In [32], the DFT, decision tree, and rule-based fuzzy logic are used for fault detection and classification for protecting DFIG-wind farm integrated UPFC-CTL. Here, the voltage and current from both ends of the TL are utilized at a sampling frequency of 4.8 kHz. In [33, 34], the adaptive impedance calculation-based boundary setting of a relay is used for protecting DFIG-wind farm integrated UPFC-CTL using voltage and current from the wind farm side. In [35], the self-tunning Kalman filter is used to extract the phasor quantities of voltage and current from both ends to utilize the features in the decision tree classifier to protect DFIG-wind farm integrated UPFC-CTL. Here, the sampling frequency is used at 1.2 kHz. In [36], the spectral energy of wavelet transform coefficient of current at 20 kHz sampling-based differential protection is reported for DFIG-wind farm integrated UPFC-CTL.

In [37], an adaptive trip boundary setting of the relay is used for protecting DFIG-wind farm integrated shunt-CTL using voltage and current from the wind farm side. Here, two shunt devices SVC and STATCOM are connected in the network. In [38], the fast discrete S-transform is used for fault detection, and traveling wave-based fault location schemes are reported for protecting DFIG-wind farm integrated TCSC-CTL. Here, the voltage and current from both ends of the TL are utilized at a sampling frequency of 200 kHz. In [39], the DFT is used to extract the fundamental components of current and voltage and then, the features are trained and tested for fault classification for DFIG-wind farm integrated STATCOM-CTL. In [40], the approximation coefficient of third-level decomposition using db4 mother wavelet-based algorithm is reported for protecting DFIG-wind farm integrated STATCOM-CTL. Here, the spectral energies are calculated at 2 kHz sampling frequency and utilized in the ANN classifier for the detection and classification of faults.

In [6], the sign of positive sequence current phasor from both ends of the TL are used for fault detection and transient monitoring index of grid side current for fault classification scheme are reported for protecting DFIG-wind farm integrated UPFC-CTL. In [7], the sign of positive sequence α arial components of current from both ends of the TL is used for fault detection and dual time transform-decision tree of grid side current for fault classification scheme are reported for protecting DFIG-wind farm integrated STATCOM-CTL. In [8], the sign of positive sequence current phasor from both ends of the TL are used for fault detection and empirical mode decomposition-random forest of grid side current for fault classification scheme are reported for protecting DFIG-wind farm integrated TCSC-CTL.

3.3 Relaying Algorithms to Protect PMSG-Wind Farm Integrated TLs

In [41], the Pearson correlation of currents from both ends of the TL is proposed at a sampling frequency of 1 kHz for protecting PMSG-wind farm integrated TL. In [42], a pilot protection scheme based on wavelet transform, structural similarity, and square error method using current from both ends at a sampling frequency of 5 kHz is proposed for protecting PMSG-wind farm integrated TL. In [43], a pilot protection scheme using circuit parameters like resistance, inductance, shunt capacitance, and current from both ends at a sampling frequency of 1.2 kHz is utilized to protect PMSG-wind farm integrated TL. The maximum fault location error is found 0.21%. In [44], a pilot protection scheme based on cosine similarity using current from both ends at the sampling frequency of 1.2 kHz is proposed for protecting PMSG-wind farm integrated TL.

4 Conclusion

This paper comprehensively reviewed the state of the art on the protection of HV-TLs connected to large-scale wind farms and compensated with/without FACTs devices. Prior to the review, the impacts of grid-integrated large-scale wind farms and compensating devices on distance relaying-based transmission line protection are investigated. For ensuring enhanced protection for such crucial TLs, a number of time–frequency-based signal processing techniques are reported in the literature. Some of them utilize both-end data for performing the protection task and thus, require dedicated communication channels. Feature-based classifiers such as ANN, fuzzy, SVMs, DTs, and random forest proposed in the literature have their own merits and demerits. On the other hand, fault location estimation algorithms based on the traveling wave approach require data at a very high sampling frequency. The above study clearly indicates a need to explore more advanced signal processing tools and intelligence classifiers that can provide very fast and reliable protection to such crucial transmission lines.

References

1. Ackerman T (2005) Wind power in power systems. Wiley, NewYork, USA
2. Chaurasiya PK, Warudkar V, Ahmed S (2019) Wind energy development and policy in India: a review. Energy Strat Rev 24:342–357
3. Xu L, Yao L, Sasse C (2007) Grid integration of large dfig-based wind farms using vsc transmission. IEEE Trans Power Syst 22(3):976–984
4. Morshed MJ, Fekih A (2019) A coordinated controller design for dfig-based multi-machine power systems. IEEE Syst J 13(3):3211–3222
5. Boyu Q, Li H, Zhou X, Li J, Liu W (2020) Low-voltage ride-through techniques in dfig-based wind turbines: a review. Appl Sci 10(6):1–25
6. Biswas S, Nayak PK (2021) A fault detection and classification scheme for unified power flow controller compensated transmission lines connecting wind farms. IEEE Syst J 15(1):297–306
7. Biswas S, Nayak PK, Pradhan G (2021) A dual-time transform assisted intelligent relaying scheme for the statcom-compensated transmission line connecting wind farm. IEEE Syst J 1–12
8. Biswas S, Nayak PK (2021) A new approach for protecting tcsc compensated transmission lines connected to dfig-based wind farm. IEEE Trans Ind Info 17(8):5282–5291
9. Hingorani NG, Gyugyi L (2000) Understanding FACTS concepts and technology of flexible AC transmission systems. IEEE Press, New York, USA
10. Padiyar KR (2007) FACTS controllers in power transmission and distribution. New Age Inter national (P) Ltd. Press
11. Commission CER (2020) Report of the expert group: review of Indian electricity grid code. Technical Report
12. Paladhi S, Pradhan AK (2021) Adaptive distance protection for lines connecting converter-interfaced renewable plants. IEEE J Emer Sel Topics Power Elect 9(6):7088–7098
13. Paladhi S, Pradhan AK (2021) Adaptive fault type classification for transmission network connecting converter- interfaced renewable plants. IEEE Syst J 15(3):4025–4036
14. Li B, Liu J, Wang X, Zhao L (2018) Fault studies and distance protection of transmission lines connected to dfig-based wind farms. Appl Sci 562(8):1–19
15. Sulla F, Svensson J, Samuelsson O (2011) Symmetrical and unsymmetrical short-circuit current of squirrel-cage and doubly-fed induction generators. Elect Power Syst Res 81(7):1610–1618
16. Chang Z, Song G, Wang T (2018) Analysis on current characteristics of pmsg under grid three-phase fault. J Eng 2018(15):785–790
17. Pradhan AK, Joos G (2007) Adaptive distance relay setting for lines connecting wind farms. IEEE Trans Energy Conv 22(1):206–213
18. Sadeghi H (2012) A novel method for adaptive distance protection of transmission line connectedto wind farms. Int J Elect Power Energy Syst 43(1):1376–1382
19. Dubey RK, Samantaray SR, Panigrahi BK (2014) Adaptive distance relaying scheme for transmission network connecting wind farms. Elect Power Comp Syst 42(11):1181–1193
20. Hooshyar A, Azzouz MA, El-Saadany EF (2014) Distance protection of lines connected to induction generator-based wind farms during balanced faults. IEEE Trans Sust Energy 5(4):1193–1203
21. Chen S, Tai N, Fan C, Liu J, Hong S (2017) Adaptive distance protection for grounded fault of lines connected with doubly-fed induction generators. IET Gen Trans Dist 11(7):1513–1520. (Apr 2017)
22. Xu C, Xianggen Y, Zhe Z (2017) Impacts of dfig-based wind farm integration on its tie line distance protection and countermeasures. IEEJ Trans Elect Electro Eng 12(4):1–12
23. Chen Y, Wen M, Yin X, Cai Y, Zheng J (2018) Distance protection for transmission lines of dfig-based wind power integration system. Int J Elect Power Energy Syst 100:438–448
24. Fan Z, Song G, Kang X, Tang J, Wang X (2019) Three-phase fault direction identification method for outgoing transmission line of dfig-based wind farms. J Mod Power Syst Clean Energy 7(5):1155–1164

25. Fang Y, Jia K, Yang Z, Li Y, Bi T (2019) Impact of inverter-interfaced renewable energy generators on distance protection and an improved scheme. IEEE Trans Ind Elect 66(9):7078–7088
26. Chen Z, Lu Y (2020) A novel pilot protection scheme for wind farm transmission line based on harmonic content ratio. In: 2020 12th IEEE PES Asia-pacific power and energy engineering conference, pp 1–6
27. Kapoor G, Purohit SD, Singh Jat D, Poonia RC, Kumar S, Hiranwal S, (eds) (2021) Springer Singapore, Singapore, pp 73–84
28. Rezaei N, Uddin MN, Amin IK, Othman ML, Marsadek MB, Hasan MM (2020) A novel hybrid machine learning classifier-based digital differential protection scheme for intertie zone of large-scale centralized dfig-based wind farms. IEEE Trans Ind Appl 56(4):3453–3465
29. Prasad C, Biswal M (2020) Swarm intelligence-based differential protection scheme for wind integrated transmission system. Comp Elect Eng 86:106709
30. Mishra S, Gupta S, Yadav A (2021) "A novel two-terminal fault location approach utilizing traveling-waves for series compensated line connected to wind farms. Elect Power Syst Res 198:107362
31. Wang X, Gao X, Liu Y, Wang Y (2021) Stockwell-transform and random forest based double-terminal fault diagnosis method for offshore wind farm transmission line. IET Renew Power Gener 15:2368–2382
32. Jena MK, Samantaray SR, Tripathy L (2014) Decision tree-induced fuzzy rule-based differential relaying for transmission line including unified power flow controller and wind- farms. IET Gen Trans Dist 8(8):2144–2152. (Dec 2014)
33. Dubey R, Samantaray S, Panigrahi B, Venkoparao G (2015) Adaptive distance relay setting for parallel transmission network connecting wind farms and upfc. Int J Elect. Power Energy Syst 65:113–123
34. Dubey R, Samantaray SR, Panigrahi BK (2014) Simultaneous impact of unified power flow controller and off-shore wind penetration on distance relay characteristics. IET Gen Trans Dist 8(11):1869–1880. (Nov 2014)
35. Jena MK, Samantaray SR (2016) Data-mining-based intelligent differential relaying for transmission lines including upfc and wind farms. IEEE Trans Neural Net Learn Syst 27(1): 8–17
36. Tripathy LN, Jena MK, Samantaray SR, Dash DR (2014) A differential protection scheme for tapped transmission line containing upfc and wind farm. In: Proceedings of the 2014 IEEE Students' Technology Symposium, pp 319–324
37. Dubey R, Samantaray SR, Panigrahi BK (2016) Adaptive distance protection scheme for shunt-facts compensated line connecting wind farm. IET Gen Trans Dist 10(9):247–256. (Jan 2016)
38. Sahoo B, Samantaray SR (2018) An enhanced fault detection and location estimation method for tcsc compensated line connecting wind farm. Int J Elect Power Energy Syst 96:432–441
39. Kumar S, Koley E (2020) A k-nearest neighbour based protection scheme for upfc compensated double circuit transmission line with wind farms. In: 2020 IEEE First International Conference on Smart Technology for Power Energy and Control, pp 1–5
40. Mishra SK, Tripathy LN (2020) "A novel relaying approach for performance enhancement in a statcom integrated wind-fed transmission line using single-terminal measurement. Iran J Sci Tech Trans Elect Eng 44:897–910
41. Jia K, Li Y, Fang Y, Zheng L, Bi T, Yang Q (2018) Transient current similarity based protection for wind farm transmission lines. Appl Energy 225:42–51
42. Zheng L, Jia K, Bi T, Yang Z, Fang Y (2020) A novel structural similarity based pilot protection for renewable power transmission line. IEEE Trans Power Del 35(6):2672–2681
43. Saber A, Zeineldin HH, El-Fouly THM, Al-Durra A (2021) Time-domain fault location algorithm for double-circuit transmission lines connected to large scale wind farms. IEEE Access 9:11 393–11 404
44. Zheng L, Jia K, Bi T, Fang Y, Yang Z (2021) Cosine similarity based line protection for large-scale wind farms. IEEE Trans Ind Elect 68(7):5990–5999

The manufacturer's authorised representative in the EU is Springer Nature Customer Service Centre GmbH, Europaplatz 3, 69115 Heidelberg, Germany. If you have any concerns regarding our products, please contact ProductSafety@springernature.com

Printed and bound by CPI Group (UK) Ltd, Croydon, CR0 4YY
30/09/2026
02225315-0010